SPACE SCIENCES

2nd Edition

SPACE SCIENCES

Second Edition
Volume 1: Space Business

MACMILLAN REFERENCE USA
A part of Gale, Cengage Learning

GALE
CENGAGE Learning·

Detroit • New York • San Francisco • New Haven, Conn • Waterville, Maine • London

Space Sciences 2ⁿᵈ Edition

Project Editor: John F. McCoy

Contributing Editors: Jason M. Everett, Debra M. Kirby

Product Manager: Douglas A. Dentino

Composition: Evi Seoud

Cover Design: Jenny Willingham, Infiniti

Imaging: John Watkins

Manufacturing: Wendy Blurton

Product Design: Kristine Julien

Rights Acquisition and Management: Sheila Rivers-Spencer

For product information and technology assistance, contact us at
Gale Customer Support, 1-800-877-4253.
For permission to use material from this text or product,
submit all requests online at **www.cengage.com/permissions.**
Further permissions questions can be emailed to
permissionrequest@cengage.com

Cover photographs courtesy of the following: Sun rising over Earth in space, © Worldspec/NASA/Alamy. Artist's rendition of Earth, © PhotoStocker/ShutterStock. com. Nebula, © NASA, ESA, and The Hubble Heritage Team (STScI/AURA). International Space Station, Shuttle launch, and Space walk all courtesy of NASA.

While every effort has been made to ensure the reliability of the information presented in this publication, Gale, a part of Cengage Learning, does not guarantee the accuracy of the data contained herein. Gale accepts no payment for listing; and inclusion in the publication of any organization, agency, institution, publication, service, or individual does not imply endorsement of the editors or publisher. Errors brought to the attention of the publisher and verified to the satisfaction of the publisher will be corrected in future editions.

LIBRARY OF CONGRESS CATALOGING-IN-PUBLICATION DATA

Space sciences. -- 2nd ed.
  p. cm.
  Includes bibliographical references and index.
  ISBN 978-0-02-866214-5 (set) -- ISBN 978-0-02-866215-2 (vol. 1, space business) -- ISBN 978-0-02-866216-9 (vol. 2, planetary science and astronomy) -- ISBN 978-0-02-866217-6 (vol. 3, humans in space) -- ISBN 978-0-02-866218-3 (vol. 4, our future in space)
   1. Space sciences. I. Gale (Firm)

QB500.S63 2012
500.5--dc23   2011048057

Gale
27500 Drake Rd.
Farmington Hills, MI, 48331-3535

ISBN-13: 978-0-02-866214-5 (set)   ISBN-10: 0-02-866214-8 (set)
ISBN-13: 978-0-02-866215-2 (vol. 1)   ISBN-10: 0-02-866215-6 (vol. 1)
ISBN-13: 978-0-02-866216-9 (vol. 2)   ISBN-10: 0-02-866216-4 (vol. 2)
ISBN-13: 978-0-02-866217-6 (vol. 3)   ISBN-10: 0-02-866217-2 (vol. 3)
ISBN-13: 978-0-02-866218-3 (vol. 4)   ISBN-10: 0-02-866218-0 (vol. 4)

This title is also available as an e-book.
ISBN-13: 978-0-02-866219-0 ISBN-10: 0-02-866219-9
Contact your Gale, a part of Cengage Learning, sales representative for ordering information.

Printed in China
1 2 3 4 5 6 7 16 15 14 13 12

Table of Contents

Table of Contents

Volume 2: Planetary Science and Astronomy

Table of Contents

Volume 4: Our Future in Space

Table of Contents

Preface

The American humorist Will Rogers frequently shared the homey folk wisdom that "It ain't what you don't know what makes you look like a fool, it's what you do know, what ain't so." In a much more scholarly style, the eminent scholar of the Library of Congress, Daniel Boorstin, writing about the great "Age of Exploration" five hundred years ago, observed that "The greatest obstacle to discovering the shape of the earth, the continents, and the oceans was not ignorance but the illusion of knowledge."

Both were saying that prior to reaping the fruits of exploration or experience, people have to revise things they believe, sometimes very fundamental things they believe. To open the way for new knowledge, old mis-knowledge—misconceptions, misinterpretations, delusions, deceptions, all sorts of variations of false information—has to be cleared out.

Most people—in fact, most human societies in history—go through their lives learning things and then stubbornly clinging to them for the rest of their lives. Learning something contrary to past lessons—requiring yourself to un-learn, or doing something differently—encounters immense resistance.

The antidote to this mental and cultural constipation is exploration. History has shown that it's the best way to shake up people's concepts, to shake up societies, to shake up the course of history.

Today's point of the spear of exploration is space exploration, and its success has so permeated our culture that we can no longer imagine an earlier time. Space exploration has provided tangible value to countries that spend the time, effort, treasure, and on occasion human lives to conduct it. National prestige directly translates into commercial, cultural, and diplomatic power. New discoveries, and new technologies developed to attain them, enhance the productivity of industry. But these things, the things we can touch, pale to insignificance when compared to the greatest treasures reaped by exploration: the things that touch us. And what touches us most intimately is our ability to be caught by surprise—and then to react positively, to our benefit. It is our capacity to be astonished, and then to strive to understand, that space exploration encourages.

Just one example: the volcanoes of Io, Jupiter's innermost large moon. The first probes to Jupiter never saw them, because they had not been explicitly instructed to directly look for a feature that nobody had seriously expected to find. Even the more sophisticated Voyager-1 probe in 1979 wasn't looking for them, and wouldn't have seen them except for a post-flyby image captured to refine the probe's navigational accuracy.

It was then that 25-year-old Linda Morabito, measuring the image for the navigation team, noticed the strange smudge on the edge of the disc that proved to be the profile of a volcanic plume. As it turned out, a handful of scientists had only recently predicted the possibility of such eruptions based on calculations of the moon's internal heating, but few had paid attention.

Morabito's discovery is a shining example of the surprises that shower down on Earth from space exploration, and the handful of scientists who had predicted the discovery show that when stretched, the human imagination is capable of properly processing them.

Compare this success to the main-boggling prediction by British biologist J.B.S. Haldane in his 1927 book, *Possible Worlds.* In an often quoted and paraphrased passage, he wrote: "I have no doubt that in reality the future will be vastly more surprising than anything I can imagine. Now my own suspicion is that the Universe is not only queerer than we suppose, but queerer than we *can* suppose."

Space exploration is the vaccine against the inability to adequately "suppose," or to "imagine." It fires our imaginations, preparing us—sometimes just barely—for the astonishments it delivers, and to other surprises throughout our civilization. Here's how it does that:

As new knowledge is accumulated by exploration, it both adds to "old knowledge" as well as often replacing it with more accurate "new knowledge." It is a constant threat to minds that are satisfied that what they already know is all they'll ever need to know.

Intelligence is often foolishly measured—and tested—as a quality of knowing what things are. Recite dates, recite city names, recite muscle groups, recite a poem. But true intelligence is in knowing how to find out what the things are and—more importantly—knowing when to change your mind. Exploration—particularly space exploration—is a mind-changing machine, an engine for the process of reminding us that we "know" things with a degree of uncertainty and must always be prepared to modify that "knowledge" with new discoveries.

These discoveries accumulate new knowledge, to be sure. But more importantly they foster new attitudes. Tomorrow is not going to be a linear outgrowth of things we know today. Today's attitudes—and prejudices—are not insurmountable barriers to different ones that could take root tomorrow. Our choices and actions can, and often do, reflect this historically rare attitude that people's thoughts can change the shape of the future.

Recent conceptual revolutions in space exploration have provided in-your-face examples of our earlier foolishness about things we once thought we "knew"—and without space exploration, would still mistakenly "know."

One of my favorites is our search for "life as we know it." Our early searches—even our fundamental questions—were misdirected because what we "knew" about life, that it needs sunlight and oxygen, wasn't true.

It turns out, wonderfully and mind-blowingly, that we were wrong about life on Earth, and that the portions we knew about—the surface sunlight-based bio-scum—make up only a small fraction of the planet's actual biomass. Most of our home planet's life forms thrive quite merrily deep underground, away from sunlight and oxygen. True, this life is microbial in nature, but in total mass, it far outweighs all of the complex biology we "knew" about and thought was all there is. And the best part is this: from the point of view of this branch of Earth's biota, the environment it now inhabits is closely duplicated elsewhere in the Solar System. And not just one or two spots—it looks like there potentially are a dozen or more ecological habitats that could readily host Earthborn species—or alien life developed along similar lines—deep underground where surface conditions (temperature, atmosphere, radiation) just don't matter.

Then there is commercial space flight. It has long been a theme in science fiction (*The Man Who Sold the Moon,* for example) and mass-access "space tourism" has been awaited impatiently for decades, but in 2011 is still not a reality. Whereas today's leading space-based business profit centers were poorly predicted, if at all. Personal navigation services, and personal communications support, are now multi-billion-dollar businesses, while the headline-grabbing rocket launch business has become an isolated sideline, almost a backwater of "space business." What we "knew" did not turn out to be "true," we had been insufficiently imaginative.

Space exploration has also taught us that the Universe beyond Earth is more violent than people had imagined. And closer to home, this violence—in energy, in solid matter, in other forces—is not insulated from Earth. In the past it has affected Earth's biosphere and has inflicted calamities upon it—and could do so again. What once had been inconceivable is now a topic for hazard assessment and countermeasure design.

Sometimes what we "know" turns out to be overoptimistic. A generation ago it was common to believe that after the Apollo moon landings, it was only a matter of faster rockets and bigger spaceships to bring Mars within range of astronauts. But the "unknown unknowns" had been lurking out of our sight and beyond our imaginations. In hindsight, most space experts now believe that no matter what the budget, such a project would not have possessed technologies that were advanced enough to make it work. It would have wound up delayed and cancelled, or worse, could have launched an expedition that ended disastrously. "Going around in circles," as critics sneered at the International Space Station, has turned out to have been the only reliable strategy for developing and verifying the life-support technologies crucial to human deep-space ventures in the years ahead. Imagine that!

Another example is the semi-mystical philosophical concept of a century ago called "panspermia." This is the notion that life could originate on one world in the Universe and then spread, naturally, to other worlds, like it has spread across the surface of the Earth from niche to niche. Just how panspermia was supposed to happen, though, those philosophers never specified. And because astronomers and geologists for most of the twentieth century could not imagine how it could happen, they concluded it could not happen. It seemed logical to do so.

But then suppose, in Haldane's use of that term, that a simple life form chose to live inside of rocks, as shelter from harsh surface conditions (as microbes do in the dry valleys of Antarctica). Further suppose occasional impacts of infalling asteroids, some of them at large angles to the target planet's surface, blast some of these rocks into space

Suppose some of the rubble later—a million years perhaps—hits another planet, this one with a thick atmosphere. The meteorite is briefly surrounded by a fiery train of flaring plasma, but it slows quickly, before the skin heating soaks too far inside. It falls to the ground. It is rained upon. The passengers, evolved to form hardy spores by their harsh world of origin, wake. And so life spreads from world to world, at least in theory.

In all of human history, scientists had never—until recent years—been able to imagine such events. Some, without knowing exactly how, had speculated it could happen. Most, not knowing how, concluded that the limits of their own imaginations were more or less the limits of the Universe's repertory. But as we explored and discovered, and recognized unusualness where we stumbled upon it, things we never imagined came into the range of our senses. We have to believe that process is nowhere near its end.

As for our future in space, the brief history of its past teaches us that imagination is the main ingredient to enable it, and imagination is the main ingredient in understanding its discoveries and bounties. If, as Jules Verne wrote, "Anything one man can imagine, other men can make real," is valid—and I think spaceflight shows that it is—then every one of us needs, every day, to exercise that imagination, to stretch our minds, and to reach out our hands among the stars.

James E. Oberg

www.jamesoberg.com

How To Use This Book

. . . .

Space Sciences, 2ⁿᵈ Edition provides a comprehensive overview of all aspects of space exploration, astronomy, planetary science, and their impact on daily life on Earth. This is a broad scope, and for ease of use entries have been grouped into four subject-based volumes.

Volume 1: Space Business This volume features entries on the business aspects of space exploration. The many different businesses focused on space are discussed, such as the launch industry and the emerging space tourism sector. Prominent businesspeople are profiled, as are a variety of space-related career paths. This volume also examines the business aspects of a variety of topics that at first glance might not seem commercial in nature, such as the International Space Station and military space programs. Likewise, the connections between space and everyday topics like advertising, education, and toys are covered here.

Volume 2: Planetary Science and Astronomy This volume is focused on presenting facts about our solar system and the universe at large, as well as explaining the science behind this information. This is where the reader will find entries on the planets and their satellites, other features of the solar system like the solar wind and the Oort cloud, and also on topics like black holes and galaxies. Different facets of astronomy are explored, as is cosmology, and the theories that underlie these sciences are explained. There is also extensive coverage of unmanned space probes and orbiting observatories. Biographies of some noteworthy space scientists complete the volume.

Volume 3: Humans in Space The drama of manned space exploration is explored in this book. An overarching history of spaceflight is supplemented by detailed entries on programs like Apollo, Soviet manned space missions, and the space shuttle. Biographies of prominent astronauts are included. The volume also features entries on a variety of concepts that are important to understanding human spaceflight, such as: how spacecraft maneuver, how life is supported in space, and the physiological effects of living in space. Coverage of the ground operations that support spaceflight are also found in this book.

Volume 4: Our Future in Space This volume explores a number of concepts associated with space exploration that range from experimental, to theoretical, to downright fanciful given current science. For example: antimatter propulsion, faster-than-light travel, lightsails, permanent settlements on other worlds and in space, space elevators, and teleportation. While some of these ideas may never come to fruition, they inspire today's scientists and space enthusiasts. This volume also covers the connection between space sciences and science fiction. Biographies of space visionaries are included as well.

All entries include a bibliography to assist in conducting additional research. Where appropriate, technical terms are defined in the margins of the entries, and photos and charts are used to illustrate most entries.

Entries are arranged in alphabetical order within each volume, with biographies listed by their last names. All four volumes include a comprehensive table of contents and cumulative index that can further assist the reader in locating the information they need. Other features found in every volume are:

- ■ **Preface** — An essay by noted space historian and journalist James Oberg discussing the importance of space exploration and science.

- ■ **For Your Reference** — This section defines scientific units of measurement and supplies conversion charts to customary units of measurement. It also provides basic statistical information on the eight planets of the solar system.

- ■ **Milestones in Space History** — This chronology lists major events in space science and exploration.

- ■ **Human Achievements in Space** — This chronology is focused on the history of human spaceflight, starting with the beginning of the space race in 1957.

- ■ **Glossary** — The glossary defines over 325 technical terms used throughout the set.

- ■ **Directory of Space Organizations** — This directory provides contact information for over fifty space-related organizations, including all major national space agencies.

For Your Reference

··

This section provides information that may be of assistance in understanding the entries that make up this book: a comparative table of data about the planets; definitions for SI terms and symbols, and; conversion tables for SI measurements to other measurement systems.

SOLAR SYSTEM PLANET DATA

	Mercury	Venus[2]	Earth	Mars	Jupiter	Saturn	Uranus	Neptune
Mean distance from the Sun (AU):[1]	0.387	0.723	1	1.524	5.202	9.555	19.218	30.109
Siderial period of orbit (years):	0.24	0.62	1	1.88	11.86	29.46	84.01	164.79
Mean orbital velocity (km/sec):	47.89	35.04	29.79	24.14	13.06	9.64	6.81	5.43
Orbital eccentricity:	0.206	0.007	0.017	0.093	0.048	0.056	0.047	0.009
Inclination to ecliptic (degrees):	7.00	3.40	0	1.85	1.3	2.49	0.77	1.77
Equatorial radius (km):	2439	6052	6378	3397	71492	60268	25559	24764
Polar radius (km):	same	same	6357	3380	66854	54360	24973	24340
Mass of planet (Earth = 1):[3]	0.06	0.82	1	0.11	317.89	95.18	14.54	17.15
Mean density (g/cm^3):	5.44	5.25	5.52	3.94	1.33	0.69	1.27	1.64
Body rotation period (hours):	1408	5832.R	23.93	24.62	9.92	10.66	17.24	16.11
Tilt of equator to orbit (degrees):	0	2.12	23.45	23.98	3.08	26.73	97.92	28.8

[1]AU indicates one astronomical unit, defined as the mean distance between Earth and the Sun (~1.495 x 10^8 km).
[2]R indicates planet rotation is retrograde (i.e., opposite to the planet's orbit).
[3]Earth's mass is approximately 5.976 x 10^{26} grams.

SI BASE AND SUPPLEMENTARY UNIT NAMES AND SYMBOLS

Physical Quality	Name	Symbol
Length	meter	m
Mass	kilogram	kg
Time	second	s
Electric current	ampere	A
Thermodynamic temperature	kelvin	K
Amount of substance	mole	mol
Luminous intensity	candela	cd
Plane angle	radian	rad
Solid angle	steradian	sr

Temperature

Scientists commonly use the Celsius system. Although not recommended for scientific and technical use, earth scientists also use the familiar Fahrenheit temperature scale (°F). 1°F = 1.8°C or K. The triple point of H_2O, where gas, liquid, and solid water coexist, is 32°F.

- To change from Fahrenheit (F) to Celsius (C): °C = (°F-32)/(1.8)
- To change from Celsius (C) to Fahrenheit (F): °F = (°C x 1.8) + 32
- To change from Celsius (C) to Kelvin (K): K = °C + 273.15
- To change from Fahrenheit (F) to Kelvin (K): K = (°F-32)/(1.8) + 273.15

UNITS DERIVED FROM SI, WITH SPECIAL NAMES AND SYMBOLS

Derived Quantity	Name of SI Unit	Symbol for SI Unit	Expression in Terms of SI Base Units
Frequency	hertz	Hz	s^{-1}
Force	newton	N	$m \cdot kg \cdot s^{-2}$
Pressure, stress	pascal	Pa	$m^{-1} \cdot kg \cdot s^{-2}$
Energy, work, heat	joule	J	$m^2 \cdot kg \cdot s^{-2}$
Power, radiant flux	watt	W	$m^2 \cdot kg \cdot s^{-3}$
Electric charge	coulomb	C	$s \cdot A$
Electric potential, electromotive force	volt	V	$m^2 \cdot kg \cdot s^{-3} \cdot A^{-1}$
Electric resistance	ohm	Ω	$m^2 \cdot kg \cdot s^{-3} \cdot A^{-2}$
Celsius temperature	degree Celsius	°C	K
Luminous flux	lumen	lm	Cd
Illuminance	lux	lx	$m^{-2} \cdot cd$

UNITS USED WITH SI, WITH NAME, SYMBOL, AND VALUES IN SI UNITS

The following units, not part of the SI, will continue to be used in appropriate contexts (e.g., angtsrom):

Physical Quantity	Name of Unit	Symbol for Unit	Value in SI Units
Time	minute	min	60 s
	hour	h	3,600 s
	day	d	86,400 s
Plane angle	degree	°	$(\pi/180)$ rad
	minute	'	$(\pi/10,800)$ rad
	second	"	$(\pi/648,000)$ rad
Length	angstrom	Å	10^{-10} m
Volume	liter	l, L	$1\ dm^3 = 10^{-3}\ m^3$
Mass	ton	t	$1\ Mg = 10^3$ kg
	unified atomic mass unit	u	$\approx 1.66054 \times 10^{-27}$ kg
Pressure	bar	bar	$10^5\ Pa = 10^5\ N\ m^{-2}$
Energy	electronvolt	eV (= e X V)	$\approx 1.60218 \times 10^{-19}$ J

CONVERSIONS FOR STANDARD, DERIVED, AND CUSTOMARY MEASUREMENTS

Length

1 angstrom (Å)	0.1 nanometer (exactly)
	0.000000004 inch
1 centimeter (cm)	0.3937 inches
1 foot (ft)	0.3048 meter (exactly)
1 inch (in)	2.54 centimeters (exactly)
1 kilometer (km)	0.621 mile
1 meter (m)	39.37 inches
	1.094 yards
1 mile (mi)	5,280 feet (exactly)
	1.609 kilometers
1 astronomical unit (AU)	1.495979×10^{13} cm
1 parsec (pc)	206,264.806 AU
	3.085678×10^{18} cm
	3.261633 light-years
1 light-year	9.460530×10^{17} cm

Area

1 acre	43,560 square feet (exactly)
	0.405 hectare
1 hectare	2.471 acres
1 square centimeter (cm^2)	0.155 square inch
1 square foot (ft^2)	929.030 square centimeters
1 square inch (in^2)	6.4516 square centimeters (exactly)
1 square kilometer (km^2)	247.104 acres
	0.386 square mile
1 square meter (m^2)	1.196 square yards
	10.764 square feet
1 square mile (mi^2)	258.999 hectares

MEASUREMENTS AND ABBREVIATIONS

Volume

1 barrel (bbl)*, liquid	31 to 42 gallons
1 cubic centimeter (cm^3)	0.061 cubic inch
1 cubic foot (ft^3)	7.481 gallons
	28.316 cubic decimeters
1 cubic inch (in^3)	0.554 fluid ounce
1 dram, fluid (or liquid)	$1/8$ fluid ounce (exactly)
	0.226 cubic inch
	3.697 milliliters
1 gallon (gal) (U.S.)	231 cubic inches (exactly)
	3.785 liters
	128 U.S. fluid ounces (exactly)
1 gallon (gal) (British Imperial)	277.42 cubic inches
	1.201 U.S. gallons
	4.546 liters
1 liter	1 cubic decimeter (exactly)
	1.057 liquid quarts
	0.908 dry quart
	61.025 cubic inches
1 ounce, fluid (or liquid)	1.805 cubic inches
	29.573 mililiters
1 ounce, fluid (fl oz) (British)	0.961 U.S. fluid ounce
	1.734 cubic inches
	28.412 milliliters
1 quart (qt), dry (U.S.)	67.201 cubic inches
	1.101 liters
1 quart (qt), liquid (U.S.)	57.75 cubic inches (exactly)
	0.946 liter

Units of mass

1 carat (ct)	200 milligrams (exactly)
	3.086 grains
1 grain	64.79891 milligrams (exactly)
1 gram (g)	15.432 grains
	0.035 ounce
1 kilogram (kg)	2.205 pounds
1 microgram (µg)	0.000001 gram (exactly)
1 milligram (mg)	0.015 grain
1 ounce (oz)	437.5 grains (exactly)
	28.350 grams
1 pound (lb)	7,000 grains (exactly)
	453.59237 grams (exactly)
1 ton, gross or long	2,240 pounds (exactly)
	1.12 net tons (exactly)
	1.016 metric tons
1 ton, metric (t)	2,204.623 pounds
	0.984 gross ton
	1.102 net tons
1 ton, net or short	2,000 pounds (exactly)
	0.893 gross ton
	0.907 metric ton

Pressure

1 kilogram/square centimeter (kg/cm^2)	0.96784 atmosphere (atm)
	14.2233 pounds/square inch (lb/in^2)
	0.98067 bar
1 bar	0.98692 atmosphere (atm)
	1.02 kilograms/square centimeter (kg/cm^2)

* There are a variety of "barrels" established by law or usage. For example, U.S. federal taxes on fermented liquors are based on a barrel of 31 gallons (141 liters); many state laws fix the "barrel for liquids" as $31 1/2$ gallons (119.2 liters); one state fixes a 36-gallon (160.5 liters) barrel for cistern measurment; federal law recognizes a 40-gallon (178 liters) barrel for "proof spirts"; by custom, 42 gallons (159 liters) comprise a barrel of crude oil or petroleum products for statistical purposes, and this equivalent is recognized "for liquids" by four states.

Milestones in Space History

c. 850	The Chinese invent a form of gunpowder for rocket propulsion.
1242	Englishman Roger Bacon develops gunpowder.
1379	Rockets are used as weapons in the Siege of Chioggia, Italy.
1804	William Congrieve develops ship-fired rockets.
1903	Konstantin Tsiolkovsky publishes *Research into Interplanetary Science by Means of Rocket Power,* a treatise on space travel.
1909	Robert H. Goddard develops designs for liquid-fueled rockets.
1917	Smithsonian Institute issues grant to Goddard for rocket research.
1918	Goddard publishes the monograph *Method of Attaining Extreme Altitudes.*
1921	Soviet Union establishes a state laboratory for solid rocket research.
1922	Hermann Oberth publishes *Die Rakete zu den Planetenräumen,* a work on rocket travel through space.
1923	Tsiolkovsky publishes work postulating multi-staged rockets.
1924	Walter Hohmann publishes work on rocket flight and orbital motion.
1927	The German Society for Space Travel holds its first meeting.
1927	Max Valier proposes rocket-powered aircraft adapted from Junkers G23.
1928	Oberth designs liquid rocket for the film *Woman in the Moon.*
1929	Goddard launches rocket carrying barometer.
1930	Soviet rocket designer Valentin Glusko designs U.S.S.R. liquid rocket engine.
1931	Eugen Sänger test fires liquid rocket engines in Vienna.
1932	German Rocket Society fires first rocket in test flight.
1933	Goddard receives grant from Guggenheim Foundation for rocket studies.
1934	Wernher von Braun, member of the German Rocket Society, test fires water-cooled rocket.
1935	Goddard fires advanced liquid rocket that reaches 700 miles per hour.
1936	Glushko publishes work on liquid rocket engines.
1937	The Rocket Research Project of the California Institute of Technology begins research program on rocket designs.

1938	von Braun's rocket researchers open center at Pennemünde.
1939	Sänger and Irene Brendt refine rocket designs and propose advanced winged suborbital bomber.
1940	Goddard develops centrifugal pumps for rocket engines.
1941	Germans test rocket-powered interceptor aircraft Me 163.
1942	V-2 rocket fired from Pennemünde enters space during ballistic flight.
1943	First operational V-2 launch.
1944	V-2 rocket launched to strike London.
1945	Arthur C. Clarke proposes geostationary satellites.
1946	Soviet Union tests version of German V-2 rocket.
1947	United States test fires Corporal missile from White Sands New Mexico.
	X-1 research rocket aircraft flies past the speed of sound.
1948	United States reveals development plan for Earth satellite adapted from RAND.
1949	Chinese rocket scientist Hsueh-Sen proposes hypersonic aircraft.
1950	United States fires Viking 4 rocket to record 106 miles from USS Norton Sound.
1951	Bell Aircraft Corporation proposes winged suborbital rocket-plane.
1952	Wernher von Braun proposes wheel-shaped Earth-orbiting space station.
1953	U.S. Navy D-558II sets world altitude record of 15 miles above Earth.
1954	Soviet Union begins design of RD-107, RD-108 ballistic missile engines.
1955	Soviet Union launches dogs aboard research rocket on suborbital flight.
1956	United States announces plan to launch Earth satellite as part of Geophysical Year program.
1957	U.S. Army Ballistic Missile Agency is formed.
	Soviet Union test fires R-7 ballistic missile.
	Soviet Union launches the world's first Earth satellite, Sputnik-1, aboard R-7.
	United States launches 3-stage Jupiter C on test flight.
	United States attempts Vanguard 1 satellite launch; rocket explodes.
1958	United States orbits Explorer-1 Earth satellite aboard Jupiter-C rocket.
	United States establishes the National Aeronautics and Space Administration (NASA) as civilian space research organization.
	NASA establishes Project Mercury manned space project.
	United States orbits Atlas rocket with Project Score.

1959   Soviet Union sends Luna 1 towards Moon; misses by 3100 miles.

NASA announces the selection of seven astronauts for Earth space missions.

Soviet Union launches Luna 2, which strikes the Moon.

1960   United States launches Echo satellite balloon.

United States launches Discoverer 14 into orbit, capsule caught in midair.

Soviet Union launches two dogs into Earth orbit.

Mercury-Redstone rocket test fired in suborbital flight test.

1961   Soviet Union tests Vostok capsule in Earth orbit with dummy passenger.

Soviet Union launches Yuri Gagarin aboard Vostok-1; he becomes the first human in space.

United States launches Alan B. Shepard on suborbital flight.

United States proposes goal of landing humans on the Moon before 1970.

Soviet Union launches Gherman Titov into Earth orbital flight for one day.

United States launches Virgil I. "Gus" Grissom on suborbital flight.

United States launches first Saturn 1 rocket in suborbital test.

1962   United States launches John H. Glenn into 3-orbit flight.

United States launches Ranger to impact Moon; craft fails.

First United States/United Kingdom international satellite launch; Ariel 1 enters orbit.

X-15 research aircraft sets new altitude record of 246,700 feet.

United States launches Scott Carpenter into 3-orbit flight.

United States orbits Telstar 1 communications satellite.

Soviet Union launches Vostok 3 and 4 into Earth orbital flight.

United States launches Mariner II toward Venus flyby.

United States launches Walter Schirra into 6-orbit flight.

Soviet Union launches Mars 1 flight; craft fails.

1963   United States launches Gordon Cooper into 22-orbit flight.

Soviet Union launches Vostok 5 into 119-hour orbital flight.

United States test fires advanced solid rockets for Titan 3C.

First Apollo Project test in Little Joe II launch.

Soviet Union orbits Vostok 6, which carries Valentina Tereshkova, the first woman into space.

Soviet Union tests advanced version of R-7 called Soyuz launcher.

1964   United States conducts first Saturn 1 launch with live second stage; enters orbit.

U.S. Ranger 6 mission launched towards Moon; craft fails.

Soviet Union launches Zond 1 to Venus; craft fails.

United States launches Ranger 7 on successful Moon impact.

United States launches Syncom 3 communications satellite.

Soviet Union launches Voshkod 1 carrying 3 cosmonauts.

United States launches Mariner 4 on Martian flyby mission.

1965   Soviet Union launches Voshkod 2; first space walk.

United States launches Gemini 3 on 3-orbit piloted test flight.

United States launches Early Bird 1 communications satellite.

United States launches Gemini 4 on 4-day flight; first U.S. space walk.

United States launches Gemini 5 on 8-day flight.

United States launches Titan 3C on maiden flight.

Europe launches Asterix 1 satellite into orbit.

United States Gemini 6/7 conduct first space rendezvous.

1966   Soviet Union launches Luna 9, which soft lands on Moon.

United States Gemini 8 conducts first space docking; flight aborted.

United States launches Surveyor 1 to Moon soft landing.

United States tests Atlas Centaur advanced launch vehicle.

Gemini 9 flight encounters space walk troubles.

Gemini 10 flight conducts double rendezvous.

United States launches Lunar Orbiter 1 to orbit Moon.

Gemini 11 tests advanced space walks.

United States launches Saturn IB on unpiloted test flight.

Soviet Union tests advanced Proton launch vehicle.

United States launches Gemini 12 to conclude 2-man missions.

1967   Apollo 1 astronauts killed in launch pad fire.

Soviet Soyuz 1 flight fails; cosmonaut killed.

Britain launches Ariel 3 communications satellite.

United States conducts test flight of M2F2 lifting body research craft.

United States sends Surveyor 3 to dig lunar soils.

Soviet Union orbits anti-satellite system.

United States conducts first flight of Saturn V rocket (Apollo 4).

1968   Yuri Gagarin killed in plane crash.

Soviet Union docks Cosmos 212 and 213 automatically in orbit.

United States conducts Apollo 6 Saturn V test flight; partial success.

Nuclear rocket engine tested in Nevada.

United States launches Apollo 7 in 3-person orbital test flight.

Soviet Union launches Soyuz 3 on 3-day piloted flight.

United States sends Apollo 8 into lunar orbit; first human flight to Moon.

1969   Soviet Union launches Soyuz 4 and 5 into orbit; craft dock.

Largest tactical communications satellite launched.

United States flies Apollo 9 on test of lunar landing craft in Earth orbit.

United States flies Apollo 10 to Moon in dress rehearsal of landing attempt.

United States cancels military space station program.

United States flies Apollo 11 to first landing on the Moon.

United States cancels production of Saturn V in budget cut.

Soviet lunar rocket N-1 fails in launch explosion.

United States sends Mariner 6 on Mars flyby.

United States flies Apollo 12 on second lunar landing mission.

Soviet Union flies Soyuz 6 and 7 missions.

United States launches Skynet military satellites for Britain.

1970   China orbits first satellite.

Japan orbits domestic satellite.

United States Apollo 13 mission suffers explosion; crew returns safely.

Soviet Union launches Venera 7 for landing on Venus.

United States launches military early warning satellite.

Soviet Union launches Luna 17 to Moon.

United States announces modifications to Apollo spacecraft.

1971   United States flies Apollo 14 to Moon landing.

Soviet Union launches Salyut 1 space station into orbit.

First crew to Salyut station, Soyuz 11, perishes.

Soviet Union launches Mars 3 to make landing on the red planet.

United States flies Apollo 15 to Moon with roving vehicle aboard.

1972   United States and the Soviet Union sign space cooperation agreement.

United States launches Pioneer 10 to Jupiter flyby.

Soviet Union launches Venera 8 to soft land on Venus.

United States launches Apollo 16 to moon.

India and Soviet Union sign agreement for launch of Indian satellite.

United States initiates space shuttle project.

United States flies Apollo 17, last lunar landing mission.

1973   United States launches Skylab space station.

United States launches first crew to Skylab station.

Soviet Union launches Soyuz 12 mission.

United States launches second crew to Skylab space station.

1974   United States launches ATS research satellite.

Soviet Union launches Salyut 3 on unpiloted test flight.

Soviet Union launches Soyuz 12, 13 and 14 flights.

Soviet Union launches Salyut 4 space station.

1975   Soviet Union launches Soyuz 17 to dock with Salyut 4 station.

Soviet Union launches Venera 9 to soft land on Venus.

United States and Soviet Union conduct Apollo-Soyuz Test Project joint flight.

China orbits large military satellite.

United States sends Viking 1, 2 towards landing on Martian surface.

Soviet Union launches unpiloted Soyuz 20.

1976   Soviet Union launches Salyut 5 space station.

First space shuttle rolls out; Enterprise prototype.

Soviet Union docks Soyuz 21 to station.

China begins tests of advanced ballistic missile.

1977   Soyuz 24 docks with station.

United States conducts atmospheric test flights of shuttle Enterprise.

United States launches Voyager 1 and 2 on deep space missions.

Soviet Union launches Salyut 6 space station.

Soviet Soyuz 25 fails to dock with station.

Soyuz 26 is launched and docks with station.

1978   Soyuz 27 is launched and docks with Salyut 6 station.

Soyuz 28 docks with Soyuz 27/Salyut complex.

United States launches Pioneer/Venus 1 mission.

Soyuz 29 docks with station.

Soviet Union launches Progress unpiloted tankers to station.

Soyuz 30 docks with station.

United States launches Pioneer/Venus 2.

Soyuz 31 docks with station.

1979   Soyuz 32 docks with Salyut station.

Voyager 1 flies past Jupiter.

Soyuz 33 fails to dock with station.

Voyager 2 flies past Jupiter.

1980   First Ariane rocket launches from French Guiana; fails.

Soviet Union begins new Soyuz T piloted missions.

STS-1 first shuttle mission moves to launching pad.

1981   Soviet Union orbits advanced Salyut stations.

STS-1 launched on first space shuttle mission.

United States launches STS-2 on second shuttle flight; mission curtailed.

1982   United States launches STS-5 first operational shuttle flight.

1983   United States launches Challenger, second orbital shuttle, on STS-6.

United States launches Sally Ride, the first American woman in space on STS-7.

United States launches Guion Bluford, the first African-American astronaut on STS-8.

United States launches first Spacelab mission aboard STS-9.

1984	Soviet Union tests advanced orbital station designs.
	Shuttle Discovery makes first flights.
	United States proposes permanent space station as goal.
1985	Space shuttle Atlantis enters service.
	United States announces policy for commercial rocket sales.
	United States flies U.S. Senator aboard space shuttle Challenger.
1986	Soviet Union launches and occupies advanced Mir space station.
	Challenger—on its tenth mission, STS-51-L—is destroyed during launch phase, 73 seconds after liftoff.
	United States restricts payloads on future shuttle missions.
	United States orders replacement shuttle for Challenger.
1987	Soviet Union flies advanced Soyuz T-2 designs.
	United States Delta, Atlas and Titan rockets grounded in launch failures.
	Soviet Union launches Energyia advanced heavy lift rocket.
1988	Soviet Union orbits unpiloted shuttle Buran.
	United States launches space shuttle Discovery on STS-26 flight.
	United States launches STS-27 military shuttle flight.
1989	United States launches STS-29 flight.
	United States launches Magellan probe from shuttle.
1990	Shuttle fleet grounded for hydrogen leaks.
	United States launches Hubble Space Telescope.
1992	Replacement shuttle Endeavour enters service.
	United States probe Mars Observer fails.
1993	United States and Russia announce space station partnership.
1994	United States shuttles begin visits to Russian space station Mir.
1995	Europe launches first Ariane 5 advanced booster; flight fails.
1996	United States announces X-33 project to replace shuttles.
1997	Mars Pathfinder lands on Mars.
1998	First elements of International Space Station (ISS) launched.

1999   First Ocean space launch of Zenit rocket in Sea Launch program.

2000   Twin United States Mars missions fail.

2001   United States cancels shuttle replacements X-33 and X-34 because of space budget cutbacks.

United States orbits Mars Odyssey probe around Mars.

2002   First launches of United States advanced Delta IV and Atlas V commercial rockets.

2003   During STS-107, on its twentieth-eighth mission, space shuttle Columbia breaks apart during reentry.

China becomes the third nation (after the U.S. and Russia) to launch a human into space.

Europe launches the Mars Express spacecraft using a Russian Soyuz/Fregat rocket.

2004   NASA robotic rovers Opportunity and Spirit land on Mars and begin traveling over the planet's surface.

SpaceShipOne becomes the first privately-owned, manned spacecraft to reach space.

NASA's Cassini spacecraft begins orbiting Saturn.

SpaceShipOne wins the $10 million Ansari X-prize space competition.

NASA announces Constellation program for shuttle replacement and return to the Moon.

2005   STS-114 is the first shuttle mission since the Columbia disaster.

China launches a capsule with two taikonauts (astronauts) onboard.

2006   United States launches New Horizons spacecraft to study dwarf planet Pluto.

NASA's Stardust spacecraft returns a capsule to Earth containing comet dust.

2007   United States sends Dawn spacecraft to study dwarf planet Ceres.

American astronaut Peggy Whitson becomes the first female space station commander.

2008   China launches a capsule with three men inside; two perform spacewalks.

SpaceX Corporation launches its Falcon 1 rocket, becoming the first privately-funded liquid fueled rocket to achieve orbit.

2009   United States launches Kepler spacecraft to search for extrasolar planets.

Europe launches the Herschel Space Observatory on an Ariane 5 rocket.

Shuttle Atlantis performs last Hubble Space Telescope servicing mission during STS-125.

2010   President Obama cancels Constellation program and lunar manned missions.

SpaceX launches and recovers its Dragon capsule designed for future manned flight. This is the first time a private company has launched and recovered its own space capsule.

2011   STS-133 marks the last flight of shuttle Discovery.

STS-134 marks the final spaceflight of shuttle Endeavor.

Space shuttle program ends with flight of Atlantis during STS-135.

United States launches Mars Science Laboratory and Curiosity rover to Mars.

Human Achievements in Space

The road to space has been neither steady nor easy, but the journey has cast humans into a new role in history. Here are some of the milestones and achievements.

Oct. 4, 1957	The Soviet Union launches the first artificial satellite, a 184-pound spacecraft named Sputnik.
Nov. 3, 1957	The Soviets continue pushing the space frontier with the launch of a dog named Laika into orbit aboard Sputnik 2. The dog lives for seven days, an indication that perhaps people may also be able to survive in space.
Jan. 31, 1958	The United States launches Explorer 1, the first U.S. satellite, and discovers that Earth is surrounded by radiation belts. James Van Allen, who instrumented the satellite, is credited with the discovery.
Apr. 12, 1961	Yuri Gagarin becomes the first person in space. He is launched by the Soviet Union aboard a Vostok rocket for a two-hour orbital flight around the planet.
May 5, 1961	Astronaut Alan Shepard becomes the first American in space. Shepard demonstrates that individuals can control a vehicle during weightlessness and high gravitational forces. During his 15-minute suborbital flight, Shepard reaches speeds of 5,100 mph.
May 24, 1961	Stung by the series of Soviet firsts in space, President John F. Kennedy announces a bold plan to land men on the Moon and bring them safely back to Earth before the end of the decade.
Feb. 20, 1962	John Glenn becomes the first American in orbit. He flies around the planet for nearly five hours in his Mercury capsule, Friendship 7.
June 16, 1963	The Soviets launch the first woman, Valentina Tereshkova, into space. She circles Earth in her Vostok spacecraft for three days.
Nov. 28, 1964	NASA launches Mariner 4 spacecraft for a flyby of Mars.
Mar. 18, 1965	Cosmonaut Alexei Leonov performs the world's first space walk outside his Voskhod 2 spacecraft. The outing lasts 10 minutes.
Mar. 23, 1965	Astronauts Virgil I. "Gus" Grissom and John Young blast off on the first Gemini mission and demonstrate for the first time how to maneuver from one orbit to another.
June 3, 1965	Astronaut Edward White becomes the first American to walk in space during a 21-minute outing outside his Gemini spacecraft.
Mar. 16, 1966	Gemini astronauts Neil Armstrong and David Scott dock their spacecraft with an unmanned target vehicle to complete the first joining of two spacecraft in orbit. A stuck thruster forces an early end to the experiment, and the crew makes America's first emergency landing from space.

Jan. 27, 1967	The Apollo 1 crew is killed when a fire breaks out in their command module during a prelaunch test. The fatalities devastate the American space community, but a subsequent spacecraft redesign helps the United States achieve its goal of sending men to the Moon.
Apr. 24, 1967	Tragedy strikes the Soviet space program also, with the death of cosmonaut Vladimir Komarov. His new Soyuz spacecraft gets tangled with parachute lines during re-entry and crashes to Earth.
Dec. 21, 1968	Apollo 8, the first manned mission to the Moon, blasts off from Cape Canaveral, Florida. Frank Borman, Jim Lovell and Bill Anders orbit the Moon ten times, coming to within 70 miles of the lunar surface.
July 20, 1969	Humans walk on another world for the first time when astronauts Neil Armstrong and Edwin "Buzz" Aldrin climb out of their spaceship and set foot on the Moon.
Apr. 13, 1970	The Apollo 13 mission to the Moon is aborted when an oxygen tank explosion cripples the spacecraft. NASA's most serious inflight emergency ends four days later when the astronauts, ill and freezing, splash down in the Pacific Ocean.
June 6, 1971	Cosmonauts blast off for the first mission in the world's first space station, the Soviet Union's Salyut 1. The crew spends twenty-two days aboard the outpost. During re-entry, however, a faulty valve leaks air from the Soyuz capsule, and the crew is killed.
Jan. 5, 1972	President Nixon announces plans to build "an entirely new type of space transportation system," pumping life into NASA's dream to build a reusable, multi-purpose space shuttle.
Dec. 7, 1972	The seventh and final Apollo mission to the Moon is launched, as public interest and political support for the Apollo program dims.
May 14, 1973	NASA launches the first U.S. space station, Skylab 1, into orbit. Three crews live on the station between May 1973 and February 1974. NASA hopes to have the shuttle flying in time to reboost and resupply Skylab, but the outpost falls from orbit on July 11, 1979.
July 17, 1975	In a momentary break from Cold War tensions, the United States and Soviet Union conduct the first linking of American and Russian spaceships in orbit. The Apollo-Soyuz mission is a harbinger of the cooperative space programs that develop between the world's two space powers twenty years later.
Apr. 12, 1981	Space shuttle Columbia blasts off with a two-man crew for the first test-flight of NASA's new reusable spaceship. After two days in orbit, the shuttle lands at Edwards Air Force Base in California.
June 18, 1983	For the first time, a space shuttle crew includes a woman. Astronaut Sally Ride becomes America's first woman in orbit.
Oct. 30, 1983	NASA's increasingly diverse astronaut corps includes an African-American for the first time. Guion Bluford, an aerospace engineer, is one of the five crewmen assigned to the STS-8 mission.

Nov. 28, 1983	NASA flies its first Spacelab mission and its first European astronaut, Ulf Merbold.
Feb. 7, 1984	Shuttle astronauts Bruce McCandless and Robert Stewart take the first untethered space walks, using a jet backpack to fly up to 320 feet from the orbiter.
Apr. 9–11, 1984	First retrieval and repair of an orbital satellite.
Jan. 28, 1986	Space shuttle Challenger explodes 73 seconds after launch, killing its seven-member crew. Aboard the shuttle was Teacher-in-Space finalist Christa McAuliffe, who was to conduct lessons from orbit. NASA grounds the shuttle fleet for two and a half years.
Feb. 20. 1986	The Soviets launch the core module of their new space station, Mir, into orbit. Mir is the first outpost designed as a module system to be expanded in orbit. The expected lifetime of the station is five years.
May 15, 1987	Soviets launch a new heavy-lift booster from the Baikonur Cosmodrome in Kazakhstan.
Oct. 1, 1987	Mir cosmonaut Yuri Romanenko breaks the record for the longest space mission, surpassing the 236-day flight by Salyut cosmonauts set in 1984.
Sept. 29, 1988	NASA launches the space shuttle Discovery on the first crewed U.S. mission since the 1986 Challenger explosion. The shuttle carries a replacement communications satellite for the one lost onboard Challenger.
Nov. 15, 1989	The Soviets launch their space shuttle Buran, which means snowstorm, on its debut flight. There is no crew onboard, and unlike the U.S. shuttle, no engines to help place it into orbit. Lofted into orbit by twin Energia heavy-lift boosters, Buran circles Earth twice and lands. Buran never flies again.
May 4, 1989	Astronauts dispatch a planetary probe from the shuttle for the first time. The Magellan radar mapper is bound for Venus.
Apr. 24, 1990	NASA launches the long-awaited Hubble Space Telescope, the cornerstone of the agency's "Great Observatory" program, aboard space shuttle Discovery. Shortly after placing the telescope in orbit, astronomers discover that the telescope's prime mirror is misshapen.
Dec. 2, 1993	Space shuttle Endeavour takes off for one of NASA's most critical shuttle missions: repairing the Hubble Space Telescope. During an unprecedented five space walks, astronauts install corrective optics. The mission is a complete success.
Feb. 3, 1994	A Russian cosmonaut, Sergei Krikalev, flies aboard a U.S. spaceship for the first time.
Mar. 16, 1995	NASA astronaut Norman Thagard begins a three and a half month mission on Mir—the first American to train and fly on a Russian spaceship. He is the first of seven Americans to live on Mir.
Mar. 22, 1995	Cosmonaut Valeri Polyakov sets a new space endurance record of 437 days, 18 hours.

June 29, 1995	The U.S. space shuttle docks at the Russian space station Mir for the first time.
Mar. 24, 1996	Shannon Lucid begins her stay aboard space aboard Mir, which lasts 188 days—a U.S. record for spaceflight endurance at that time.
Feb. 24, 1997	An oxygen canister on Mir bursts into flames, cutting off the route to one of the station's emergency escape vehicles. The fire only goes out when the canister's oxygen is used up. The six crewmembers onboard, including U.S. astronaut Jerry Linenger, are unharmed.
June 27, 1997	During a practice of a new docking technique, Mir commander Vasily Tsibliyev loses control of an unpiloted cargo ship and it plows into the station. The Spektr module is punctured, and the crew hurriedly seals off the compartment to save the ship.
Oct. 29, 1998	Senator John Glenn, one of the original Mercury astronauts, returns to space aboard the shuttle.
Nov. 20, 1998	A Russian Proton rocket hurls the first piece of the International Space Station into orbit.
Aug. 27, 1999	Cosmonauts Viktor Afanasyev, Sergei Avdeyev, and Jean-Pierre Haignere leave Mir. The station is unoccupied for the first time in almost a decade.
Oct. 31, 2000	The first joint American-Russian crew is launched to the International Space Station. Commander Bill Shepherd requests the radio call sign "Alpha" for the station and the name sticks.
Mar. 23, 2001	Russian ground control manuevers the Mir space station out of orbit, and it burns up in Earth's atmosphere.
Apr. 28, 2001	Russia launches the world's first space tourist for a weeklong stay at the International Space Station. NASA objects to the flight, but is powerless to stop it.
Mar. 1, 2002	Space shuttle Columbia lifts off to the Hubble Space Telescope on a repair and servicing mission. New solar arrays are installed on Hubble, increasing power by 30 percent. New optics and guidance units are also installed.
Feb. 1, 2003	Space shuttle Columbia breaks up during reentry, killing all seven crewmembers onboard. NASA grounds the shuttle fleet for nearly two and a half years. A subsequent investigation finds that foam shed from the shuttle's External Tank punctured Columbia's thermal protection system, which protects the shuttle from the intense heat produced during reentry.
Oct. 15, 2003	China launches a taikonaut (astronaut) into Earth orbit, thus becoming the third nation in the world—after Russia and the United States—to possess a manned space program of its own. The space capsule is named Shenzhou 5, and is launched atop a Long March IIF rocket.
Jan. 14, 2004	President George W. Bush announces the Vision for Space Exploration. In response, NASA creates the Constellation program to develop the rockets, spacecraft, and

other system elements needed to replace the space shuttle fleet and return astronauts to the Moon.

June 21, 2004 SpaceShipOne, a vehicle built by Scaled Composites, becomes the first manned spacecraft developed with private funds to reach outer space—100 kilometers (62 miles) above Earth's surface—during a suborbital flight.

Nov. 6, 2004 SpaceShipOne is officially declared winner of the $10 million Ansari X-prize when it performs two suborbital flights within two weeks.

July 26, 2005 Discovery begins mission STS-114, becoming the first shuttle to launch following the Columbia disaster. In addition to delivering supplies to the International Space Station, Discovery's crew also implements new safety procedures, including remote viewing of the shuttle's thermal protection system.

Jan. 18, 2006 NASA announces the Commercial Orbital Transportation Services (COTS) program. The COTS program is created to fund private space companies to ferry cargo and crewmembers from Earth to the International Space Station.

Sept. 18, 2006 Iranian-American Anousheh Ansari becomes the fourth space tourist overall—and first female space tourist—to live aboard the International Space Station. She arrives aboard a Russian-built Soyuz TMA spacecraft.

Oct. 24, 2007 China begins its Lunar Exploration Program with the launch of the Chang'e-1 lunar orbiter. Chang'e-1 is only the first in a series of planned Chinese missions to the Moon, including landing a rover there and bringing lunar soil to Earth, culminating with Chinese taikonauts landing on the Moon by 2030.

Sept. 25, 2008 China launches three taikonauts aboard the Shenzhou 7 space capsule. The spacewalks performed by two of the taikonauts are a first for the Chinese space program.

Dec. 23, 2008 NASA awards its Cargo Resupply Services (CRS) contract to SpaceX to ferry supplies to the International Space Station. The CRS is worth $1.6 billion to SpaceX for a dozen supply missions to the Space Station. SpaceX will use its planned Dragon capsule and Falcon 9 rocket to fulfill the CRS contract.

May 11, 2009 Space shuttle Atlantis travels to the Hubble Space Telescope to perform its last servicing and repair mission. Sensors, gyroscopes, and batteries are replaced, while two new optical instruments are installed on Hubble.

Feb. 1, 2010 President Obama's budget proposal eliminates funding for NASA's Constellation program that, among other goals, planned on returning astronauts to the Moon by 2020. The President instead directs NASA to send astronauts to nearby asteroids, followed by a mission to Mars.

June 4, 2010 The first launch of SpaceX's Falcon 9 rocket into low Earth orbit.

Dec. 8, 2010 The second launch of SpaceX's Falcon 9 rocket, this time carrying the Dragon capsule. After two orbits around Earth, Dragon reenters the atmosphere and splashes down in the Pacific Ocean. This is the first time a private firm successfully launched and recovered a space capsule.

Feb. 24, 2011   The last mission of space shuttle Discovery launches to the International Space Station.

May 16, 2011   The Endeavor space shuttle travels to the International Space Station on its last mission, delivering equipment that includes scientific instruments.

July 8, 2011   The space shuttle programs ends with the final launch of Atlantis to the International Space Station. Atlantis brings equipment and supplies to the Space Station.

Nov. 3, 2011   Two unmanned Chinese spacecraft—Shenzhou 8 and Tiangong 1—dock together. The Chinese government claims that this docking demonstration constitutes a major step towards constructing its own space station.

A

Accessing Space

The task of placing humans, satellites, and other payloads into orbit* about Earth or to send such human or nonhuman cargo to another celestial body has proven formidable. Current technology dictates that rockets be used to access space. A rocket is a cylindrical metal object containing inflammable material, which, when ignited, propels the rocket to a significant height or distance. Rocket-powered vehicles are quite different from jet aircraft, in that jets use the atmosphere as a source of oxidizer (oxygen in the air) with which to burn the fuel. Rocket-propelled vehicles must carry along all propellants (both fuel and oxidizer) because once they reach a certain height above Earth's surface, little oxygen is available.

* **orbit** the circular or elliptical path of an object around a much larger object, governed by the gravitational field of the larger object

Pre-Space Age Rocketry Developments

Many centuries ago the Chinese first employed crude rockets using solidified propellants to scare their enemies with the resulting loud noises and flashing overhead lights. Later, rockets became popular for displays and celebrations. Early devices, however, were crude, used low-energy propellants, and were largely uncontrollable. It was not until the 1900s that major technological advances in rocketry were realized.

Around the turn of the twentieth century, Konstantin Tsiolkovsky, a Russian schoolteacher, discovered the fundamental relationship between the amount of propellant needed in a rocket and the resulting change in speed. This remains the most fundamental relationship of rocketry, and it is referred to as the "Tsiolkovsky rocket equation," or simply as the "rocket equation." In the 1920s, American physicist Robert H. Goddard designed and built the first liquid-propelled rocket motor and demonstrated its use in flight. This was a significant step toward the development of modern missiles and space launchers.

The onset of World War II (1939–1945) created a sense of urgency in advancing the development and deployment of long-range, rocket-propelled artillery projectiles and bombs. In both Germany and the United States, major efforts were begun to create rocket-propelled guided bombs, that is, missiles. By 1944, operational V-2 missiles were being launched militarily by Germany toward England. Although these were the first successfully guided bombs, they lacked good terminal guidance and usually missed their primary targets. They were, however, very effective as instruments of mass intimidation.

* **trajectories** paths followed
through space by missiles and
spacecraft moving under the
influence of gravity

* **expendable launch vehicles**
launch vehicles, such as rockets,
not intended to be reused

After the war, one group of German rocket engineers and scientists defected to the United States and another to the Soviet Union. Wernher von Braun led the group that went to the United States. The mission of these scientists was to continue work on missile technology and, by the 1950s, the V-2 had been improved and transformed into a variety of missiles. In order to create an intercontinental ballistic missile (ICBM) that could travel several thousand miles, however, improved guidance systems and multistage booster designs were needed, and these two areas of technology became the focus of 1950s rocketry research. Precise guidance systems ensure accurate trajectories* and precision targeting, while two-stage vehicles can overcome the pull of Earth's strong gravity, along with their low-propellant energies, to achieve great distances. These same technologies were needed for orbital launcher vehicles.

Rocketry Advances During the Late Twentieth Century Space Age

When the Soviets launched the first artificial satellite in 1957 (*Sputnik 1*) in an orbit about Earth, the United States quickly followed by modifying its ICBM inventory to create orbital launchers, and the "Space Race" between the two countries was on in earnest. By 1960, both the United States and the Soviet Union were producing launch vehicles at will. In 1961, President John F. Kennedy challenged America to send humans, before the end of the decade, to the surface of the Moon and return them safely. As amazing as it seemed at the time, two American astronauts, Neil Armstrong and Edwin (Buzz) Aldrin, walked on the Moon in July 1969. By the end of Project Apollo from the National Aeronautics and Space Administration (NASA), in 1972, a total of twelve astronauts had walked on the lunar soil and returned safely to Earth, and the United States was well established as the dominant spacefaring nation.

During the 1960s and 1970s, both the Soviet Union and the United States developed several families of space launchers. The Soviet inventory included the Kosmos, Proton, Soyuz, and Molniya, and the U.S. inventory included the Titan, Atlas, and Delta. In terms of the number and frequency of satellite launches, the Soviets were far more prolific; that is, until the breakup of the Soviet Union in 1991—which formed Russia and many other new countries. Whereas the Soviets (and later Russia) focused on putting large numbers of relatively crude satellites in orbit, the United States focused on sophistication and reliability. Thus, the West was very successful in collecting a good deal more science data with fewer satellites.

Early in the 1970s, U.S. President Richard Nixon approved the development of the Space Transportation System (STS), better known as the space shuttle fleet. This was to be a partially reusable system to replace all U.S. expendable launch vehicles*. Thus, when the space shuttle *Columbia*, the first operational NASA shuttle, started flying in 1981, production lines for the Delta and Atlas boosters were shut down. They stayed shut down until the 1986 space shuttle *Challenger* disaster. At that point, it became clear that expendables were still needed and would be needed for a long time to come. By 1989, the shuttle fleet and several expendables were back in business. The three-year U.S. launch hiatus, however, permitted other countries to enter the commercial launcher business. The most prominent of these is the European Space Agency's launcher family, Ariane, which, in the past, launched (by the French company Arianespace) roughly 40% of the world's largest communications satellites and, in 2008, launched about 50% of them to geostationary transfer orbit. Other competitors in the marketplace include Russia, India, China,

◀

The European Space Agency's fifth-generation Ariane rocket, shown here launching in October 1997, was capable of delivering an 18-ton satellite into a low Earth orbit. © *AP Images*.

* **reusable vehicles** launchers that
can be used many times before
discarding

and Japan. Even Israel, the Ukraine, the United Kingdom, South Korea,
Romania, and Brazil have been active in developing and launching small-
capacity booster vehicles.

The Future of Rocketry in the Twenty-First Century

During the first decade of the twenty-first century, there were more
than twenty families of expendable launchers from within Europe and
eight countries outside Europe. Only one reusable vehicle group was
available during the same period, the U.S. space shuttle fleet. That program ended in August 2011, leaving no operational non-expendable
rocket launch system in existence. In many respects, the space shuttle
program was a great success. In the thirty year period between 1981 and
2011, there were 135 successful missions, marred by two major disasters, the loss of *Challenger* in 1986 and *Columbia* in 2003. The shuttles
completed missions of anywhere from five to seventeen days, carrying
crews of two to eight astronauts and payloads of up to 24,000 kilograms (54,000 pounds). Although criticism has been leveled at NASA's
decision to invest so much of its resources in a single type of launch
vehicle, it is difficult to argue with much of the success of the program.
Following termination of the space shuttle program, there has been only
one major governmental effort to develop a new reusable launch vehicle
system, the Russian Kliper (English: Clipper) program. Russian officials
first announced the existence of the Kliper program at a press conference
in 2004, explaining the the Kliper was intended as a partially reusable
substitute for the aging Soyuz spacecraft for ferrying humans and cargo
back and forth between Earth and the International Space Station (ISS).
First projections were that the Kliper would be operational as early as
2009. In fact, such was not to be the case. A number of technical, financial, and political problems arose that required Russia to seek an alliance
first with Europe, then with Japan, in an effort to complete the program. Neither effort was successful and in July 2006, the Russian space
agency, Roscosmos, announced that development of the program had
been deferred. In 2009, Roscosmos announced another effort to replace
the Soyuz with a partially reusable spacecraft called Perspektivnaya
Pilotiruemaya Transportnaya Sistema (Prospective Piloted Transport
System or simply "Rus"). As of late 2011, the date of the first flight of
Rus is projected to be in 2017.

As the second decade of the twenty-first century begins, the outlook
for space vehicles appears to be one in which the emphasis is on reusable vehicles*, like the space shuttle, with a smaller number of expendables under development. While the new expendables should offer some
relief in terms of launch prices, reusable vehicles hold the promise for
the significant cost reductions that are needed for extensive expansion
of applications, such as space tourism. In pursuit of this objective, over
twenty private companies (non-governmental entities) in the United
States are trying to develop robotic or manned vehicles, with some

of them being partially or fully reusable. Some of the companies propose to build a single-stage system in which the entire vehicle travels from the launch pad all the way to a suborbital altitude (and later to orbit), separates from the carrier vehicle, and returns to the launch site. Others propose two-stage vehicles in which a booster orbiter* combination leaves the launch pad together and returns separately. One of the first companies to experience success in this challenge was Scaled Composites, a subsidiary of Northrup Grumman. In 2004, the company won the $10,000,000 Ansari X Prize for being the first nongovernmental entity to launch a reusable manned spacecraft into space twice within two weeks. The successful spacecraft, SpaceShipOne, was launched from an aircraft known as White Knight that takes off and lands conventionally and rises to an altitude of about 15 kilometers (9 miles). At that height, it releases the space ship, which then initiates its own propulsion system, before eventually returning to Earth by way of a conventional landing path. Shortly after the successful SpaceShipOne flight, Scaled Composites formed a joint venture with Virgin Galactic, a subsidiary of Richard Branson's Virgin Group. The new company, called the Spaceship Company, is at work on the development of a modification of SpaceShipOne, called SpaceShipTwo, to be launched from a modified White Knight aircraft, called White Knight Two. The company's immediate goal is to build five of the first aircraft and two of the second, to be followed, if work progresses successfully, by a third generation of aircraft.

A new motivation for the transport of humans into space has come with the rise of so-called space tourism. The term *space tourism* refers to trips made into space by private individuals primarily for recreational purposes. The cost of such trips to an individual prior to the 2010s, was prohibitive except for the wealthiest individuals. The first such trip occurred in 2001 when American businessman Dennis Tito paid $20 million to travel to the International Space Station for a period of eight days. Between 2001 and 2009, six other individuals paid up to $35 million for a similar experience, before the work schedule at the ISS itself made further such trips impossible. A number of private space companies have now taken up the challenge of building spacecraft that will allow humans to travel into orbital or sub-orbital space at more reasonable costs. Perhaps the best known of these companies is Virgin Galactic, which is currently selling tickets for its first sub-orbital flights at some as-yet unannounced date in the future. Tickets cost $200,000 a piece, with a 10 percent down payment required at time of reservation. As of late 2011, about 430 people had paid the necessary down payment and made their reservation.

* **orbiter** spacecraft that uses engines and/or aerobraking, and whose main function is to orbit about a planet or natural satellite

▶ *See also* **Aldrin, Buzz (Volume 1)** • **Apollo (Volume 3)** • **Armstrong, Neil (Volume 3)** • **Goddard, Robert Hutchings (Volume 1)** • **Launch Vehicles, Expendable (Volume 1)** • **Launch Vehicles, Reusable (Volume 1)** • **Reusable Launch Vehicles (Volume 4)** • **Space Shuttle (Volume 3)** • **Tsiolkovsky, Konstantin (Volume 3)** • **von Braun, Wernher (Volume 3)**

Resources

Books and Articles

Anderson, John D., Jr. *Introduction to Flight.* New York: McGraw-Hill, 2008.

Bizony, Piers. *The Space Shuttle: Celebrating Thirty Years of NASA's First Space Plane.* Minneapolis: Zenith Press, 2011.

Dubbs, Chris, and Emeline Paat-Dahlstrom. *Realizing Tomorrow: The Path to Private Spaceflight.* Lincoln: University of Nebraska Press, 2011.

Isakowitz, Steven J., Joseph P. Hopkins Jr., and Joshua B. Hopkins. *International Reference Guide to Space Launch Systems,* 4th ed. Reston, VA: American Institute of Aeronautics and Astronautics, 2004.

Linehan, Dan. *Burt Rutan's Race to Space: The Magician of Mojave and His Flying Innovations.* Minneapolis: Zenith Press, 2011.

Websites

Brief History of Rockets. <http://quest.nasa.gov/space/teachers/rockets/history.html> (accessed October 23, 2011).

Future of Space tourism: Who's Offering What? Space.com. <http://www.space.com/11477-space-tourism-options-private-spaceships.html> (accessed October 23, 2011).

History of the Space Shuttle. <http://history.nasa.gov/shuttlehistory.html> (accessed October 23, 2011).

Space Adventures. <http://www.spaceadventures.com/index.cfm> (accessed October 23, 2011).

A Timeline of Rocket History. <http://history.msfc.nasa.gov/rocketry/> (accessed October 23, 2011).

Virgin Galactic Website. <http://www.virgingalactic.com/> (accessed October 23, 2011).

Advertising

In the early days of human spaceflight in the 1960s, public curiosity about astronauts was fueled by regular headlines in the media. Products selected for the space program were perceived to be exceptional, and promoters were quick to exploit this by playing up the fascination and mystery surrounding spaceflight. A crystallized, dehydrated, orange-flavored beverage called Tang® was touted as "what the astronauts drink," and sales skyrocketed as the public clamored to have what the astronauts had. With the space age came space-themed advertising, the use of advertised messages in outer space, and ads related to space-based activities.

Consumers are bombarded daily with multimedia advertisements, coaxing people to buy, choose, or react to a myriad of products and services. Advertisers are hired to promote these products and services to specific markets based on a careful calculation of a target population's propensity to consume. To appeal to this possibility, advertisers strive to stay in the mainstream of the target audience in fashion, entertainment, food, and new technology by implanting a brand with a message that is crafted to be remembered by the recipient. Over the years, space themes have been used as a backdrop for many new products.

Before humans were orbiting Earth, space-themed advertisements were uncommon because the general public did not connect them with outer space. In the early twenty-first century, with discussion of futuristic orbiting hotels and launch adventure trips within the realm of technological possibility, space as a backdrop or theme for advertising is well-established.

Some fantastic concepts have been considered for advertising in space. For example, one firm considered using an Earth-based laser to beam messages onto the Moon. They soon realized this was impractical, however, because the images needed to be about the size of the state of Texas to be visible to earthlings!

What was not impractical was to film a commercial in space. In 1997, the Israeli-based company Tnuva filmed the first commercial in space, an advertisement for milk that was filmed inside the Mir space station. Pizza Hut Inc., contracted with a Russian launch firm to affix a nine-meter-high (30-foot-high) new corporate logo on a Proton rocket carrying aloft a service module to the International Space Station and scheduled for launch in November 1999. Advertising the event prior to the launch date gave Pizza Hut international recognition, and the company expected 500 million people to watch the live televised event. The launch was planned to coincide with a release of Pizza Hut's transformed millennium image; the launch, however, was postponed for eight months because of technical problems. The rocket was finally launched on July 12, 2000, but not without controversy as part of the Russian space program sued Pizza Hut for improperly associating the Russian space agency with Pizza Hut. Then in 2001, Pizza Hut became the first pizza delivery company to deliver pizzas to outer space. They delivered several vacuum-packed pizzas to the crewmembers aboard the International Space Station.

Pepsi, the soft drink company, paid a large sum of money so that Russian cosmonauts would unveil a newly designed brand logo on a simulated "can" during missions to the Russian space station Mir in May 1996. The company has also pursued smaller scale promotional ventures in the U.S. space program, through the National Aeronautics and Space Administration (NASA), since 1984.

NASA and other outer space agencies have researched the profitability of permitting advertising through the display of logos on space hardware,

Pepsi Cola launched a new advertising campaign from the Mir space station on April 25, 1996. This particular campaign unveiled a new version of Pepsi's logo. © *AP Images.*

such as the International Space Station. While there is a market for such advertising, studies suggest that demand would not necessarily be sustained beyond the novelty of the first few paying customers.

Space.com, one of several space-related web sites that appeared in the dot-com boom of the late 1990s, derived significant revenue from advertising banners at the web site. Interestingly, the advertising content tended to relate to very down-to-Earth necessities—credit cards, cars, goods and services—and not space merchandise or otherworldly creations.

Manned space travel has been the exclusive domain of governments since the first astronauts and cosmonauts were launched into outer space beginning in the early 1960s. However, several commercial space companies have entered the business of sending paying customers into outer space. Initially, these private enterprises would be limited to suborbital flights of limited duration. Virgin Galactic is a subsidiary of the Virgin Group founded by Richard Branson. As of late 2011, some 450 people had reserved spots to make suborbital journeys with Virgin Galactic from a spaceport in New Mexico, with flights projected to begin sometime in 2013. Other commercial spaceflight companies are working to launch paying customers into space as well. If these private space ventures are a success, they will undoubtedly increase their advertising budgets in search of potential customers for their spaceflight services, especially as ticket prices come down and competition increases. One or more of these private spaceflight companies might also choose to boost revenues by selling advertising space on their launch vehicles and/or spacecraft.

Privately-sponsored space prizes are also sparking interest in the possibilities of space advertising. The Google Lunar X-Prize is a competition funded by Internet search-engine giant Google Inc. The Lunar X-Prize is endowed with a purse of 30 million dollars to the non-government team or teams that can land a rover on the Moon and successfully perform certain tasks, such as beaming high resolution video of the lunar surface back to Earth. One company in particular, White Label Space, founded in 2008, is hoping to fund its mission to the Moon by selling advertising at each stage of the mission, from launch to lunar landing. As of 2011 the company had at least a dozen partners from around the world contributing their expertise in all areas of spaceflight, from software development, to the electronics for the necessary space hardware.

In 1993 the U.S. Congress passed a bill banning all U.S. advertisements in space. It was later amended to include only "obtrusive advertising," which means advertising can be placed on certain items such as launch vehicles and clothing. The Federal Aviation Administration has the authority in the United States to enforce the regulations banning obtrusive space advertising; but as of 2011 it had yet to encounter a need to invoke those regulations. Also, the international scientific community has considered recommending a limitation on obtrusive advertising in space. But at the start of the 2010s, no such international regulations had been established.

 See also **Commercialization (Volume 1) • International Space Station (Volumes 1 and 3) • Mir (Volume 3)**

Resources

Books and Articles

National Aeronautics and Space Administration. Office of Advanced Concepts and Technology. *Spinoff 93.* Washington, DC: U.S. Government Printing Office, 1994.

Solomon, Lewis D. *Rocketeers: How a Visionary Band of Business Leaders, Engineers, and Pilots Is Boldly Privatizing Space.* New York: Smithsonian Books, 2007.

Solomon, Lewis. *The Privatization of Space Exploration: Business, Technology, Law and Policy.* New Brunswick, N.J.: Transaction Publishers, 2008.

Weil, Elizabeth. "American Megamillionaire Gets Russki Space Heap!" *New York Times Magazine,* August 23, 2000.

Websites

Obtrusive space advertising and astronomical research. United Nations General Assembly. <http://www.unoosa.org/pdf/reports/ac105/AC105_777E.pdf> (accessed October 21, 2011).

Pizza Hut Pioneers Space Commercialization with Successful Rocket Launch. SpaceRef.com. July 12, 2000, <http://www.spaceref.com/news/viewpr.html?pid=2202> (accessed October 21, 2011).

* **in situ** in the natural or original location

* **meteorites** any part of a meteoroid that survives passage through Earth's atmosphere

The Daimler AG automotive group contributes to the European Aeronautic Defence and Space Company's development by manufacturing Cluster II satellites. © *AP Images.*

Aerospace Corporations

For most of history, humankind has had to study space from on or near the surface of Earth. This meant that most of our knowledge was limited to what could be deduced from observations conducted through dust and light pollution and the distorting and degrading effects of Earth's atmosphere. No in situ* study or direct analysis of materials from space (except for studies of meteorites*) was possible. These conditions changed drastically with the development of space technology. First machines, then humans, were able to enter outer space, beginning a new era in space study and exploration. This era has grown to include the exploitation of space for public and private purposes. Designing, building, and operating the systems that make this possible is the role of aerospace corporations of the twenty-first century.

Historical Overview

The characteristics of aerospace corporations and the current structure of the aerospace industry result from the numerous political and economic forces that have created, shaped, and reshaped it. The first of these forces,

and the one responsible in large part for creating the aerospace industry, was the Cold War between the United States and the former Soviet Union. As World War II (1939–1945) came to a close, the uneasy alliance between the Soviet Union (now, Russia) and the United States began to disintegrate. Leaders on both sides sought to achieve a military advantage by capturing advanced German technology and the scientists and engineers who developed it. This included the German rocket technology that created the V-2 missile, the first vehicle to enter the realm of space.

This competition was a precursor to the space race between the two superpowers, the United States and the Soviet Union. That competition began in earnest on October 4, 1957, when the Soviet Union launched the 184-pound *Sputnik 1* space probe, Earth's first artificial satellite, into an orbit* approximately 800 kilometers (500 miles) above Earth. This demanded a response from the U.S. Department of Defense (DoD) and intelligence and scientific communities within the United States.

To develop the systems needed to engage in this competition, the U.S. government established contracts with existing aircraft and aeronautics companies. Martin Aircraft was the manufacturer of the B-26 Marauder, a World War II bomber. Its corporate successor, Martin Marietta, developed the Titan rocket that was used first as an intercontinental ballistic missile (ICBM) during the Cold War. The rocket was later modified to boost two astronauts in Gemini* capsules into orbit during the space race. (The last of 368 Titan launches took place on October 19, 2005. The rocket was retired for a number of reasons, one of which was the problem involved in dealing with its toxic hydrazine-nitrogen tetroxide propellant.) In 1959, the U.S. government turned to Pratt & Whitney, an aircraft engine manufacturer, to develop the first liquid hydrogen-fueled engine to operate successfully in space. It was used on the Surveyor lunar landers, the Viking Mars landers, and the Voyager outer-planet flyby* missions. A derivative of this engine is used in the second stage of the Delta III satellite launch rockets.

The intelligence community was also interested in using space technology. The United States' first space-based overhead reconnaissance* program, CORONA, began flight in 1959. It, too, relied on established companies. Lockheed (now Lockheed Martin), a prominent aircraft manufacturer, developed the launch vehicle's upper stage. Eastman Kodak (now Kodak) produced special film that would function properly in space and low-Earth-orbit (LEO) environments. General Electric designed and manufactured the recovery capsule to protect exposed film as it was de-orbited and re-entered Earth's atmosphere for airborne capture and recovery.

The government's interest in, and contracts for, space systems also created new companies. TRW (now part of Northrop Grumman) resulted from efforts to build the *Atlas* missile (what is today a successful family of Atlas rockets) and the *Pioneer 1* spacecraft, the first U.S. ICBM and satellite, respectively. Currently, commercial involvement in

* **orbit** the circular or elliptical path of an object around a much larger object, governed by the gravitational field of the larger object

* **Gemini** the second series of American-piloted spacecraft, crewed by two astronauts; the Gemini missions were rehearsals of the spaceflight techniques needed to go to the Moon

* **flyby** flight path that takes the spacecraft close enough to a planet to obtain good observations; the spacecraft then continues on a path away from the planet but may make multiple passes

* **reconnaissance** a survey or preliminary exploration of a region of interest

the aerospace industry is growing, but government involvement continues to be significant.

Space Systems Overview

Each space system is composed of a collection of subsystems, often grouped into segments. Typical groupings are the launch segment; the space segment; the ground, or control, segment; and the user segment. The launch segment includes the equipment, facilities, and personnel needed to place elements of the system into space. The space segment includes the spacecraft, other equipment, and personnel that are placed into space. The ground, or control, segment includes the equipment, facilities, and personnel that control and operate the spacecraft as it performs its mission. The user segment includes the equipment, facilities, and personnel using the products of the space system to accomplish other purposes. Aerospace corporations are the source of virtually all of the equipment and facilities that these segments require. Moreover, these corporations frequently train or provide personnel to operate and maintain them.

The Launch Segment The most visible activity associated with space missions is usually the launch of the space elements of the system. Television and film coverage has often featured footage of the U.S. space shuttle with its large boosters gushing fire and smoke as it rises slowly into the sky. The launch vehicle and upper stages, along with the facilities, equipment, and team at the launch site and associated range, are part of the launch segment. The two U.S. aerospace corporations that provide the most frequently used large launch vehicles are the Boeing Company (Delta family of rockets) and Lockheed Martin Corporation (Atlas families of rockets). In 1996, Boeing and Lockheed Martin formed a joint venture, United Space Alliance (USA) to coordinate their activities on more than 30 contracts held by the two companies in connection with the National Aeronautics and Space Administration's (NASA) space shuttle program. In December 2006, the same two companies formed a second joint venture, United Launch Alliance (ULA), to combine their launch programs for the Atlas and Delta rockets. As of late 2011, Boeing and Lockheed Martin produced four rocket launchers: Atlas V, Delta II, Delta IV, and Delta IVH. With the completion of the space shuttle program in 2011, USA transfered its attention to the future NASA Constellation Program, the successor to the Space Transportation System (STS— commonly called NASA's space shuttle fleet), and International Space Station (ISS) activities. The company also works with space programs located in the U.S. Department of Energy and Department of Defense. It also hopes to expand its activities in future commercial space programs operated by private companies.

In addition to the Atlas and Delta rockets, the U.S. Federal Aviation Administration (FAA) licenses five expendable vehicle launch systems, the Falcon 1 and 9 rockets, produced by Space Exploration Technologies Corporation (better known simply as SpaceX); Pegaus XL and Taurus, built by Orbital Sciences Corporation; and Zenit 3SL, built by the Sea Launch Corporation. Sea Launch is a consortium of four companies from Norway, Russia, Ukraine, and the United States, managed by Boeing, with the goal of launching rockets from sea platforms at equatorial latitudes. The company filed for Chapter 11 bankruptcy protection in June 2009, and emerged from bankruptcy as of October 27, 2010. Three other launch vehicles built by Orbital Sciences, the Minotaur I, IV, and V, are used for government-only programs and do not require FAA licensing. The Falcon rockets are two-stage systems powered by oxygen-kerosene engines designed to deliver payloads to mid-level orbits. On July 14, 2009, a Falcon 1 rocket was used to deliver the Malaysian RazakSAT satellite to orbit. After that flight, the rocket was retired and replaced by a somewhat larger and more powerful rocket, the Falcon 1e. The first successful launch of the Falcon 9 took place on June 4, 2010. SpaceX has since received a contract from NASA for the resupply of the International Space Station. In April 2011 the company announced plans for an even more powerful rocket, the Falcon Heavy, which will have the capability of delivering payloads to low- or medium-level orbit or to more distant destinations that include the Moon and Mars. Orbital Sciences has experienced a long record of success with its Pegasus rocket, capable of placing more than 1,000 pounds in lower-Earth orbit. It recorded 34 successful flights between 1994 and 2008, the last 26 in a row. The company's Taurus rocket, launched from a "bare" pad to minimize operating costs, has been less successful. After five successful launches between 1994 and 2000, it experienced three failures out of four launches between 2001 and 2011. The last of these failures occurred on March 4, 2011, in an attempt to launch four satellites. The last two failures have been said to cost NASA in excess of $700 million in lost payloads alone. The troubled Sea Launch program is managed by Boeing Commercial Space, a 40% partner in the consortium. It manufactures the payload fairing*, performs spacecraft integration, and manages overall mission operations. S.P. Korolev Rocket and Space Corporation Energia (Russia), commonly called Energia or RKK Energiya, provides the third stage, launch vehicle integration, and mission operations. SDO Yuzhnoye/PO Yuzhmash (Ukraine) provides the first two Zenit stages, launch vehicle integration support, and mission operations. Aker Solutions (Norway) provides the launch platform and the command ship.

In addition to launch rockets produced in the United States, a number of other nations have produced similar rockets, often with the use of U.S. or Russian technology. These rockets include India's Polar Satellite Launch Vehicle and Polar Satellite Space Vehicle; Japan's H-II and H-IIa rockets, based on the U.S. Delta rocket; China's Chang Zheng series of rockets, dating to 1970 and now led by the CZ-5, or "Long March" rocket; the European Space Agency's Ariane rocket, developed in France, and its smaller cousin, the Vega rocket; and a number of rockets produced first by the Soviet Union and now by Russia, especially the famous Proton rocket. Russia's Zenit rocket is also a workhorse for the nation's space program, for the delivery of military payloads and satellites to low-Earth orbit.

* **payload fairing** structure surrounding a payload; it is designed to reduce drag

Liquid vs. Solid Fuels

Liquid fuels generally provide more energy than solid fuels and are easier to control. Liquid fuel engines can be throttled up and down during a flight. Solid fuels are easier to handle. They do not give off toxic vapors or require extreme cooling during storage and pre-launch operations.

* **interplanetary trajectories** the solar orbits followed by space-craft moving from one planet in the solar system to another

* **Cassini-Huygens mission** a robotic spacecraft mission to the planet Saturn that arrived in July 2004 and dropped its *Huygens* probe into Titan's atmosphere while the *Cassini* spacecraft studied the planet

* **orbiter** spacecraft that uses engines and/or aerobraking, and whose main function is to orbit about a planet or natural satellite

Since the mid-1990s, the industry has seen a number of newcomers, many with partially or fully reusable systems but so far without any space launches. Some of these companies have provided useful technological developments for rocket launch, but have been unable to survive commercially. One exception to that trend has been Kelly Space and Technology (KS&T), located in San Bernadino, California. KS&T has designed a new approach to rocket launch called the tow-launch system. The rocket is attached to a Boeing 747 airliner, which drags the rocket into the air to a height of about 20,000 feet. At that altitude, the tow line is released and the rocket's three engines fire, raising the rocket to an altitude of about 400,000 feet, the top of Earth's atmosphere. As of late 2011, KS&T had not yet fully implemented its tow-launch system in the release of a payload into orbit.

Most launch vehicles use liquid propellants, but some use motors with solid fuels. The large, white strap-on boosters straddling the rust-orange main fuel tank of the space shuttles are solid-fuel boosters, as are the strap-on motors used with the Atlas, Delta, and Titan. In addition, most upper stages that are used to propel systems to high orbits or even into interplanetary trajectories* are also solid-fuel systems. Thiokol Corporation, Pratt & Whitney, and others make many of these motors.

The Space Segment The space segment consists of all the hardware, software, and other elements placed into space. Examples include spacecraft that orbit Earth, such as NASA's Tracking and Data Relay Satellite (TDRS), or an interplanetary probe, such as the Cassini-Huygens* spacecraft, whose mission was to explore the planet Saturn and its moons. Spacecraft used by humans, such as the space shuttle orbiter*, and human-occupied space facilities such as the International Space Station, are also included. Even the smallest spacecraft are complex machines. They must operate with limited human interaction for long periods, in a very hostile environment, and at great distances. Designing, manufacturing, and testing these spacecraft can be very demanding and requires many specialized facilities and an experienced staff.

Some of the established leaders in this segment include Boeing Satellite Development Center, Lockheed Martin, Thales Alenia Space, and EADS Astrium. All of these corporations, like many aerospace corporations, have complex histories that involve a number of mergers and acquisitions. For example, Boeing Satellite Development Center is a business unit of Boeing Integrated Defense Systems. It was created as the result of a merger of Boeing's satellite operations with GM Hughes Electronics' Space and Communications division. The Boeing Corporation itself dates to the latter years of World War I when the company began making aircraft for the U.S. Army. It expanded its operations in satellite production in 2000 when it purchased Hughes Space and Communications, at the time, the world's largest producer of earth-orbiting satellites. Hughes Space and Communications was a company that also had a long and complex corporate history dating to its founding by the legendary aviator and entrepeneur, Howard Hughes, in 1948. Lockheed Martin was also formed as the result of a merger between two giant aerospace corporations, Lockheed Corporation and Martin Marietta, in 1995. Lockheed Martin's complex operations are divided into four major areas: aeronautics, electronic systems, information systems and global solutions, and space, each of which, in turn, consists of multiple divisions. The company's space activities are located within its Lockheed Martin Space Systems division whose work is focused on strategic and missile defense systems, surveillance and navigation systems, global communications systems, sensing and exploration systems, and human space flight. Thales Alenia Space was created in 2005 by the merger of two European aerospace companies, Alcatel-Lucent, of France, and the Italian conglomerate Finmecchanica. The company is the largest satellite manufacturer in Europe with contracts for work with the International Space Station, as well as satellite manufacture for France, Germany, Italy, Brazil, South Korea, and United Arab Emirates. EADS Astrium is a division of the European Aeronautic Defence and Space Company (EADS), created in 2000 by the merger of four aerospace companies from Britain, France, Germany, and Spain. In 2003, EADS acquired British Aerospace's Astrium division, which employs 15,000 men and women in France, Germany, the Netherlands, Spain, and the United Kingdom. Its activities were originally distributed among three divisions, Astrium Satellites (spacecraft and ground segments), Astrium Space Transportation (launchers and orbital infrastructure), and Astrium Services (development and delivery of satellite services). In 2007, the company announced the creation of a fourth division, Space Tourism, to develop programs for taking four passengers into sub-orbital flights into the atmosphere.

The list of aerospace corporations is frequently augmented by the creation of new companies focusing on one or another aspect of space flight. Most of these companies survive for only a short period of time before going out of business or being absorbed by larger existing corporations. A few new companies survive and begin to make a name for themselves in the field. One of the most successful of these newcomers has been Orbital Sciences, a company formed in 1982 by three Harvard Business School graduates, David Thompson, Bruce Ferguson, and Scott Webster. The company was created to respond to the demands by NASA for ways of developing commercial space applications. As of 2011, the company had designed, developed, and deployed about 1,000 satellites, launch vehicles, and other space-related systems, and had a backlog of $5.6 billion in long-term contracts. Their workforce included about 3,700 employees, about half of whom were scientists and engineers. The company significantly increased its outlook in 2010 with the purchase of the General Dynamics Advanced Information Systems, which itself had begun as a satellite-production company, Spectrum Astro, in 1988.

The Ground, or Control, Segment The ground, or control, segment is probably the least glamorous and least public element of any space system. Although it lacks the showmanship of a launch or the mystique of traveling through space, it is critical to mission success. This segment consists of all the hardware, software, and other elements used to command the spacecraft and to downlink*, distribute, and archive science and spacecraft systems status data. This segment serves as a combined control center and management information system for the mission. Aerospace corporations build and often operate these systems.

Most aerospace corporations include ground control systems as part of their missions. An example is the Raytheon corporation, which has been involved in the design, development, and deployment of ground control systems for more than 40 years. During that time, the company has deployed more than 110 unique stations as part of a variety of ground control systems. In 2009, Raytheon announced that it had been chosen by the U.S. Air Force to design a new ground control system for the next generation of global positioning satellites that will include new anti-jamming technologies, more advanced predictive algorithms, and more frequent clock and ephemeris updates. The system will satisfy the demands of both military and civilian needs with better anti-jamming technology and increased accuracy and reliability. As with most aerospace projects, Raytheon will be developing its new ground control system in conjunction with a number of other aerospace and related corporations, including ITT Technologies, Boeing, Braxton Technologies, Infinity Systems Engineering, SRI International, and NASA's Jet Propulsion Laboratory.

The User Segment Although all segments of a space system are necessary, the user segment is the most important. It is here that the mission of a space program is achieved. The user segment consists of all the hardware, software, and other elements required to make use of the data. A very public example of user segment equipment is the GPS receiver. Many of these units are sold to drivers, campers, hikers, boaters, and others who desire an easy and accurate means of determining their location. The user segment is, also, where science data are processed, formatted, and delivered to the scientists and other investigators for study and analysis. The U.S. Geological Survey's Earth Resources Observation Systems (EROS) Data Center near Sioux Falls, South Dakota, is a major scientific data processing, archive, and product distribution center for spacecraft, shuttle, and aerial land sciences data and imagery. Data are processed into usable formats and then made available to researchers and other users. Many aerospace corporations perform further processing and formatting of EROS data to generate information for sale.

Cross-Segment Approaches

Most aerospace corporations design, develop, and operate facilities and equipment in more than one segment. Often a specific space program will have one aerospace corporation serve as a lead or prime contractor, managing or integrating the work of many other companies. The International

Space Station provides an excellent example. Boeing is the prime contractor of a space station team that includes a number of partners. Major U.S. teammates and some of their contributions include: Lockheed Martin, providing solar arrays* and communications systems; Space Systems/Loral, providing batteries and electronics; Honeywell (formerly Allied Signal), providing gyroscopes* and other navigational gear; Honeywell, providing command and data systems as well as gimbal motors*; and United Technologies, providing pumps and control valve assemblies.

One of the more unusual competitors spanning all segments is Sierra Nevada Corporation's (SNC) Space Systems. Established in 2009 through the merger of SNC subsidiary SpaceDev Inc., its subsidiary Starsys Research, and Microsat Systems, Inc., SNC is a commercial space exploration and development company for small, low-cost, commercial space missions, space products, and affordable space services. In October 2011, it was announced that SNC's "Dream Chaser" space plane would make a high-altitude test flight during the summer of 2012. The Dream Chaser is being developed as a reusable spacecraft capable of carrying up to seven crew members and cargo to and from the ISS.

Other Industry Roles

In addition to designing, developing, and operating space systems in and across the various segments, some aerospace corporations perform more focused roles, such as providing systems engineering and other technical assistance or producing subsystems, components, and parts for systems. Many of these corporations are not as readily recognized as other members of the aerospace industry.

Systems Engineering and Technical Assistance Systems engineering and technical assistance (SETA) is a role performed by a number of aerospace corporations. As a SETA contractor, an aerospace corporation may develop, review, analyze, or assess concepts and designs for space missions, programs, and systems. Typically, SETA contractors do not provide hardware for the programs they support. Instead, they provide valuable expertise and a viewpoint independent of those manufacturing the system's components. For example, the Raytheon Information Technology and Scientific Services (RITSS) company has been the technical support contractor to the U.S. Geological Survey's EROS Data Center discussed previously. Science Applications International Corporation (SAIC) provides a variety of SETA services to NASA, the DoD, and some commercial space programs. Dynamics Research Corporation and OAO Corporation (part of Lockheed Martin) are examples of other companies that provide SETA support to NASA, the DoD, and aerospace prime contractors.

Parts, Components, or Subsystems Providers Another role for an aerospace corporation is that of parts, components, or subsystems provider. This category encompasses the greatest number of aerospace corporations. Many of these corporations provide a broad range of subsystems and

* **solar arrays** groups of solar cells or other solar power collectors arranged to capture energy from the Sun and use it to generate electrical power

* **gyroscope** a spinning disk mounted so that its axis can turn freely and maintain a constant orientation in space

* **gimbal motors** motors that direct the nozzle of a rocket engine to provide steering

components and may also manufacture complete spacecraft. Others specialize in a specific type of space hardware, software, or service. Space products from Ball Aerospace and Technologies Corporation include antennas, fuel cell systems, mirrors, pointing and tracking components (such as star trackers), and reaction/momentum wheels. Malin Space Science Systems designs, develops, and operates instruments to fly on unmanned spacecraft. Thermacore International works on heat pipes for space applications. These pipes are used to move heat from one location to another with little loss in temperature. Analytical Graphics, Inc., produces a commercial computer program, Satellite Tool Kit, which possesses extensive space mission and system analysis and modeling capabilities.

Non-aerospace Corporations

Many corporations that support aerospace programs are not commonly recognized as members of the aerospace industry. Kodak, a world-famous film and camera manufacturer, has been involved with aerospace almost since the beginning. Kodak developed the special film used in CORONA's orbiting cameras to photograph former Soviet missile sites, air and naval bases, and weapons storage facilities. Its charge-coupled device (CCD) image sensors were used on NASA's *Mars Pathfinder* spacecraft, which visited Mars in 1997 and explored the planet with its *Sojourner* rover. Today, Kodak manufactures digital cameras used from space to capture images of Earth's surface. These images are of value to scientists, farmers, and many others. International Business Machines Corporation (IBM), another well-known corporation, supports many aerospace programs. During the 1999 space shuttle mission that returned U.S. astronaut and politician John Glenn to space, twenty IBM ThinkPads (branded notebook computers) were onboard. (Since 2005, ThinkPad notebooks are manufactured and marketed by Lenova.)

Other "unsung heroes" of aerospace include insurance and finance companies that are growing in importance as the primary revenue source of aerospace corporations shifts from government to the commercial sector.

 See also **Emerging Space Businesses (Volume 1) • Getting to Space Cheaply (Volume 1) • Insurance (Volume 1) • Launch Vehicles, Expendable (Volume 1) • Launch Vehicles, Reusable (Volume 1) • Navigation from Space (Volume 1) • Reusable Launch Vehicles (Volume 4) • Satellite Industry (Volume 1)**

Resources

Books and Articles

Bauer, E. E. *Boeing: The First Century & Beyond*, 2nd ed. Issaquah, WA: TABA Publishing, 2006.

Brennan, Louis, and Alessandra Vecchi. *The Business of Space: The Next Frontier of International Competition*. Houndmills, UK: Palgrave Macmillan, 2011.

Cliff, Roger, Chad J. R. Ohlandt, and David Yang. *Ready for Takeoff: China's Advancing Aerospace Industry.* Santa Monica, CA: Rand Corporation, 2011.

Dubbs, Chris, and Emeline Paat-Dahlstrom. *Realizing Tomorrow: The Path to Private Spaceflight.* Lincoln: University of Nebraska Press, 2011.

Federal Aviation Administration. *2011 U.S. Commercial Space Transportation Developments and Concepts: Vehicles, Technologies, and Spaceports.* Washington, DC: Federal Aviation Administration, 2011.

Industrial College of the Armed Forces. *Space 2009.* Fort McNair, Washington, DC: National Defense University, 2009.

Isakowitz, Steven J., Joseph P. Hopkins, Jr., and Joshua B. Hopkins. *International Reference Guide to Space Launch Systems,* 4th ed. Reston, VA: American Institute of Aeronautics and Astronautics, 2004.

Morris, Langdon, and Kenneth J. Cox, eds. *Space Commerce: The Inside Story by the People Who Are Making It Happen.* Washington, DC: Aerospace Technology Working Group, 2010.

Websites

Defense Industry Projects a Good 2011 but 2012 and Out Look Worse. <http://www.defenseprocurementnews.com/2011/08/22/defense-industry-projects-a-good-2011-but-2012-and-out-look-worse/> (accessed October 3, 2011).

Launch Vehicle. Encyclopedia Britannica. <http://www.britannica.com/EBchecked/topic/332323/launch-vehicle/272735/Russia-and-Ukraine> (accessed October 3, 2011).

Reusable Launch Vehicle Technology. <http://www.kellyspace.com/launchvehicle2/> (accessed October 3, 2011).

Space News. <http://www.spacenews.com/> (accessed October 3, 2011).

Aging Studies

Soon after entry into outer space, outside the envelope of Earth's atmosphere, physical changes occur within the human body, and these changes become more severe and diverse as flight duration increases. When in space (under the influence of microgravity), humans experience early signs of a decrease in blood volume (that is, less liquid as it is distributed more evenly through the body) and red cell mass (thus, less oxygen being delivered throughout the body), aerobic capacity, endurance, strength, and muscle mass. Moreover, there is a reduction of bone density in the lower limbs, hips, and spine, and in the absorption of calcium through the gut. Visual-spatial orientation and eye-hand coordination are also affected.

When humans return to Earth's gravity, this reduction in physical fitness manifests itself through the body's inability to maintain the blood pressure control necessary to prevent fainting. This inability occurs because the heart and blood vessels are less responsive. Balance, gait, and motor coordination are also severely affected. Similar but less-intense symptoms occur during and after complete bed rest.

Bed Rest Studies

Researchers study bed rest to understand the mechanisms that bring about these symptoms, and to develop preventive treatments. Both astronauts and volunteers for these bed rest studies recover, with the speed of recovery being proportional to the duration of the flight or the bed rest. Research into the mechanisms that contribute to these symptoms has pointed to the variety of ways humans use Earth's gravity to promote stimuli the body needs to maintain normal physiology.

While orbiting above Earth or another celestial body (where weightlessness makes gravity seem small) or while at large distances away from a massive celestial body (where the influence of gravity is negligible), the load normally felt on Earth from human mass is much reduced. Exercise, such as walking, is ineffective, because a person is not working against the force of gravity. Signals to the parts of our nervous system that control blood pressure from changing position (such as standing or lying down) are also absent, and humans can no longer sense what is up and what is down.

Bed rest studies have been conducted to determine the minimum daily exposure to gravity's stimuli needed to maintain normal physiology. This research indicated that a change in posture from lying down to upright—a total of two to four hours of being upright so gravity pulled maximally

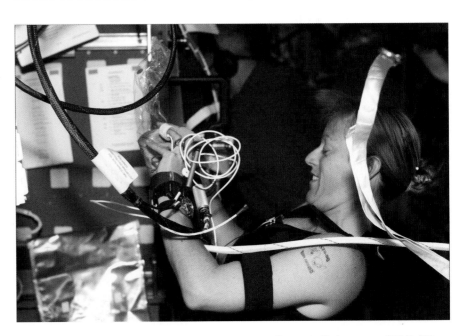

Astronaut Kathryn P. Hire undergoes a sleep study experiment in the Neurolab (STS-90, April 17 to May 3, 1998) on the space shuttle *Columbia.* Bed rest studies are used to examine the physical changes that occur during human spaceflight. *NASA.*

in the head-to-toe direction—at least eight to sixteen times a day could prevent the decline in blood pressure control, aerobic capacity, blood volume, and muscle strength. As little as thirty minutes a day of walking at a pace of three miles per hour prevented the increased loss of calcium produced by bed rest. These results suggested that intermittent exposure to gravity in space, as provided by a centrifuge*, may be an effective way to keep astronauts healthy on long trips.

Are the Effects of Aging Irreversible?

The similarity of the set of symptoms resulting from going into space to those associated with aging is striking. In the elderly, these symptoms have been assumed to be part of the normal course of aging and therefore irreversible. Space and bed rest research argues against this assumption. Research indicates that the symptoms of aging are due to an increasingly sedentary lifestyle* rather than aging and are, thus, also reversible in the elderly. In the mid-1990s, Maria Fiatarone and her coworkers reported that weight training and nutritional supplements reversed muscle and bone atrophy* of aging in people aged seventy-three to ninety-eight years. American astronaut and politician John Glenn's second trip to space, in 1998, at the age of seventy-seven years on shuttle mission STS-95 helped increase understanding of the effects of a healthy lifestyle on aging.

Because of the similarities between the effects of space travel and aging, the National Aeronautics and Space Administration (NASA) and the National Institute on Aging (NIA) are jointly studying the two, especially with regards to their effects on bones and muscles, along with the quality and quantity of sleep. For instance, their studies on osteoporosis (the weakening of bones) both in space and on Earth may one day help to eliminate the problem. However, it still remains to be seen if space research will also help scientists understand the more fundamental mechanisms of aging. Preliminary research suggests cell cycle and cell death may be affected. But, it will only be through the conducting of experiments long enough to explore life span and chromosomal and genetic mechanisms that these questions will be answered.

 See also **Careers in Space Medicine (Volume 1)** • **Glenn, John (Volume 3)**

Resources

Books and Articles

Barratt, Michael R., and Sam Pool, eds. *Principles of Clinical Medicine for Space Flight.* New York: Springer, 2008.

Fiatarone, Maria Antoinette, et al. "Exercise Training and Nutritional Supplementation for Physical Frailty in Very Elderly People." *New England Journal of Medicine* 330 (1994):1,769–1,775.

* **centrifuge** a device that uses centrifugal force caused by spinning to simulate gravity

* **sedentary lifestyle** a lifestyle characterized by little movement or exercise

* **atrophy** condition that involves withering, shrinking, or wasting away

* **microgravity** the condition experienced in freefall as a spacecraft orbits Earth or another body; commonly called weightlessness; only very small forces are perceived in freefall, on the order of one-millionth the force of gravity on Earth's surface

* **crystallography** the study of the internal structure of crystals

McPhee, Jancy C., and John B. Charles, eds. *Human Health and Performance Risks of Space Exploration Missions: Evidence Reviewed by the NASA Human Research Program.* Houston, TX: National Aeronautics and Space Administration, 2009.

Vernikos, Joan. "Human Physiology in Space." *Bioessays* 18 (1996): 1,029–1,037.

Websites

Aging and Space Flight—In Orbit and Over the Hill?. National Aeronautics and Space Administration. <http://weboflife.nasa.gov/currentResearch/currentResearchGeneralArchives/agingSpaceflight.htm> (accessed October 4, 2011).

Synergistic Research. National Aeronautics and Space Administration. <http://spaceflight.nasa.gov/shuttle/archives/sts-95/aging.html> (accessed October 4, 2011).

AIDS Research

Since 1985, the National Aeronautics and Space Administration (NASA) has supported fundamental studies on the various factors that affect protein crystal growth processes. More than thirty principal investigators from universities throughout the United States have investigated questions such as why crystals stop growing, what factors cause defect formations in growing crystals, and what influence parameters such as protein purity, temperature, pH (a solution's degree of acid or base properties), protein concentration, and fluid flows exert around growing crystals.

The majority of these studies were conducted in Earth-based laboratories, with a limited number of experiments performed on U.S. space shuttle flights. The purpose of the space experiments is to determine the effect that a microgravity* environment has on the ultimate size and quality of protein crystals. This research was propelled by the need to improve success rates in producing high-quality crystals to be used for x-ray crystallography* structure determinations. X-ray crystallography involves exposing a protein crystal to powerful x-ray radiation. When this occurs, the crystal produces a pattern of diffracted spots that can be used to mathematically determine (using computers) the structure of the protein (i.e., the positions of all the atoms that comprise the protein molecule).

Structure-Based Drug Design

The three-dimensional structure of a protein is useful because it helps scientists understand the protein's function in biological systems. In addition, most known diseases are based on proteins that are not working properly within the human body or on foreign proteins that enter the body as part of

◀

* **quiescent** inactive

harmful bacteria, viruses, or other pathogens. The three-dimensional structure of these disease-related proteins can aid scientists in designing new pharmaceutical agents (drugs) that specifically interact with the protein, thereby alleviating or lessening the harmful effects of the associated diseases.

This method of designing new and more effective pharmaceuticals, known as structure-based drug design, was used to develop many of the new-generation AIDS (acquired immune deficiency syndrome) drugs. These drugs were developed using Earth-grown crystals. There are, however, a number of other protein targets in HIV (human immunodeficiency virus), as well as in most other pathogens, that have yet to be crystallized. Attempts to grow crystals large enough and of sufficient quality are often unsuccessful, thereby preventing the use of structure-based drug design.

On Earth, when crystals begin to grow, lighter molecules float upward in the protein solution while heavier molecules are pulled down by gravity's forces (a process known as buoyancy-induced fluid flow). This flow of solution causes the protein to be swept to the surface of the crystal where it must align in a near perfect arrangement with other protein molecules. It is believed that the rapid flow of solution causes the protein molecules to become trapped in misalignments, thereby affecting the quality of the crystal and, eventually, even terminating crystal growth.

Crystal Growth in Microgravity

In a microgravity environment, these harmful flows are nonexistent because gravity's influence is minimal. Thus, the movement of protein molecules in microgravity is much slower, caused only by a process known as random diffusion (the inherent vibration of individual molecules). It is thought by scientists that the lack of buoyancy-induced fluid flows (as occurs on Earth) creates a more quiescent* environment for growing crystals. The protein molecules have sufficient time to become more perfectly ordered in the crystal before being trapped by additional incoming molecules.

The scientific community is divided about the role that microgravity can play in improving the size and quality of protein crystals. In addition, the excessive cost of performing experiments in space has caused scientists to question the value of these experiments. Proponents of the space protein crystal growth program are optimistic that the longer growth times that will be available on the International Space Station will significantly improve microgravity success rates for producing crystals of significantly higher quality. In October 2007, a Malaysian astronaut, Sheikh Muszaphar Shukor, also an orthopedic surgeon, was sent to the Space Station (onboard the Russian Soyuz *TMA-11* spacecraft) for the purpose of conducting AIDS protein experiments, along with other medical experiments regarding liver cancer and leukemia cells and microbes. With the expansion of the International Space Station to six crew members (from its previous capacity of three members), it is hoped that the additional manpower will increase space-based scientific and medical research, including AIDS research, onboard.

Since 2009, research on protein crystal growth has shifted to include venues other than the International Space Station. Some of that research has taken place on the Japanese spacecraft *Kibo* (Hope, in Japanese) in a joint program with Russia, which provides the rockets that have placed *Kibo* in orbit. The Malaysian National Space Agency has also been involved in the program, providing materials and experience gained in research on the International Space Station. Officials at the Japanese Aerospace Exploration Agency (JAXA) have laid out an extended research program for *Kibo* with the goal of finding applications for protein crystal growth experiments in the development of breakthrough medicines.

 See also **Careers in Space Medicine (Volume 1) • Crystal Growth (Volume 3) • Microgravity (Volume 2) • Zero Gravity (Volume 3)**

Resources

Books and Articles

National Research Council. *Future Biotechnology Research on the International Space Station.* Washington, DC, 2000.

Pletser, Robert Bosch, Lothar Potthast, Peter Lautenschlager, and Ronald Kassel. "The Protein Crystallisation Diagnostics Facility (PCDF) on Board ESA Columbus Laboratory." *Microgravity Science and Technology* 21 (2009): 269-277.

Sadhal, S. S., ed. *Transport Phenomena in Microgravity.* New York: New York Academy of Sciences, 2004.

Task Group for the Evaluation of NASA's Biotechnology Facility for the International Space Station, et al. *Future Biotechnology Research on the International Space Station.* Washington, DC: National Academy Press, 2000.

Takahashi, S., et al. "High-quality Crystals of Human Haemato-poietic Prostaglandin D Synthase with Novel Inhibitors." *Acta Crystallographica. Section F, Structural Biology and Crystallization Communications.* 66 (2010): 846-50.

Websites

Commercial Protein Crystal Growth-High Density (CPCG-H). Pro-ach Models. <http://www.spaceref.com/iss/payloads/cpcg.html> (accessed September 27, 2011).

The Era of International Space Station Utilization. National Aero-nautics and Space Administration, et al. <http://www.nasa.gov/pdf/506512main_Summary_PSF_ISS_Utilization_2010.pdf> (accessed September 27, 2011).

High Quality Protein Crystallization Research (HQPC). Japan Aerospace Exploration Agency. <http://kibo.jaxa.jp/en/experiment/theme/first/hqpc/index.html> (accessed September 27, 2011).

Uhran, Mark L. *Positioning the International Space Station for the Utili-zation Era.* National Aeronautics and Space Administration. <http://www.nasa.gov/pdf/508618main_Uhran_2011_AIAA_Paper.pdf> (accessed September 27, 2011).

Buzz Aldrin, along with Neil Armstrong, became the first humans to land on the Moon. *Photograph by Kipp Teague. NASA.*

Aldrin, Buzz

American Astronaut and Engineer
1930–

On July 20, 1969, Edwin "Buzz" Aldrin and his fellow astronaut Neil Armstrong became the first humans to land on another world: Earth's Moon. This achievement is arguably the technological high-water mark of the twentieth century.

Aldrin's passion for exploration and quest for excellence and achieve-ment began early in his life. Born on January 20, 1930, in Montclair, New Jersey, Aldrin received a bachelor's of science degree from the U.S. Military Academy at West Point in 1951, graduating third in his class. After entering the U.S. Air Force, Aldrin earned his pilot wings in 1952.

As an F-86 fighter pilot in the Korean War, Aldrin flew sixty-six com-bat missions. He later attended the Massachusetts Institute of Technology (MIT), where he wrote a thesis titled "Guidance for Manned Orbital Rendezvous." After his doctoral studies, Aldrin was assigned to the Air Force Systems Command in Los Angeles.

Aldrin's interest in space exploration led him to apply for a National Aeronautics and Space Administration (NASA) tour of duty as an astro-naut. Aldrin was selected as an astronaut in October 1963—actually

* **Agena** a multipurpose rocket designed to perform ascent, precision orbit injection, and missions from low Earth orbit (LEO) to interplanetary space; also served as a docking target for the Gemini spacecraft

within the third group of astronauts for the United States. The research expertise in the new field of space rendezvous he had acquired during his studies at MIT were applied in the U.S. Gemini and Apollo programs. In fact, the techniques he pioneered in rendezvous and docking helped to make these NASA programs successful. During those early days of manned space exploration, Aldrin was often called "Dr. Rendezvous" by the other astronauts because of his advanced degree that he applied in the critical development of those missions.

On November 11, 1966, Aldrin, with James Lovell, flew into space aboard the two-seater *Gemini 12* spacecraft. On that mission the Gemini astronauts rendezvoused and docked with an Agena* target stage. During the link-up, Aldrin carried out a then-record 5.5-hour space walk. Using hand-holds and foot restraints while carefully pacing himself, Aldrin achieved a pioneering extravehicular feat in light of the many difficulties experienced by earlier space walkers. In fact, his ability to adapt to extravehicular activities (EVAs) in space proved to NASA that astronauts could work effectively outside of a spacecraft in the challenging volume of outer space.

Aldrin's unique skills in developing rendezvous techniques were tested again in July 1969. Aldrin and his fellow *Apollo 11* astronauts, Neil Armstrong and Mike Collins, were the first crew to attempt a human landing on the Moon. Once in lunar orbit, Armstrong and Aldrin piloted a landing craft, the *Eagle,* to a safe touchdown on the Moon's Sea of Tranquility. After joining Armstrong on the lunar surface, Aldrin described the scene as "magnificent desolation." They explored the landing area for two hours, setting up science gear and gathering rocks and soil samples. The two astronauts then rejoined Collins in the command module *Columbia* for the voyage back to Earth.

Aldrin returned to active military duty in 1971 and was assigned to Edwards Air Force Base in California as commander of the Test Pilots School. He retired from the U.S. Air Force as a colonel in 1972. Aldrin is an active spokesperson for a stronger and greatly expanded space program. He still advances new ideas for low-cost space transportation and promotes public space travel. Aldrin continues to spark new ideas for accessing the inner solar system. One of his concepts is the creation of a reusable cycling spaceship transportation system linking Earth and Mars for the routine movement of people and cargo.

Aldrin has written several books, sharing with readers his experiences in space. Among his books are the autobiography *Return to Earth* (1973) and the historical documentary *Men from Earth* (1991). As a co-writer, Aldrin has also authored science fiction novels that depict the evolution of space exploration in the far future. For instance, Aldrin co-authored, with author John Barnes, the science fiction books *Encounter With Tiber* (1997) and *The Return* (2001). For children, Aldrin wrote the illustrated book *Reaching for the Moon* in 2008. Aldrin has written an autobiography called *Magnificent Desolation: The Long Journey Home from the Moon,* which was released in 2009. Also released that same year is another children's book written by Aldrin entitled *Look to the Stars.*

Three U.S. patents are also credited to Aldrin in the areas of reusable rockets, modular space stations, and multi-crew space modules. In addition, he founded Starcraft Boosters, Inc., which is a rocket design company, and ShareSpace Foundation, a non-profit organization that promotes space education and exploration for all people. His company StarBuzz LLC, formed in 2008, promotes the "Rocket Hero" brand.

Aldrin was awarded a honorary doctorate of science from Gustavus Adolphus College in 1967. Numerous awards have been presented to him over the years—including the General James E. Hill Lifetime Space Achievement Award by Space Foundation in 2006. Aldrin is also a member of the Astronaut Hall of Fame. In the movies, the space character "Buzz Lightyear," from the movie *Toy Story* was named after Buzz Aldrin.

Aldrin has continued to stay active. In 2010, at the age of 80, Aldrin performed with his professional female dance partner Ashley Costa on the TV show "Dancing with the Stars." He was voted off the show in the second week of competition. Aldrin, along with his two crewmates from the *Apollo 11* mission, and John Glenn, first American to orbit the Earth, were awarded the Congressional Gold Medal in the rotunda of the U.S. Capitol on November 16, 2011. The Gold Medal is in recognition of their contributions to society through their accomplishments in space and on Earth, and is the highest civilian honor awarded by the U.S. Congress.

 See also **Apollo (Volume 3)** • **Armstrong, Neil (Volume 3)** • **Gemini (Volume 3)** • **NASA (Volume 3)** • **Space Walks (Volume 3)**

Resources

Books and Articles

Aldrin, Buzz. *Magnificent Desolation: The Long Journey Home from the Moon.* New York: Harmony Books, 2009.

Aldrin, Buzz. *Reaching for the Moon.* New York: HarperCollins, 2008.

Raum, Elizabeth. *Buzz Aldrin.* Chicago: Heinemann Library, 2006.

Websites

Buzz Aldrin. <http://www.buzzaldrin.com> (accessed October 23, 2011).

"Buzz Aldrin's Moon Landing Memories." BBC News. <http://news.bbc.co.uk/2/hi/7910275.stm> (accessed October 23, 2011).

Buzz Aldrin: Biographical Data. National Aeronautics and Space Administration. <http://www.jsc.nasa.gov/Bios/htmlbios/aldrin-b.html> (accessed October 23, 2011).

Speculative artwork, such as this depiction of spacecraft assembly in Mars orbit, can make the theoretical seem possible. *Illustration by Bonestell Space Art.* © *Bonestell Space Art.*

Artwork

Astronautics owes much of its existence to the arts. On the one hand, literary works by authors such as Jules Verne (1828–1905) were directly responsible for inspiring the founders of modern spaceflight; on the other hand, artists such as Chesley Bonestell (1888–1986) made spaceflight seem possible. When Bonestell's space art was first published in the 1940s and early 1950s, spaceflight to most people still belonged in the realm of comic books and pulp fiction. Bonestell—working with such great space scientists as Wernher von Braun—depicted space travel with such vivid reality that it suddenly no longer seemed so fantastic. Emerging as it did when the United States was first taking an interest in astronautics, these paintings went a long way toward encouraging both public and government support.

Space Art Comes of Age

Since Bonestell's time, there have been many other artists who have specialized in space art, though even in the early twenty-first century there are probably fewer than a hundred who work at it full-time. Some have been able to develop specialties within the field. Robert McCall (1919–2010) and Pat Rawlings, for example, devoted themselves to rendering spacecraft, while others, such as Michael Carroll and Ron Miller (1947–), concentrate on astronomical scenes, including views of the surface of Mars or the moons of Saturn. Some artists are interested in how we are going to explore space, while others are more interested in what we are going to find once we get there.

Although most space artists have a background in art, either as gallery artists or commercial artists, there are some notable exceptions. William (Bill) K. Hartmann (1939–), for example, is a professional astronomer who happens to also be an excellent painter. He is able to combine his artistic talent with his expert knowledge of astronomy. Only a very few space artists have ever flown in space. Alexei Leonov (1934–), a Russian cosmonaut who was also the first man to walk in space, is a very fine painter who took drawing supplies with him into orbit. Vladimir Dzhanibekov (1942–) is another cosmonaut who has translated his experiences in space onto canvas. Of the American astronauts, only one has had a serious interest in art. Alan Bean (1932–), who walked on the Moon in 1969, has devoted himself since retiring from the astronaut corps to painting and has become extraordinarily successful depicting scenes from his experiences in the Apollo program.

Beyond Aesthetics

The artists who specialize in astronomical scenes perform a very valuable service. In one sense, they are like the artists who re-create dinosaurs. By taking astronomical information and combining this with their knowledge of geology, meteorology*, and other sciences, as well as their expertise in light, shadow, perspective, and color, they can create a realistic landscape of some other world. Most often these are places that have never been visited by human beings or unmanned probes. In other instances, an orbiter* or a spacecraft on a flyby* mission may have already taken photographs of a moon or planet. In this case, the artist can use these photos to create an impression of what it might look like to stand on the surface. Since it can be very difficult to interpret orbital photos—which look down on their subjects—paintings like these are very useful in helping to understand what the features actually look like.

Into the Twenty-First Century

Until recently, most space artists worked in the same traditional materials, such as oil paint, acrylics, and watercolors, as other artists and illustrators. Many space artists now also use the computer to enhance their traditionally rendered work or to generate artwork from scratch. Don Davis (1952–), one of the twenty-first century's best astronomical artists, no longer works with brushes at all, choosing instead to work exclusively on a computer. There are advantages, both technically and aesthetically, to both methods but it is very unlikely that the computer will ever entirely replace traditional tools. It is the wise artist, however, who is at least familiar with computer techniques.

An International Genre

There are space artists, both professional and amateur and both women and men, working in almost every nation. Indeed, one of the first great artists to specialize in the field, and who helped create it, was a French artist named Lucien Rudaux (1874–1947), who created beautiful space paintings in the 1920s and 1930s. Rudaux set a standard not exceeded

▲

Saturn's rings, as seen from the upper reaches of the planet's atmosphere. *Illustration by Bonestell Space Art. © Bonestell Space Art.*

* **meteorology** the study of atmospheric phenomena or weather

* **orbiter** spacecraft that uses engines and/or aerobraking, and whose main function is to orbit about a planet or natural satellite

* **flyby** flight path that takes the spacecraft close enough to a planet to obtain good observations; the spacecraft then continues on a path away from the planet but may make multiple passes

until Bonestell published his first space art in the 1940s. Ludek Pešek (1919–1999), a Czechoslovakian expatriate who lived in Switzerland, was probably the best and most influential space artist to follow Bonestell. Pešek illustrated a dozen books with paintings of the planets that looked so natural and realistic it seemed as though they must have been done on location. David A. Hardy (1936–) of Great Britain is as adept at depicting spacecraft as he is landscapes of other worlds.

Notable women artists include Pamela Lee, who is highly regarded for her meticulously rendered depictions of astronauts at work, and MariLynn Flynn, who creates planetary landscapes in the tradition of Bonestell and Pešek. The membership roster of the International Association of Astronomical Artists, an organization of space artists, includes people from Germany, Armenia, Sweden, Japan, Russia, Canada, Belgium, and many other countries, all united by their mutual interest in space travel, astronomy, and art.

 See also **Bonestell, Chesley (Volume 4) • Literature (Volume 1) • Mccall, Robert (Volume 1) • Rawlings, Pat (Volume 4) • Verne, Jules (Volume 1) • von Braun, Wernher (Volume 3)**

Resources

Books and Articles

Carroll, Michael W. *Space Art: How to Draw and Paint Planets, Moons, and Landscapes of Alien Worlds.* New York: Watson-Guptill, 2007.

DiFate, Vincent. *Infinite Worlds: The Fantastic Visions of Science Fiction Art.* New York: Wonderland, 1997.

Hardy, David A. *Visions of Space: Artists Journey through the Cosmos.* Limpsfield, UK: Paper Tiger, 1989.

Launius, Roger D., and Bertram Ulrich. *NASA and the Exploration of Space.* New York: Stewart, Tabori & Chang, 1998.

Miller, Ron, and Frederick C. Durant III. *The Art of Chesley Bonestell.* London: Paper Tiger, 2001.

Villard, Ray, and Lynette R. Cook. *Infinite Worlds: An Illustrated Voyage to Planets beyond Our Sun.* Berkeley: University of California Press, 2005.

Websites

A. Leonov, A. Sakharov. (In Russian.) <http://scifiart.narod.ru/> (accessed October 5, 2011).

Alan Bean Gallery. <http://www.alanbeangallery.com/> (accessed October 5, 2011).

The Art of Chesley Bonestell. Bonestell LLC. <http://www.bonestell.org/> (accessed October 5, 2011).

Bill Hartmann. <http://www.psi.edu/hartmann/index.html> (accessed October 5, 2011).

Black Cat Studios: Ron Miller. <http://www.black-cat-studios.com/> (accessed October 5, 2011).

Don Davis: Space Artist and Animator. <http://www.donaldedavis.com/> (accessed October 5, 2011).

International Association of Astronomical Artists. <http://iaaa.org/> (accessed October 5, 2011).

Michael Carroll's Space Art and Writing. <http://stock-space-images.com/> (accessed October 5, 2011).

Pat Rawlings. <http://www.patrawlings.com/> (accessed October 5, 2011).

Robert McCall Studios. <http://www.mccallstudios.com/> (accessed October 5, 2011).

Augustine, Norman
Space Industry Leader
1935–

Norman R. Augustine was chairman and chief executive officer (CEO) of Lockheed Martin Corporation prior to his retirement in 1997. Augustine was undersecretary of the army in the administration of President Gerald R. Ford in 1975, having previously served as assistant secretary of the army for President Richard M. Nixon in 1973. Augustine held a variety of engineering assignments during his career. In 1958, following his graduation from Princeton University with both bachelor's and master's degrees in aeronautical engineering, Augustine joined Douglas Aircraft Company as program manager and chief engineer. In 1965, he was appointed assistant director for defense research and engineering in the Office of Secretary of Defense and then was named vice president for advanced programs and marketing for LTV Missiles and Space Company.

After joining Martin Marietta Corporation in 1977, Augustine was promoted to CEO and then chairman in 1987 and 1988, respectively. Following the formation of Lockheed Martin from the 1995 merger of Martin Marietta and Lockheed Corporation, he initially served as Lockheed Martin's president before becoming CEO and chairman in 1996.

In 1990, Augustine played a major role in defining the issues facing the U.S. space industry as head of President George H. W. Bush's Space Task Force. The group's report called for substantial increases in U.S. space spending, as well as setting new national goals in space exploration. The administration responded to the report in part by announcing a series of advanced space goals. However, funding for the projects

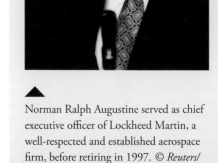

Norman Ralph Augustine served as chief executive officer of Lockheed Martin, a well-respected and established aerospace firm, before retiring in 1997. © *Reuters/ Mike Theiler/Archive Photos/Getty Images.*

was not supported by Congress, and the initiatives were abandoned. Augustine would later chair President Obama's 2009 Review of United States Human Space Flight Plans Committee.

Augustine retired from Lockheed Martin in August 1997. He then became a professor at Princeton University, serving in this capacity until July 1999.

Augustine has written several books, including *Augustine's Laws* (1990), a humorous chronicle of his experiences in defense contracting. He received the Distinguished Service Medal of the Department of Defense five times, the Defense Meritorious Service Medal, the Army Distinguished Service Medal, the Air Force Exceptional Service medal, and at least 23 honorary degrees.

In addition, Augustine received the National Space Club Goddard Award (1991), the NASA Distinguished Public Service Medal (1997), the Space Foundation General James E. Hill Lifetime Space Achievement Award (2002), the National Aeronautic Association Wright Brothers Memorial Trophy (2008), and the National Science Foundation Vannevar Bush Award (2008).

 See also **Launch Vehicles, Expendable (Volume 1) • Market Share (Volume 1)**

Resources

Books and Articles

Augustine, Norman R. *Augustine's Laws.* 6th ed. Reston, VA: American Institute of Aeronautics and Astronautics, 1997.

Augustine, Norman R. *Augustine's Travels: A World-Class Leader Looks at Life, Business, and What It Takes to Succeed at Both.* New York: AMACOM, 1998.

Websites

Crock, Stan. "CEO Chuckles." *Business Week.* <http://www.businessweek.com/1997/52/b3559124.htm> (accessed October 13, 2011).

Norman Augustine to Join the Faculty of Princeton University's School of Engineering and Applied Science. Princeton University. <http://www.princeton.edu/pr/news/97/q2/0418augu.html> (accessed October 13, 2011).

Norman R. Augustine to Receive 2006 Public Welfare Medal, Academy's Highest Honor. EurekAlert.com. <http://www.eurekalert.org/pub_releases/2006-01/tna-nra011106.php> (accessed October 13, 2011).

B

Barriers to Space Commerce

Space commerce exists in the early twenty-first century as over a $100-plus billion industry. It consists primarily of firms providing commercial telecommunication and remote sensing* services using satellites, as well as the manufacture and launch of those satellites. Space commerce also includes many organizations that provide products and services (including satellites, satellite services, ground-support services, launch, operations, and research) to government agencies in support of national civil and military space programs. Finally, a small number of firms provide other space services on a commercial basis, including space station access, on-orbit experimentation using a commercial module carried on rockets, and the launch of ashes for "burial" in space. Several small private organizations are also emerging in the area of space tourism, offering suborbital space flights beginning after 2013.

* **remote sensing** the act of observing from orbit what may be seen or sensed below on Earth or another planetary body

Efforts by space advocates, aerospace firms, and government agencies to further expand space commerce generally focus on extending the scope of commercial space activities beyond today's space telecommunications by fostering new space industries. Ongoing ventures propose expanding space services into new realms: high bandwidth Internet connectivity, on-orbit research and manufacturing, entertainment, education, power, and even routine space tourism. It is often asserted that the development of these industries is hindered by economic and policy barriers, and that these barriers can be overcome with appropriate government policy or industry initiatives.

Economic Barriers to Entry

There are three major economic barriers to the growth of space commerce: the cost of entry, the risk associated with space activities, and the cost of transportation. These factors are closely interrelated.

The Cost of Entry Space is an expensive business. The cost of manufacturing and launching a routine telecommunications satellite exceeds $150 million. The cost to establish a new capability, such as a reusable launch vehicle (RLV) or an on-orbit manufacturing facility, is likely to be in the multi-billion dollar range. The high level of capital required to establish a new capability is needed to build facilities, develop and test hardware and software, staff up with a specialized engineering team, and ultimately get to orbit. This high level of required start-up capital creates a barrier to entry into the space industry, especially for small and/or start-up firms.

After key barriers to space commerce have been overcome, profit can be made through the development of new spacecraft, such as this theoretical depiction of a lunar freighter, which would be used in transport. *NASA*.

▶

* **orbit** the circular or elliptical path of an object around a much larger object, governed by the gravitational field of the larger object

High Risk Factors The risks associated with space activities also increase the difficulty of entering the business. Risks arise from both technical factors and market factors. Technical risks exist because space systems are complex, often requiring new technology, and because space activities occur in a hazardous, challenging, and distant environment where maintenance and repair are expensive and may not be possible.

Market risks arise because in many cases the services being offered are new and it is difficult to predict what will be the customer's response. In addition, complex systems and new technology make managing costs a challenge, which can negatively affect prices. Finally, for some space services, cheaper terrestrial alternatives may be developed. These risks are exacerbated by the timing and schedule associated with space projects. Major system expenditures have to be made years prior to the beginning of operations and, as a result, financing costs are high and the time frame for achieving a return on investment is fairly long.

The Cost of Transportation Many in the industry characterize the high cost of transportation—typically expressed as the price per pound or kilogram to orbit—as the primary economic barrier, based on the premise that significantly reduced transportation costs to orbit* would make new space business activities financially feasible. In theory, this would then lead to increased launch rates, which would further reduce launch costs, and this cycle would help reduce costs of entry. This basic logic—reduce the cost of getting to space and space commerce will grow rapidly—underlies many government and industry efforts to foster space commerce. The development of reusable commercial launch vehicles, for example, is generally supported by the contention that reusing vehicles (as opposed to using a vehicle only once, as is the case with today's commercial rockets) will ultimately provide lower costs to orbit. Reusable launch vehicles* will, however, be expensive to develop. The costs and benefits of reusable launch vehicles will be a major issue for space commerce in the coming years.

Government Policy

Government policies affect space commerce and, in the minds of many industry observers, create the greatest barriers. Government barriers to commercial space come in two varieties: areas where government regulation and oversight are perceived as restrictive or inappropriately competitive (i.e., the government should do less in order to foster space commerce) and areas where government policies and actions are perceived as insufficiently supportive (i.e., the government should do more in order to foster space commerce).

Export/import restrictions, safety and licensing regulations, and launch range use policies are examples of areas that have been criticized as too restrictive. This has led to some reforms. For example, in 1984 a single licensing entity for commercial launch services was created in the U.S. Department of Transportation, so that commercial launch service providers were relieved of the requirement to interact with more than a dozen government agencies in order get permission to launch. However, other policy barriers still exist. In the United States, for instance, export/import controls aimed at limiting the transfer of valuable or sensitive technology to other countries affect U.S. commercial satellite and launch firms competing in international markets.

Sometimes government agencies, in their conduct of space activities, are viewed as competing with industry. These concerns typically arise from government operation of systems or programs for which there is a commercial demand. For example, the space shuttle fleet launched commercial satellites in the early 1980s, but stopped this activity after the space shuttle *Challenger* was destroyed in 1986. In addition, the cost of taking a satellite into space for corporate, military, and governmental customers with a space shuttle was much more costly than with the use of expendable rockets. Concerns may also emerge regarding systems or programs operated directly by government agencies, when they could be operated by industry. From the mid-1990s to the early 2010s, the daily operation and management of the space shuttle were contracted out to the industry consortium United Space Alliance (USA), which was a joint effort of The Boeing Company and Lockheed-Martin. From September 1996 to September 2006, USA ran the space shuttle program under the Space Flight Operations Contract (SFOC). From 1996 to the end of the shuttle program on August 31, 2011, USA operated the program under the Space Program Operations Contract (SPOC). In September

* **reusable launch vehicles** launch vehicles, such as the space shuttle, designed to be recovered and reused many times

Secondary Industries Enabled by Space

Many secondary industries are enabled by space assets and capabilities, such as television broadcasting, weather forecasting, tracking and navigation using satellites, and many types of voice and data communication.

2011, the company began a two-year extension on the SPOC contract to cover shuttle transition and retirement after the Space Transportation System (STS) program (commonly called the space shuttle program) was terminated.

Government policies generally express the intent to support commercial space. However, space advocates often criticize the implementation of this intent as inadequate. They seek government policies that will support commercial space, such as the government procuring commercial launch services rather than conducting government launches (an area in which the United States has made significant progress), undertaking technology development programs to reduce the risks to industry associated with advanced space concepts, serving as an anchor customer for new ventures, and providing loan guarantees, tax credits, and other financial incentives to space firms.

The impact of government policy and activities on space commerce should be viewed in a balanced way. All spacefaring nations have implemented some level of supportive government policy. While efforts to eliminate barriers to space commerce result in media attention and high-visibility policy discussion, it is important to note that many government policies have in fact been enabling space commerce. For example, government agencies have borne a significant proportion of the development costs of the major commercial launch vehicles worldwide. In the United States, Russia, and China, many vehicle families began as government launch systems that were eventually privatized; while in Europe and Japan, commercial launchers were developed as government activities. Government agencies provide access to launch facilities and support new technology development and programs to reduce technology risks. Government acquisition of satellites and launch vehicles provides important economies of scale to manufacturing and launch firms. Intense international competition in space commerce has raised the issue of the fairness of different levels of government support for commercial space activities and given rise to international agreements aimed at leveling the playing field.

Conclusions

Barriers to space commerce are both economic and policy-based. The costs and risks of space activities create barriers to entry and limit the viability of new space industries. Despite the increasing commercial focus of space activities, government expenditures and policies will continue to have a major impact on space commerce. The greatest potential impact of government policies will arise from expenditures to reduce the costs of access to space, most likely through the development of reusable launch vehicles. The magnitude of this impact, even if launch costs drop dramatically, is difficult to predict. This uncertainty about potential benefits may inhibit government and industry willingness to commit significant resources to fostering new space markets. Finally, decision-making in both government and industry regarding space commerce will be increasingly shaped by international competition.

WILL LOWER LAUNCH COSTS LEAD TO A SPACE BOOM?

Is it true that space commerce will boom if launch costs are significantly reduced? No one knows for sure. There have been several analyses that have assessed the demand for launch services by proposed space businesses (on-orbit manufacturing, space tourism, and so on) as a function of lower launch prices.

The most widely used data come from a study conducted in the mid–1990s by six large aerospace companies in conjunction with the National Aeronautics and Space Administration. The "Commercial Space Transportation Study" projected that if launch prices were reduced from modern levels of $4,000 to $10,000 per pound (depending on the destination orbit) to less than $500 to $1,000 per pound, there could be significant growth in launch activity.

In 1984, the Commercial Space Launch Act was enacted by the U.S. Congress to encourage the development of new and emerging companies within the commercial space launch industry. It then passed the Commercial Space Launch Amendments Act of 2004, which made a significant number of improvements to the previous Act. Currently, the Office of Commercial Space Transportation within the Federal Aviation Administration (FAA) oversees this Act and Amendment and authorizes all commercial launches from within the United States.

However, barriers to space commerce continue to exist. As more companies move to space-based or space-related operations, the legal, bureaucratic, and financial barriers that hinder space expansion should be lessened while, at the same time, effective standards set in place to help regulate and oversee the space industry. One legal barrier that is slowly being eroded away is the lack of lawyers concentrating on space law. In May 2008, law student Michael Dodge was awarded a juris doctor degree. What makes him special is that his law degree from the National Center for Remote Sensing, Air and Space Law at the University of Mississippi School of Law included a special space law certificate—the first one awarded in the United States.

Bureaucratic barriers are being taken away, too. The National Aeronautics and Space Administration (NASA) has been directed by the Barack Obama administration to help private industry take over the ferrying of humans and cargo into low-Earth orbit, such as trips to and from the International Space Station (ISS). The Commercial Orbital Transportation Services (COTS) program by NASA helps to coordinate the delivery of humans and cargo to the ISS by private companies. With the termination of the space shuttle fleet, NASA needs a reliable means to get astronauts and cargo up into space so it can concentrate on the development of a deep-space propulsion system and capsule to get astronauts to asteroids, Mars, and other bodies in the inner solar system. On December 23, 2008, NASA announced that Orbital Sciences Corporation (Orbital Sciences, based in Dulles, Virginia) and Space Exploration Technologies Corporation (SpaceX, headquartered in Hawthorne, California) were awarded contracts under a NASA Commercial Resupply Services (CRS) program. The contracts stipulate the delivery of a minimum of eight unmanned missions for Orbital Sciences (for $1.9 billion) and 12 unmanned missions for SpaceX (for $1.6 billion) between 2009 and 2016, to the International Space Station. Although both companies have encountered delays in their test flights, they are both proceeding with the hope of beginning cargo missions to the ISS in 2012.

▶ *See also* **Accessing Space (Volume 1) • Burial (Volume 1) • Launch Vehicles, Expendable (Volume 1) • Launch Vehicles, Reusable (Volume 1) • Legislative Environment (Volume 1) • Remote Sensing Systems (Volume 1) • Reusable Launch Vehicles (Volume 4) • Space Tourism, Evolution of (Volume 4) • Tourism (Volume 1)**

Resources

Books and Articles

Dubbs, Chris, and Emeline Paat-Dahlstrom. *Realizing Tomorrow: The Path to Private Spaceflight.* Lincoln: University of Nebraska Press, 2011.

Handberg, Roger. *International Space Commerce: Building from Scratch.* Gainesville: University Press of Florida, 2006.

Harris, Robert. *Space Enterprise: Living and Working Offworld in the 21st Century.* Berlin: Praxis, 2009.

McCurdy, Howard E. *Space and the American Imagination.* Baltimore: Johns Hopkins University Press, 2011.

Websites

Commercial Space Transportation Study. National Aeronautics and Space Administration. <http://www.hq.nasa.gov/webaccess/CommSpaceTrans/> (accessed September 27, 2011).

Dinkin, Sam. *Property rights and space commercialization.* The Space Review. <http://www.thespacereview.com/article/141/1> (accessed September 27, 2011).

Law Graduation Includes First-Time Certificate in Space Law. Newswise. <http://www.newswise.com/articles/view/540528/> (accessed October 4, 2011).

NASA Awards Space Station Commercial Resupply Services Contracts. National Aerospace and Space Administration. <http://www.nasa.gov/home/hqnews/2008/dec/HQ_C08-069_ISS_Resupply.html> (accessed September 27, 2011).

Office of Space Commercialization. <http://www.space.commerce.gov/> (accessed September 27, 2011).

Updates on SpaceX and Orbital's COTS progress. NewSpace Journal. <http://www.newspacejournal.com/2011/07/27/updates-on-spacex-and-orbitals-cots-progress/> (accessed September 27, 2011).

U.S. Congress, PUBLIC LAW 108–492—DEC. 23, 2004. *Commercial Space Launch Amendments Act of 2004: Commercial Space Transportation.* <http://www.faa.gov/about/office_org/headquarters_offices/ast/mcdia/PL108-492.pdf> (accessed September 27, 2011).

Burial

The first space burial took place on April 21, 1997, when the cremated remains (or ashes) of twenty-four people were launched into Earth orbit. Among the remains were from *Star Trek* television series creator Gene Roddenberry, rocket scientist Krafft Ehricke, physicist Gerald O'Neill, and writer Timothy Leary. The Houston, Texas-based company Celestis, Inc. performed this historic space memorial service. Approximately seven grams of ashes from each individual were placed into a lipstick-sized flight capsule. Each capsule was inscribed with the person's name and a personal message. The capsules were then placed in the memorial satellite—a small satellite about the size of a coffee can. The memorial satellite was launched into space aboard a commercial rocket and placed into Earth orbit.

Celestis continued to launch a memorial satellites through 2001. That year, Space Services Inc. (SSI) took over the assets of Celestis and the company is continuing to provide cremation memorial spaceflights. The name of their service is called Celestis Memorial Spaceflight Service, with Celestis an affiliate company of SSI.

In 2006, some of the ashes from astronomer Clyde Tombaugh, the discover of the dwarf planet Pluto, was placed onboard the NASA (National Aeronautics and Space Administration) *New Horizons* space probe. Its long-duration mission is to explore Pluto.

On May 2, 2009, the eighth Celestis Memorial Spaceflight was launched from New Mexico. However, a problem developed during ascent and the spacecraft did not reach space. Among the remains onboard was undersea explorer Ralph White, who was also the co-discoverer of the RMS *Titanic* wreckage.

As of 2011, only one company, SSI, offered space burials. SSI has facilities to service people in the United States, Canada, Japan, Germany, France, Russia, the Netherlands, and Australia. Because of the expense, only cremated remains have so far been launched into space. Other people whose ashes have been launched into space include: space illustrator

A True Space Burial

On July 31, 1999, the *Lunar Prospector* spacecraft finished its mission of mapping the Moon and was directed by NASA scientists to impact the Moon's surface. Aboard the spacecraft were the ashes of noted planetary geologist Eugene Shoemaker, the co-discoverer of the comet Shoemaker-Levy 9. The crash of the *Lunar Prospector* essentially buried Shoemaker's ashes on the Moon, where they remain today.

These lipstick-sized capsules contain cremated remains. Celestis, Inc. pioneered spaceflight memorials whereby a commercial rocket would carry and deposit this small payload in outer space. © *AP Images.*

▶

* **orbit** the circular or elliptical path of an object around a much larger object, governed by the gravitational field of the larger object

Charles O. Bennett, inventor Momofuku Ando, actor James Doohan, and astronaut Gordon Cooper.

As of late 2011, there had been ten space burials; an eleventh burial flight was planned for December 2011 or January 2012. Additional flights are planned for 2012, 2013, and 2014. Of the ten flights that have been completed, three did not achieve their intended orbit. Memorial flights now carry the ashes of more than a hundred individuals. Some of these individuals are famous for their space-related or other accomplishments, although the vast majority are ordinary individuals whose families chose the space burial because their loved ones had wanted to travel in space in their lifetimes. Each successive satellite has included more individuals as news has spread of this unique space-age service. As science progresses it is expected that the cost and difficulties of space burials will be reduced, and other companies may enter the market. However, the high cost of getting goods into Earth orbit (thus the small amount of ashes actually launched) and the strict regulations and permits necessary to conduct this novel business have helped to limit competition. In addition, as the space memorial service itself is new and unusual, it requires increased public knowledge and acceptance for the industry to grow.

One might wonder about the space environment and all the memorial satellites in orbit—are they a type of orbital debris cluttering space? The memorial satellites do not remain in orbit* forever. They are eventually drawn by gravity back to Earth, where they burn up harmlessly in the atmosphere.

▶ *See also* **Roddenberry, Gene (Volume 1)** • **Space Debris (Volume 2)**

Resources

Websites

Celestis, Inc. Web Site. <http://www.celestis.com/> (accessed November 2, 2011).

How Final Is a Space Burial? <http://discovermagazine.com/2007/sep/how-final-is-a-space-burial> (accessed November 2, 2011).

What It Costs for a Space Burial. <http://outthere.whatitcosts.com/ashes-in-space.htm> (accessed November 2, 2011).

Business Failures

Many companies have built successful businesses sending television programs, computer data, long-distance phone calls, and other information around the world via satellite. Satellites are able to send information to people across an entire continent and around the world (and even from far out in outer space). Orbiting high above Earth, a satellite can simultaneously send the same information to a vast number of users within its coverage area. In addition, satellites are not affected by geography or topography and can transmit information beyond the reach of ground-based antennas. This characteristic, in particular, attracted entrepreneurs eager to make a profit by using satellites to provide telephone service for people who live or work in remote locations or in developing nations that have underdeveloped terrestrial communications systems.

Failures in the Satellite Telephone Industry

While some satellite systems have proven successful by offering advantages not matched by ground-based systems, the satellite telephone business has had a more difficult time. This difficulty is in part because cellular telephone companies greatly expanded their reach while the satellite systems were being built. Cellular systems use ground-based antennas, and it is generally much less expensive to use a cellular phone than to place a call with a satellite phone.

Two satellite telephone companies were forced into bankruptcy in mid-1999 because of limitations in their business plans, and because the communications business evolved while the companies were still in development. One of the companies, Iridium SSC, began service in 1998. The firm, however, could not attract enough customers to pay back the $5 billion it had borrowed to place 66 satellites in orbit* to provide a global satellite phone system that would work anywhere on Earth. In late 2000, the newly formed Iridium Satellite LLC, headquartered in Bethesda, Maryland, purchased the Iridium satellite system and associated ground systems for $25 million, a fraction of its original cost. Iridium Satellite soon began selling Iridium phone service at much lower prices than those

▲

A Chinese Long March 2C rocket carrying two U.S.-made Iridium satellites launches on March 26, 1998. © *AP Images.*

* **orbit** the circular or elliptical path of an object around a much larger object, governed by the gravitational field of the larger object

of its predecessor. In February 2009, the company was in the news because its Iridium 33 satellite collided with an obsolete Russian Cosmos-2251 satellite over Siberia, creating two large debris clouds in space.

By the middle of 2010, Iridium had recovered from its previous poor condition to a remarkable extent. The company claimed to have more than half a million subscribers as of June 2010. At the time, it announced a new phase in its operations to be known as Iridium NEXT. That program will involve the design, construction, and launch of 72 new satellites to replace those already in orbit. An additional nine back-up satellites will also be constructed as potential replacements for active units. The total cost of Iridium NEXT was estimated at $2.9 billion, of which $1.8 billion was to be financed through an agreement with the French export credit agency Coface. The company selected Thales Alenia Space, a French-Italian corporation, to design and build its new satellites. The first satellites in the new system are expected to be launched in 2015.

Another satellite phone company, ICO Global Communications Ltd., headquartered in Reston, Virginia, had to reorganize in May 1999 and accept new owners, who bought the company at a large discount. ICO found it difficult to raise money from lenders after the failure of Iridium. ICO evolved into New ICO, in August 2000, and developed a new, more diversified business plan that was not limited to satellite phone service. In 2008, it launched ICO G1, a geosynchronous satellite, which, at the time, was considered the largest commercial satellite in space. Continued problems in the United States and Europe forced the company to file for bankruptcy in May 2009. One factor in that move was the decision by the European Commission to refuse a license for the company's European arm, ICO Satellite, to use the S band for signal transmission, a decision ICO challenged in the courts. At the time, the company also reorganized the North American arm of its operations and renamed those operations DBSD North America. Still unable to make a profit in early 2011, the company sold DBSD North America to Dish Network for $1.4 billion dollars. Also in 2011, ICO Global Communications (Holding) began the process of divesting itself of all international satellite assets. In June 2011, it also sold all of its non-satellite assets to the Pendrell Corporation, a company that specializes in intellectual property issues.

A third satellite phone venture, Globalstar LP, also ran into financial difficulty shortly after beginning commercial operations in 2000, and the company filed for bankruptcy protection in February 2002. In December 2003, controlling interest in the company was purchased by Thermo Capital Partners LLC, and the company was reincorporated in Delaware as Globalstar LLC. Three years later it was reorganized as Globalstar, Inc. The company's goal is to provide low-earth orbit (LEO) satellites for satellite phone services and low-speed data communications, similar to the program being pursued by Iridium. In 2007, it launched eight satellites into space. The company's resurgence appeared to be complete in 2010 when it announced the second phase of its satellite operations,

2010 Globalstar. The first two of the second generation satellites were launched in October 2010, and the next group of six satellites launched in July 2011. Original plans for the launch of two more sets of six satellites each before the end of 2011 had to be changed when problems at the Russian launch site developed. The company still plans to have those satellites in orbit, however, "in the near future." Two other firms, Constellation Communications Inc. and Ellipso Inc., were unable to raise enough money to build their planned satellite phone systems. As of 2011, they are both non-operational.

The Reasons for the Difficulties

Iridium, ICO, and Globalstar ran into trouble because of the high cost of building a satellite system compared to the relatively low cost of expanded multi-continent cellular service, which relies on less expensive ground-based antennas. Satellites are expensive to build, and a complete satellite system can take years to complete. In addition, rockets can cost tens of millions of dollars to launch, and they sometimes fail, requiring companies to buy insurance in case a rocket fails and destroys the spacecraft it was supposed to take into orbit.

These costs helped make satellite telephone systems much more expensive to use than cellular systems, while at the same time cellular networks were rapidly expanding the amount of territory for which they provided coverage. The cost difference, combined with the rapid growth of cellular networks, helped reduce the size of the potential market for satellite phone service during the very time that systems such as Iridium's were being developed.

In addition, satellite telephones are bigger and costlier to buy than cellular phones, and they must have an unobstructed view of the sky in order to work. Cellular phones, by contrast, work even indoors. These disadvantages further hurt the satellite phone industry.

Not all satellite phone companies have been unsuccessful. One, Inmarsat PLC., originally formed in 1979 as the International Maritime Satellite Organization (Inmarsat) at the request of the United Nations, became a private company in 1999 and has run a strong business providing mobile voice and data services for more than twenty years. But Inmarsat uses just a few satellites in geostationary orbit* to serve most of the world, whereas Iridium, ICO, and Globalstar designed their systems around relatively large fleets of spacecraft located much closer to Earth. Such low-or medium-Earth orbit systems are intended to reduce the satellite delay associated with geostationary satellites, but they are also more costly and complex to build and operate.

The Related Failures of Launch Vehicle Makers

In addition to costing investors billions of dollars, the satellite phone industry's difficulties also deflated the hopes of several companies hoping to build a new series of launch vehicles designed to carry satellites into space. For example, Iridium and Globalstar each have several dozen

* **geostationary orbit** a specific altitude of an equatorial orbit where the time required to circle the planet matches the time it takes the planet to rotate on its axis. An object in geostationary orbit will always remain over the same geographic location on the equator of the planet it orbits

* **payload** any cargo launched aboard a rocket that is destined for space, including communications satellites or modules, supplies, equipment, and astronauts; does not include the vehicle used to move the cargo or the propellant that powers the vehicle

satellites in their systems, and the expectation that the companies would have to replenish those spacecraft after several years helped inspire several firms to propose reusable rocket systems to launch new satellites.

The satellite phone systems in service in the early twenty-first century were launched using conventional rockets, which carry their payloads* into space and then are discarded. Reusable rockets are intended to save money by returning to Earth after transporting a load into orbit and embarking on additional missions. However, uncertainty about the satellite phone industry's future hurt the prospects of companies such as Rotary Rocket Co., Kistler Aerospace Corp., and Kelly Space & Technology Inc., which had looked to the satellite phone industry as a key source of business. Rotary Rocket ceased operations in 2001. Kistler Aerospace survived in various forms for 17 years, but finally filed for bankruptcy protection in June 2010. At that point it had assets of just over $100,000 and liabilities of more than $7 million. Kelly Space & Technology Inc. has survived, as of late 2011, by diversifying into a number of fields other than satellite launching, including environmental testing, biohazard elimination, jet and rocket testing, and wireless products. The company also continues to work on the development of its EXPRESS Spaceplane, also known as the Astroliner.

 See also **Communications Satellite Industry (Volume 1) • Financial Markets, Impacts on (Volume 1) • Insurance (Volume 1) • Launch Vehicles, Reusable (Volume 1) • Reusable Launch Vehicles (Volume 4)**

Resources

Books and Articles

Grant, August E., and Jennifer Harman Meadows. *Communication Technology Update and Fundamentals..* Amsterdam: Focal Press/Elsevier, 2010.

Plunkett, Jack W. *Plunkett's Telecommunications Industry Almanac 2012: Telecommunications Industry Market Research, Statistics, Trends & Leading Companies..* Houston, TX: Plunkett Research, Ltd., 2011.

U.S. Satellite Communication Companies Directory.. Washington, DC: International Business Publications, 2009.

Websites

Communication Satellites. National Aeronautics and Space Administration. <http://www.hq.nasa.gov/office/pao/History/commsat.html> (accessed October 11, 2011).

Globalcom Satellite Phone Articles Home. <http://www. globalcomsatphone.com/articles/> (accessed October 11, 2011).

Globalstar. <http://www.globalstar.com/en/> (accessed October 11, 2011).

ICO Global and DISH Networks Enter into Implementation and Restructuring Support Agreements Regarding DBSD North America. <http://www.fiercetelecom.com/press_releases/ico-global-and-dish-networks-enter-implementation-and-restructuring-support> (accessed October 11, 2011).

Iridium Everywhere. <http://www.iridium.com/default.aspx> (accessed October 11, 2011).

Kelly Space & Technology. <http://www.kellyspace.com/index.php> (accessed October 11, 2011).

Whalen, David J. *Communications Satellites: Making the Global Village Possible.* National Aeronautics and Space Administration (NASA). <http://www.hq.nasa.gov/office/pao/History/satcomhistory.html> (accessed October 11, 2011).

Business Parks

Humans have been doing business in orbit about the Earth since the early 1960s, with "business" loosely defined in this context as any useful activity. Trained specialists, within the safety of small orbiting spacecraft, studied the Earth below and the rest of the universe above. They conducted medical tests to see how their bodies responded to weightlessness. Others conducted experiments on various materials to see how microgravity (or the apparent lack of gravity) affected their interactions. Some studied the growth and behavior of plants and small animals. This early orbital activity would become the seed of today's International Space Station, which in turn may lead to tomorrow's truly defined space business parks.

Precursors to Space-Based Business Parks

For serious work, more spacious, dedicated orbiting laboratories were needed. The Soviets launched a series of Salyut stations beginning in 1971.

The American Skylab, built from the casing of a leftover Saturn I booster, was launched in 1973 and was staffed in three missions of twenty-eight, fifty-nine, and eighty-four days.

Through the 1980s and 1990s, the National Aeronautics and Space Administration (NASA) used the space shuttle orbiter's payload bay* to conduct orbital observations and experiments for periods of up to two weeks. Commercially built SpaceHab modules and the European-built Spacelab, both riding in the payload bay, allowed scientists to conduct serious work. Meanwhile, the Soviets' historic *Mir* space station grew from a single module to an ungainly but productive complex. In was deorbited from space on March 23, 2001.

* **payload bay** the area in the shuttle or other spacecraft designed to carry cargo

U.S. astronaut Charles "Pete" Conrad Jr. completes an experiment activity check-list during training for a Skylab mission. Space business parks may include modules the size of Skylab. *NASA.* ▶

* **synthesis** the act of combining different things so as to form new and different products or ideas

Scientists experimented, synthesizing* chemicals, crystals, and proteins impossible to produce in the full gravity environment on Earth. Such experiments may allow scientists to discover products valuable enough to warrant orbiting factories.

Earth orbit is a good environment in which to conduct useful research, development, and possibly manufacturing with proven economic impact.

For example, such activities can: create global communications grids; monitor the atmosphere, weather, oceans, and changing land-use patterns; and search for otherwise hidden submarine and subterranean features. Much of this has been accomplished with automated satellites. Human activities in orbit have been directed at investigating the effects of microgravity* on humans, animals, plants, and inanimate substances. Astronauts have also launched, retrieved, and repaired satellites and deep-space probes. The International Space Station (ISS) was built to ramp up all these activities to the next level. Open for business with a crew of three, ISS' final design configuration now allows for a crew of six people. (The first six-person crew of the ISS began on May 29, 2009, with the ISS Expedition 20.) Further large expansion of the current design will not be possible now that the NASA space shuttle fleet has been retired (which happened on August 31, 2011) unless other launch vehicles are built that can lift equally large massive bodies as could the shuttles. (The Russians are adding one more module to the International Space Station in 2012. At that point, the ISS will be structurally completed.)

Elements of Business Parks in Space

Orbit is also an ideal place to service satellites, refuel probes on deep-space missions, and assemble large spacecraft and platforms too big to launch whole from Earth. For these activities humans will need: a fuel depot with the capacity to scavenge residual fuel, robotic tugs to take satellites to higher orbits and fetch them for servicing, a well-equipped hangar bay in which to perform such services safely under optimum lighting, and a personnel taxi for spacecraft parking nearby.

Supporting humans in space requires a growing mix of services to keep labs busy, bring people to orbit and back, supplying equipment and parts, resupplying consumables, and handling wastes. These are all businesses that can be provided more effectively and at less cost by for-profit enterprises, given proper incentives. Physically, the ISS is similar to *Mir* in being a government-built metal maze. Operationally, the ISS has the potential to host considerable entrepreneurial activity.

A favorable climate of legislation, regulation, and taxation will foster such development. Privatizing American contributions could lead the way in this international endeavor. NASA could be mandated to use commercial providers for additional modules not in the original final design of the ISS as well as for the transfer of cargoes between orbits. In fact, NASA is doing part of this already. As part of the Commercial Orbital Transportation Services (COTS) contract by NASA, two private companies—Orbital Sciences Corporation (Orbital) and Space Exploration Technologies Corporation (SpaceX)—were awarded, on February 2008, contracts for a minimum of eight Orbital missions and twelve SpaceX missions to the International Space Station. Orbital will launch its Cygnus spacecraft from the Taurus II rocket at a launch pad on Wallops Island (Virginia), while SpaceX will launch its Dragon spacecraft from its Falcon 9 rocket at Cape Canaveral

* **microgravity** the condition experienced in freefall as a spacecraft orbits Earth or another body; commonly called weightlessness; only very small forces are perceived in freefall, on the order of one-millionth the force of gravity on Earth's surface

The International Space Station, where astronauts live and conduct research for significant stretches of time, is the next step towards realizing integrated business parks in orbit. *NASA.*

▶

* **berth space** the human accommodations needed by a space station, cargo ship, or other vessel

Air Force Station (Florida). The first test flight for Orbital's Cygnus is tentatively scheduled for late 2011, while SpaceX's Dragon began its first test flight in December 2010, with a second scheduled for November 2011.

The ISS could evolve into a business park. As more people live and work aboard the ISS, additional quarters will be needed for visiting scientists, policymakers, and journalists. A modular six-berth "hotel" could grow with demand. An open-door policy would welcome private individuals who had paid for or won their fares and who had passed physical and psychological tests.

Along with berth space*, demand will grow for adequate recreation and relaxation "commons." A voluminous sports center, built in a large inflatable structure (sphere, cylinder, or torus), could be subsidized by naming rights and paid for by television advertising. Zero-G (zero-gravity) space soccer, handball, wrestling, ballet, and gymnastics might command respectable audiences on Earth. The first players will be regular staff, but telecasting these events would feed the demand among would-be tourists.

As on-location service providers join the action, they too will need residential and recreational space. As the population grows, treating and recycling wastes in-orbit will become more attractive. This will help build the know-how needed for future Moon and Mars outposts.

Managing a Growing Space Business Park

The operating agency should be a marina/port authority-type entity, acting as park "developer," anticipating growth and demand. The expansion of the station's skeletal structure and power grid must be planned, along with additional piers, slips, and docking ports. Haphazard growth can lead to an early dead end!

Clustering activities—scientific, commercial, industrial, and tourist—will aim at creating a critical mass of goods and services providers, including frequently scheduled transportation. If collective station keeping proves feasible, a station-business park could expand to include co-orbiting satellite clusters, orbiting Earth in formation. Laboratories may choose to relocate to such parks for isolation from unwanted vibrations.

If microgravity experiments identify products that could be profitably mass-produced in orbit, manufacturing complexes may develop. Raw materials mined on the Moon or asteroids could be processed at orbiting industrial parks into building materials and manufactured goods. These products will not be aimed at consumers on Earth, where they would not be able to compete on price. Instead, they will feed the construction and furnishing of ever-more orbiting labs, business parks, factories, and tourist resorts. If Earth-to-orbit transportation costs remain high, space-sourced goods could hold a cost advantage for orbital markets. This need to support increasing activities in Earth orbit will in turn support mining settlements on the Moon and elsewhere.

The *Mir* space station, deactivated and then recommissioned, had taken the lead in encouraging for-profit space activities, including space tourism. However, the Russian Federal Space Agency decided to end its operation, and the aging space station re-entered Earth's atmosphere in 2001.

Given favorable changes in climate, bursts of commercial and business activity could vitalize the ISS core complex. The ISS is now in a high-inclination Earth orbit, an orbit chosen for its ease of access as well as to allow observation of most populated landmasses on Earth. Such orbits, however, are not optimum as staging points for higher geosynchronous orbit* or deep space. Rising demand for such crewed services should lead to another depot-business park in equatorial orbit*. Business and tourist activities in orbit are emerging from these tiny seeds.

Orbiting business parks are a possibility. They may come from private investments rather than from the governmental sector. For instance, Bigelow Aerospace is developing an inflatable habitat module that could become the basis for its expandable space habitat module program. The company is currently testing the *Genesis I* and *Genesis II* prototypes. As of October 1, 2011, both modules are currently in orbit about Earth, sending information on pressure, temperature, and radiation in and around their shells to mission control personnel in Las Vegas. The company is hoping to build larger, more complex structures in the future. It is also expecting to launch its BA 330 (formerly called Nautilus space complex module) in 2014 or 2015. The space habitation module, expandable to

* **geosynchronous** remaining fixed in an orbit 35,786 kilometers (22,300 miles) above Earth's surface

* **equatorial orbit** an orbit parallel to a body's geographic equator

330 cubic meters (12,000 cubic feet), will directly support zero-gravity scientific and manufacturing research, along with indirect activities involving space tourism.

Bigelow is also designing a BA 2100 that would require a heavy-lift rocket to send it to low-Earth orbit. If manufactured, the module would be six times larger than the BA 330—at 2,200 cubic meters (77,690 cubic feet). In addition, Bigelow Aerospace is developing a series of inflatable modules, under the name of Bigelow Next-Generation Commercial Space Station (also called *Space Complex Alpha*). The complex will include spacecraft modules, along with a central docking node, solar arrays, propulsion, and crew capsules. The company hopes to launch the components in 2014, with leasing options starting in 2015. So far, Bigelow has contracts with seven countries interested in leasing its space station modules. The company also has plans for a second orbital space station called *Space Complex Bravo*, with launches scheduled to begin in 2016 and possible commercial operations commencing in 2017.

In addition, China is developing a manned space station. A small space station module called *Tiangong 1* (in English: Heavenly Palace 1) was launched on September 29, 2011, via a Long March 2F/G rocket. This flight is considered an evolving stage to eventually develop the rendezvous and docking procedures necessary to orbit a larger semi-permanent manned space station. According to the Chinese, the 8.5-metric-ton (19,000-pound) module is capable of docking with manned and robotic spacecraft. Further, the *Tiangong 2* space station is scheduled to be launched between 2013 and 2015, and an even larger space station called *Tiangong 3* in the 2014 to 2016 timeframe.

Russia is also considering the construction of an orbital construction facility for spacecraft too massive to launch from Earth's surface. Between 2015 and 2020, the Russians have indicated that they could possibly take one or all of its Russian modules from the International Space Station for the start of its new space station called Orbital Piloted Assembly and Experiment Complex (OPSEK). One of the modules the Russian Federal Space Agency (Roscosmos) hopes to use for the new OPSEK space station is the Nauka Multipurpose Laboratory Module (MLM). This module is the last major component to be attached to the ISS, and it is expected to be launched in mid-2012. The main purpose for the OPSEK is to support deep-space missions by the Russians. Roscosmos expects to first dock at its space station before starting any manned missions to the Moon and later to Mars.

All such space construction may eventually lead to a space business park. In 2011, the only structure that might be vaguely called a space business park is the International Space Station, a project primarily of the United States, Russia, Japan, Canada, and the European Space Agency. It is the largest artificial structure in orbit about Earth. It has financial support to continue operations through 2020, and possibly through 2028.

See also **Commercialization (Volume 1) • Habitats (Volume 3) • Hotels (Volume 4) • TransHab (Volume 4)**

Resources

Books and Articles

Catchpole, John E. *The International Space Station: Building for the Future.* Berlin: Springer, 2008.

Jamison, David E., editor. *The International Space Station.* New York: Nova Science, 2010.

Reference Guide to the International Space Station. Washington, D.C.: National Aeronautics and Space Administration, 2010.

Websites

Bigelow Aerospace. <http://www.bigelowaerospace.com/> (accessed September 30, 2011).

Bigelow Aerospace's Space Station. YouTube.com. <http://www.youtube.com/watch?v=OqsHK2vxyzo> (accessed October 31, 2011).

China National Space Administration. <http://www.cnsa.gov.cn/n615709/cindex.html> (accessed September 30, 2011).

China space station signals shift in space race. Guardian.co.uk. <http://www.guardian.co.uk/world/2011/sep/30/china-space-station> (accessed September 30, 2011).

A concept of the Russian successor to the ISS. RussianSpaceWeb.com. <http://www.russianspaceweb.com/opsek.html> (accessed September 30, 2011).

International Space Station. National Aeronautics and Space Administration. <http://www.nasa.gov/mission_pages/station/main/index.html> (accessed September 28, 2011).

Project 921-2. Astronautix.com. <http://www.astronautix.com/craft/prot9212.htm> (accessed September 30, 2011).

Russian Federal Space Agency. <http://www.federalspace.ru/?lang=en> (accessed September 28, 2011).

Six-Person ISS Crew Gets To Work. AviationWeek.com. <http://www.aviationweek.com/aw/generic/story_channel.jsp?channel=space&id=news/Six060309.xml> (accessed October 31, 2011).

TransHab Concept. National Aeronautics and Space Administration. <http://spaceflight.nasa.gov/history/station/transhab/> (accessed September 30, 2011).

C

Cancer Research

The potential use of the microgravity* environment for inroads in cancer research is both important and promising. Research opportunities are broad and will include many areas of examination for investigators who are trained in both basic and clinical sciences. As one example, studies have shown that mammalian cell culture* conducted in a manner that does not allow cell settling as a result of gravitational forces holds promise in the propagation of three-dimensional tissue cellular arrays* much like those that normally comprise tissue specimens in the intact body. Because of its position in orbit about Earth, the International Space Station (ISS) has only a minute fraction of the gravitational force that is present on the surface of Earth. Culture of tissues with a three-dimensional architecture on the ISS provides a unique and powerful opportunity for studies of anti-cancer drug action with a more complex and natural tissue ultrastructure than can be attained in terrestrial laboratories.

Even the production and analysis of new anti-cancer drugs may be conducted in a superior manner in microgravity. Studies already conducted on the space shuttle have shown that, in at least some instances, superior crystal growth can be achieved in microgravity when compared to crystals grown on Earth. This success primarily is the result of a lack of liquid convection currents* in microgravity that subsequently leads to a quieter liquid environment for a gradual and more orderly growth of crystals. The quality of crystal products is an important feature in the determination of the three-dimensional structure of the molecules by x-ray* diffraction analysis. Until the three-dimensional structures of new and existing anti-cancer compounds are established, the design of superior candidates for cancer treatment is severely hampered.

It is generally recognized that cancers arise in the body more often than clinically troublesome cancer diseases occur. In many cases, the primary cancer growth is restricted in further development and the victim's immune system plays an important role in limiting cancer progression, sometimes even eradicating the cancer cells altogether. It seems that the mammalian immune system may not function as efficiently in the microgravity environment when compared to Earth.

On one hand, a weakened immune response to infectious diseases and cancer could present a serious obstacle for space travelers of the future. On the other hand, a compromised immune system in microgravity, and a subsequent increased efficiency of tumor progression, may provide a valuable test bed for research on the immune system with regard

* **microgravity** the condition experienced in freefall as a spacecraft orbits Earth or another body; commonly called weightlessness; only very small forces are perceived in freefall, on the order of one-millionth the force of gravity on Earth's surface

* **cell culture** a means of growing mammalian (including human) cells in the research laboratory under defined experimental conditions

* **cellular array** the three-dimensional placement of cells within a tissue

* **convection** the movement of heated fluid caused by a variation in density; hot fluid rises while cool fluid sinks

* **x-ray diffraction analysis** a method to determine the three-dimensional structure of molecules

NASA's light-emitting diode (LED) developed out of a need for a light that could promote plant growth in space. Surgeons now use LED technology in cancer treatments involving surgery. *NASA.*

to cancer development. The microgravity environment, where immune function is less efficient, may also provide an excellent opportunity to develop and assess new chemotherapeutic measures that can strengthen the host's immune response. Of course, there are biomedical applications well beyond cancer research since the progression of many diseases may reflect a compromised immune function.

The life-threatening radiation exposure away from the protective atmosphere of Earth, and the ensuing increase in cancer cell development, is more than a casual concern for long-distance space travel. The means to protect space travelers from increased radiation will be necessary before such adventures are common. One experiment on the effects of radiation on the human body was launched to the International Space Station in May 2001. The "subject" of that experiment was the model of a human torso name "Fred, the Phantom Torso," that looked much like an Egyptian mummy. This "mummy," however, lacked arms and legs. Inside its body were 416 dosimeters placed in strategic locations simulating the presence of human organs. After Fred completed his four-month trip on the ISS, NASA scientists sliced his body into thin segments to collect dosimeter readings on the radiation reaching each body parts. The results of that study suggested that humans might not experience harmful effects from relatively short trips into space (such as a trip to the Moon), but might face more serious challenges on longer trips (such as a visit to Mars). In 2009, the "phantom torso" experiment was repeated, this time by the European Space Agency (ESA), which called its mummy-model Matroshka. Matroshka visited the space station for four months and was analyzed by ESA researchers upon its return to Earth, with much the same results as those obtained for Fred seven years earlier.

In October 2007, Malaysian astronaut Sheikh Muszaphar Shukor was sent to the International Space Station to conduct scientific research involving the growth of cancer cells in outer space. The research of the cancer cells, from the liver, will benefit science and medical research in his country. In addition, in 2009, research by NASA scientists discovered a way to identify cancerous brain cells by having a sensor sniff them from the air. They developed an 'electric nose' that can differentiate between normal brain cells and cancerous ones. The electric nose contains chemical sensors that analyze electronically the air around a human.

In March 2009, NASA released an announcement outlining its overall objectives for research projects on the International Space Station for projects to begin after July 1, 2010. Some of the topics in which the National Cancer Institute (NCI) was interested during that time period included studies on the biology of metastasis, the role of gravity on radiation sensitivity and immune cells, the effect of gravity and radiation on cancer stem cell function, and the preparation of new therapeutic compounds for the treatment of cancer.

An example of the type of cancer research now underway at the space station is the study of tissue engineering under microgravity conditions. The term *tissue engineering* refers to the use of cells, engineering technology, and biochemical procedures to improve or replace biological structures and functions. The development of various forms of artificial skin is an example of tissue engineering. Tissue engineering experiments carried out in microgravity avoid the confounding effects of settling in solutions that occur in Earth-bound experiments. In such experiments, researchers have found that cells self-assembly into tissues in an extraordinarily perfect way, producing structures that are very similar to those found in the human body. In one such experiment, carried out by astronaut David Wolf, cells self-assembled into an almost perfect colon cancer tumor that provided an ideal model for the study of the disease. This line of research had led, as of late 2011, to more than 25 patents for the production and study of tissue cultures and created a new paradigm for the way such studies can be done on Earth. The technology is expected to be useful in studying virtually every tissue in the human body, along with their cancerous variations.

 See also **AIDS Research (Volume 1)** • **Living in Space (Volume 3)** • **Made in Space (Volume 1)** • **Medicine (Volume 3)** • **Microgravity (Volume 2)**

Resources

Books and Articles

Calhoun, Nicole, M. *International Space Station Research: Accomplishments and Challenges.* New York: Nova Science Publishers, 2010.

National Aeronautics and Space Administration, et al. *Review of NASA Plans for the International Space Station.* Washington, DC: National Academies Press, 2006.

commercial astronaut a person trained to go into space as part of a privately funded operation

Mercury the first American piloted spacecraft, which carried a single astronaut into space; six Mercury missions took place between 1961 and 1963

Gemini the second series of American-piloted spacecraft, crewed by two astronauts; the Gemini missions were rehearsals of the spaceflight techniques needed to go to the Moon

Apollo American program to land men on the Moon; *Apollo 11, Apollo 12, Apollo 14, Apollo 15, Apollo 16,* and *Apollo 17* delivered twelve men to the lunar surface between 1969 and 1972 and returned them safely back to Earth

National Research Council, Committee for the Decadal Survey on Biological and Physical Sciences in Space. *Life and Physical Sciences Research for a New Era of Space Exploration: An Interim Report.* Washington, DC: National Academies Press, 2010.

Websites

Cooney, Michael. *NASA's Electronic Nose Can Sniff Out Cancer, Space Stench.* NetworkWorld.com. <http://www.networkworld.com/community/node/41467> (accessed October 27, 2011).

The Phantom Torso Returns. National Aeronautics and Space Administration. <http://science.nasa.gov/headlines/y2009/27may_phantomtorso.htm?list970856> (accessed October 27, 2011).

Tapping into Space Research. <http://thestar.com.my/news/story.asp?file=/2007/9/22/nation/18514133&sec=nation> (accessed October 27, 2011).

Tissue Engineering and the International Space Station. <http://www.comspacewatch.com/news/viewsr.html?pid=36118> (accessed October 27, 2011).

Career Astronauts

Sometime in the future passenger-paying flights into space may likely become as routine as air travel. The fledgling space tourism industry is slowly developing with the promise of commercial suborbital flights in the very near future. Private enterprises, as of 2011, are in various stages of providing still-expensive commercial suborbital flights into space, with orbital space flights as their next step. For instance, Virgin Galactic is hoping to begin sub-orbital flights for paying customers sometime in the first half of the 2010s. However, in the early twenty-first century, opening up the frontier of is still the duty of a select cadre of highly trained individuals. These are career astronauts, people whose professional job it is to work and live in space in such roles as commander, pilot, or crewmember of a spacecraft. Other names are sometimes applied to such professional space travelers when in other countries, such as career cosmonauts in Russia, or when performed for privately held companies, such as career commercial astronauts*.

In the United States, the early pioneering days of human spaceflight gave rise to individuals with what American author Tom Wolfe called the "right stuff." These individuals were tough-as-nails experimental aircraft test pilots. They were critical in getting America's human spaceflight program, quite literally, off the ground. During the 1960s, and continuing through the 1970s, a unique corps of astronauts flew in the U.S. Mercury*, Gemini*, Apollo*, and Skylab programs.

Today, after nearly sixty years of human sojourns into suborbital trajectories, low Earth orbits*, and to the Moon, 518 people (from 38 countries), according to Fédêation Aéonautique Internationale (FAI) guidelines, have departed Earth and attained at least a minimum altitude of 100 kilometers (62.1 miles) from Earth's surface (as of August 31, 2011).

In addition, seven U.S. pilots reached an altitude of at least 80 kilometers (50 miles) into space (according to the Department of Defense definition of space) when they flew sub-orbital flights on the X-15 space plane. Since 1981, a majority of these space-faring individuals have been boosted into space courtesy of the U.S. space shuttle fleet. For instance, the first space shuttle flight was launched on April 12, 1981, when the space shuttle *Columbia* went into space commanded by John Young and piloted by Robert Crippen, the first two career astronauts for the space shuttle program.

Indeed, space travel has come a long way, from the early single-person "capsule" to the winged flight of a space shuttle, and now to pending

* **low Earth orbit** an orbit between 300 and 800 kilometers (185 and 500 miles) above Earth's surface

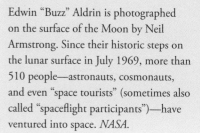

Edwin "Buzz" Aldrin is photographed on the surface of the Moon by Neil Armstrong. Since their historic steps on the lunar surface in July 1969, more than 510 people—astronauts, cosmonauts, and even "space tourists" (sometimes also called "spaceflight participants")—have ventured into space. *NASA.*

payload any cargo launched aboard a rocket Ihat is destined for space, including communications satellites or modules, supplies, equipment, and astronauts; does not include the vehicle used to move the cargo or the propellant that powers the vehicle

spacewalking moving around outside a spaceship or space station, also known as extravehicular activity

remote manipulator system a system, such as the external Canadarm2 on the International Space Station, designed to be operated from a remote location inside the space station

payload operations experiments or procedures involving cargo or "payload" carried into orbit

privately funded trips to outer space piloted by commercial astronauts and still developing government sponsored trips to asteroids, the planet Mars, and other bodies in the Solar System.

Types and Duties of NASA Astronauts

The National Aeronautics and Space Administration (NASA) recruited pilot astronaut candidates and mission specialist astronaut candidates to support the space shuttle–formally called the Space Transportation System (STS) program—which began in 1981 and ended in 2011.

Since August 2011, NASA recruits astronaut candidates only for its International Space Station (ISS) program, now that its STS program has been completed. As of August 31, 2011, the ISS has been continually staffed with crewmembers since November 2, 2000. Persons from both the civilian sector and the military services are considered for ISS duty. Applicants for the NASA Astronaut Candidate Program must be citizens of the United States. However, international astronauts are accepted by NASA from countries that have their own space agencies, such as Japan's Japan Aerospace Exploration Agency (JAXA), Canada's Canadian Space Agency (CSA), Russia's Russian Federal Space Agency (RKA, or Roscosmos), and the European Union's European Space Agency (ESA).

The crewmembers onboard the International Space Station are called flight engineers (similar to mission specialists on the space shuttle) and commanders (similar to shuttle pilots and commanders). Once a flight engineer is a member of an ISS mission, which is called an Expedition, then they may be promoted to commander on a future mission. For instance, Peggy Whitson was the first female commander of an ISS Expedition mission; specifically, Expedition 16 in 2007. However, her first stint on the International Space Station was in 2002 where she was a flight engineer on the Expedition 5 mission, which gave her the needed experience to later become an ISS commander.

Flight engineer astronauts, working under the direction of the ISS commander, have overall responsibility for the coordination of space station operations in the areas of crew activity planning, consumables usage, and payload operations. Flight engineers are required to have detailed knowledge of space station systems, as well as detailed knowledge of the operational characteristics, mission requirements and objectives, and supporting systems and equipment for each payload* element on their assigned missions. Flight engineers perform spacewalks* (extravehicular activities [EVAs]), use remote manipulator systems* to handle payloads, and performed or assisted in specific experiments. Flight engineers also perform payload operations* and science experiments on the ISS. They sometimes perform spacewalks outside of the station when repair or maintenance work or additions to the structure are needed.

Commanders in the International Space Station are responsible for the overall success of their mission, along with the safety of their Expedition crew and the space station as a whole. All ISS commanders must have prior spaceflight experience, especially many previous hours working aboard the space station on past ISS missions.

As of August 31, 2011, the Expedition 28 crew is onboard the ISS. They include commander Andrey Borisenko of Russia, and flight engineers Alexander Samokutyaev, Mike Fossum and Ron Garan (both from the United States), Sergei Volkov (Russia), and Satoshi Furukawa (Japan).

An exciting new era of space exploration is underway with the completion of the International Space Station (ISS), which is scheduled to be completed in 2012, with the addition of a Russian laboratory module. The development of this orbital facility has been called the largest international scientific and technological endeavor ever undertaken. The ISS will likely be used as a starting off point for future NASA explorations of the Solar System that includes a manned visit to an asteroid and eventually a trip to the planet Mars.

The ISS is designed to house six people, and a permanent laboratory has been established within it in a realm where gravity, temperature, and pressure can be used in a variety of scientific and engineering pursuits that are more difficult to recreate in ground-based laboratories. The ISS is a test bed for the technologies of the future and a laboratory for research on new, advanced industrial materials, communications technology, and medical research. For all of its promise for the future, the people aboard the ISS are still the most important aspect of the International Space Station. These explorers continue to lead the way out into space for the rest of the human race.

Requirements for Applicants

What minimum requirements must an individual meet prior to submitting an application for astronaut status at NASA? All candidates to be NASA astronauts must be U.S. citizens and pass a comprehensive physical examination.

For a flight engineer astronaut candidate, an individual must have a bachelor's degree from an accredited institution in engineering, a biological or physical science, or mathematics. The degree must be followed by at least three years of related, progressively responsible, professional experience. An advanced degree is desirable and may be substituted for part or the entire experience requirement (a master's degree is considered equivalent to one year of experience, while a doctoral degree equals three years of experience). The quality of the academic preparation is important. Individuals must also pass a NASA Class II space physical, which is similar to a military or civilian Class II flight physical, and includes the following specific standards:

- Distance visual acuity: 20/200 or better uncorrected, correctable to 20/20, each eye

- Blood pressure: not greater than 140/90 measured in a sitting position

- Height: between 157.5 and 193 centimeters (58.5 and 76 inches)

The minimum requirement for a commander/pilot astronaut candidate is a bachelor's degree from an accredited institution in engineering, a biological or physical science, or mathematics. However, service in the U.S. Air Force can exempt this requirement. An advanced degree is desirable. The quality of the academic preparation is important. At least 1,000 hours of pilot-in-command time in jet aircraft are necessary. Flight test experience is highly desirable. Applicants must pass a NASA Class I space physical, which is similar to a military or civilian Class I flight physical, and includes the following specific standards:

- Distant visual acuity: 20/70 or better uncorrected, correctable to 20/20, each eye
- Blood pressure: not greater than 140/90 measured in a sitting position
- Height: between 162.6 and 193 centimeters (62 and 76 inches)
- Eye surgical procedures are allowed, but at least one year must have passed from the date of application without any permanent problems from the procedure

Screening and Training

Beyond the initial application requirements, NASA's astronaut selection involves a rigorous physical and mental screening process designed to cull the best and brightest from those who are applying. Part of the screening process is psychological evaluation of the candidates. Two hours of interviews are required of the candidates from psychiatrists and psychologists. In fact, in July 1999, a NASA call for astronauts produced more than 4,000 applicants. A mere 3% made the first cut. From there, further screening by the Astronaut Selection Board led to a final twenty candidates. In 2009, NASA finalized the latest round of astronaut applicants—the last class in five years. The space agency had about 3,500 applicants and selected only nine candidates from the group, less than 0.3% of the original applicants. This 2009 class of astronauts is the twentieth group since the "Original Seven" Mercury astronauts were selected in 1959. With the retirement of the space shuttle fleet in 2011, these astronauts will be training to work onboard the International Space Station and to fly on the Russian Soyuz space capsule (and possibly some not-yet-operational private spacecraft). As NASA develops its next-generation manned space program, these astronaut trainees will likely also train within mock-ups of an advanced designed space capsule—the replacement for NASA's shuttle.

Those who make the grade as astronaut trainees are trained at NASA's Lyndon B. Johnson Space Center just outside of Houston, Texas. The selected applicants are designated astronaut candidates and undergo a one- to two-year training and evaluation period during which time they participate in the basic astronaut-training program. This effort is designed to develop the knowledge and skills required for formal mission training upon selection for a flight or mission. During their candidate period, pilot astronaut candidates must maintain proficiency in NASA aircraft.

As part of the astronaut candidate training program, trainees are required to complete military water survival exercises prior to beginning their flying studies and become scuba qualified to prepare them for space-walking training. Consequently, all astronaut candidates are required to pass a swimming test during their first month of training. They must swim three lengths of a 25-meter (82-foot) pool without stopping, and then swim three lengths of the pool in a flight suit and tennis shoes. The strokes allowed are freestyle, breaststroke, and sidestroke. There is no time limit. The candidates must also tread water continuously for ten minutes.

To simulate microgravity*, astronaut candidates have previously boarded the infamous "Vomit Comet," a converted Boeing KC-135 Stratotanker jet aircraft. In 2005, the two airplanes were replaced with a McDonnell Douglas C-9B Skytrain II airplane and given the official nickname: the "Weightless Wonder." Flown on a parabolic trajectory*, this airplane can produce periods of microgravity for some twenty seconds. Akin to an airborne version of a roller coaster, the parabolic maneuvers are repeated up to forty times a day. Those riding inside the aircraft experience microgravity similar to that felt in orbital flight, although in short bursts.

One very important note: Selection as a candidate does not ensure selection as an astronaut. Final selection is based on the satisfactory completion of the one-year program. As of August 31, 2011, NASA is not accepting further applications for astronaut candidates. However, further classes will be generated as NASA continues to ramp up its new manned space program for the 2010s and beyond.

Salaries

Salaries for civilian astronaut candidates are based on the federal government's general schedule pay scales for grades GS-11 through GS-14 and are set in accordance with each individual's academic achievements and experience. Selected military personnel are assigned to the Johnson Space Center but remain in an active duty status for pay, benefits, leave, and other similar military matters.

The latest groups of astronauts selected by NASA will likely participate in its new manned spaceflight program, which is expected to become operational in the mid- to late 2010s. They will be trained to deal with long missions to the International Space Station and with even longer missions to asteroids, the Moon, Mars, and other possible destinations as the United States continues to explore the inner Solar System with its career astronauts. NASA states, "The astronauts of the 21st century will help lead NASA through the next steps of its Vision for Space Exploration as we explore the Moon, Mars, and beyond."

Other career astronauts

Besides U.S. astronauts who perform space-related work for NASA, several other countries also have career astronauts. For many decades, Soviet cosmonauts plied their trade, first as competitors with the United States and now as Russian allies to America's exploration of space. The Russian

* **microgravity** the condition experienced in freefall as a spacecraft orbits Earth or another body; commonly called weightlessness; only very small forces are perceived in freefall, on the order of one-millionth the force of gravity on Earth's surface

* **parabolic trajectory** trajectory followed by an object with velocity equal to escape velocity

equivalent to NASA, the Russian Federal Space Agency (Roscosmos), coordinates the work of these cosmonauts as they work alongside U.S. astronauts (and other countries' astronauts) aboard the International Space Station. These cosmonauts go into space and return to Earth through Russia's reliable Soyuz space capsule and rocket system.

In addition, the Chinese manned space program recently sent its first career astronaut (or taikonaut) into space when Yang Liwei was launched aboard *Shenzhou 5* in October 2003. Liwei became the first Chinese to be sent into space directly by its Chinese space program, the China National Space Administration (CNSA). Since then, two taikonauts were sent aboard *Shenzhou 6* on October 2005, and Zhai Zhigang became the first Chinese astronaut to walk in space during his *Shenzhou 7* mission in 2008, which also involved two other Chinese astronauts. As of September 2011, six Chinese astronauts have gone into space. China is now the third country to be able to send people directly into space.

Private organizations are also gearing up to send paying customers into space. Consequently, pilots are needed for these pioneering flights. A new type of astronaut has resulted, what is being called a commercial astronaut. These commercial (or professional) astronauts will be trained to command, pilot, or serve as crewmembers on privately funded space-craft. For instance, the first commercial astronaut to go into space was Michael "Mike" Melvill, who piloted the experimental spaceplane called *SpaceShipOne* for Scaled Composites. On June 21, 2004, Melvill piloted flight 15P on the first privately funded (non-government) trip into space as part of the Ansari X Prize competition. On October 4, 2004, Brian Binnie, also for Scaled Composites, became the second commercial astro-naut to go into space when he piloted *SpaceShipOne* on flight 17P.

Currently, several private spaceflight organizations are attempting to develop spacecraft to deliver cargo and/or humans into space. Virgin Galactic, using spacecraft developed by Scaled Composites (in a partner-ship called The Spaceship Company), seems to have the best chance to become the first privately funded company to send paying customers into space with the use of its career commercial astronauts. Richard Branson heads Virgin Galactic, while Burt Rutan leads Scaled Composites. Although a firm date has yet to be set, the first commercial flight into space by this organization is hoped for sometime in the mid-2010s.

 See also **Astronauts, Types of (Volume 3) • Microgravity Research Aircraft (Volume 3) • Mission Specialists and Flight Engineers (Volume 3) • NASA (Volume 3) • Payload Specialists (Volume 3)**

Resources

Books and Articles

Mari, Christopher, editor. *U.S. National Debate Topic 2011-2012: American Space Exploration and Development.* New York: H. W. Wilson, 2011.

McCurdy, Howard E. *Space and the American Imagination.* Baltimore: Johns Hopkins University Press, 2011.

National Aeronautics and Space Administration. *Astronaut Fact Book.* Washington, D.C.: NASA, 2005.

Spires, David N, et al. *Beyond Horizons: A Half Century of Air Force Space Leadership.* Peterson Air Force Base, CO: Air Force Space Command (Air University Press), 2011.

Websites

Astronaut Candidate Program. NASA. <http://astronauts.nasa.gov/content/broch00.htm> (accessed August 31, 2011).

Astronaut Selection. National Aeronautics and Space Administration. <http://www.nasajobs.nasa.gov/astronauts/default.htm> (accessed August 31, 2011).

Astronaut Biographies. NASA. <http://history.nasa.gov/nauts.html> (accessed August 31, 2011).

Expedition 28. NASA. <http://www.nasa.gov/mission_pages/station/expeditions/expedition28/index.html> (accessed August 31, 2011).

First Chinese Astronaut Back Home Safe. China.org. <http://www.china.org.cn/english/2003/Oct/77449.htm> (accessed September 1, 2011).

Malik, Tariq. *New NASA Astronauts Will Never Fly on Shuttle.* MSNBC. <http://www.msnbc.msn.com/id/31623240/ns/technology_and_science-space/t/new-nasa-astronauts-will-never-fly-shuttle/> (accessed August 31, 2011).

News. Virgin Galactic. <http://www.virgingalactic.com/news/> (accessed September 1, 2011).

Space Station Gets Its First Woman Commander. ABC News. <http://abcnews.go.com/Technology/TenWays/story?id=3751344&page=1> (accessed September 1, 2011).

Careers in Business and Management

One of the most interesting and potentially exciting trends in space exploration in the late twentieth and early twenty-first centuries has been the move towards the privatization and commercial exploitation of space. Privatization refers to the transfer of operations from the government or public agencies to private sector management. Several organizations have suggested that many aspects of the U.S. space program's involvement in the International Space Station (ISS) and all Space Transportation System (STS, or space shuttle) operations should be privatized. Such suggestions will likely be continued with the next manned U.S. space program. The commercial exploitation of space has been a key topic of interest since the space program began. Commercialization and privatization of space go hand-in-hand, but the words have somewhat different meanings.

"Commercialization of space" is the term used by the National Aeronautics and Space Administration (NASA) and the U.S. Department of Commerce to describe the technology transfer program, where technologies developed by NASA are transferred to the private sector. The term is also used to describe purely private ventures that seek to use space as a resource for making a profit. This includes satellite delivery systems, asteroid mining, space-waste disposal, space tourism, and medical or commercial uses of the ISS. One of the earliest satellites launched was a giant balloon named *Echo*, which was used as a test of satellite communications. Now, in the early twenty-first century, the space around Earth is filled with orbiting communications satellites, mostly owned and operated by private industry.

The commercialization of space offers new opportunities for private enterprise. While large aerospace corporations continue to dominate the industry, several small companies have been formed in recent years with the intention and stated goal of commercially exploiting space. Many former astronauts, and former NASA scientists and engineers, have moved on to these companies, suggesting that people with knowledge of space exploration consider privatization and commercial development of space enterprise the way of the future.

Careers in Aerospace

Individuals well suited for a career in aerospace tend to enjoy figuring out how things work; mathematics and science; solving puzzles, especially mechanical puzzles; building flying model rockets, model airplanes, or trains; learning new things; and working with computers. There are several different ways to prepare for a career in aerospace, including taking plenty of math and science courses in high school. For those interested in design, research, or development of new aerospace systems, a college degree is generally required, preferably in engineering or science, but not necessarily aerospace engineering. After completing a degree, many seek a job in the aerospace industry or a related field and immediately apply for on-the-job training for specialized aerospace fields. Because jobs in the aerospace industries are very competitive, enlisting in one of the armed forces and applying for specialized training or even flight school are also recommended.

For the most part, everything that flies in the air or orbits Earth is made by a company within the aerospace industry. As of July 1, 2011, the industry employed 479,900 workers in the United States. Almost

60 percent of that number (277,800) were production workers, accounting for about 2.8 percent of all manufacturing jobs in the United States (2008 data). The number of employees in the aerospace industry declined substantially from its peak of 578,600 in 1998 to its current level. The U.S. Bureau of Labor Statistics (BLS) has predicted that employment levels in the industry have leveled off and will remain close to their current levels until at least 2018. BLS predicts a total decrease in employment in the industry of 1.5 percent between 2008 and 2018 levels.

The economic downturn in the United States and Europe beginning in 2008 had a dramatic effect on midterm forecasts for the future of the aerospace industry around the world. Optimism about significant long-term growth has been replaced by caution resulting from the decrease in investment in aerospace projects in these parts of the world. Analysts now suggest that the next spurt in aerospace growth is likely to occur in China, India, and South American countries such as Brazil and Argentina.

Perhaps the most significant change in the aerospace industry in the United States in the foreseeable future is the reduced commitment to some important space programs, including the end of the space shuttle program in 2011 and President Barack Obama's (1961–) 2010 decision to cancel the U.S. Constellation project, designed to return a human to the Moon in the 2010s. As of 2011, the new vision for NASA includes an emphasis on basic research not directly tied to specific engineering projects and increased support for the privatization of space and aerospace projects, a change in course for which the economic and employment consequences are not yet clear.

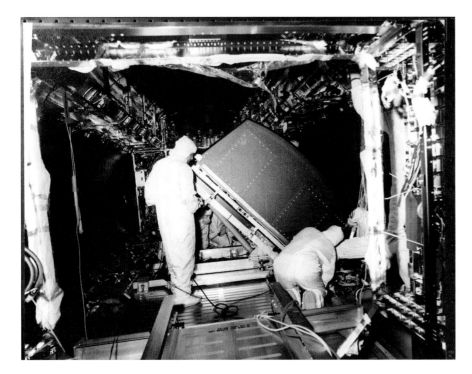

The aerospace industry is one of the largest employers in the United States. Here, Boeing technicians install the first of 24 system racks into the U.S. laboratory module for the International Space Station at the NASA Marshall Space Flight Center in Huntsville, Alabama. *NASA.*

It is almost impossible to get a job in the aerospace industry without a high school diploma; a bachelor's degree is desired, at a minimum. However, there are many different opportunities for employment in the aerospace industry, at many different levels from high school graduates to persons with advanced degrees in science, mathematics, and engineering. At whatever level a person is employed, special training or skill preparation is required. Administrative assistants working in the aerospace industry must be able to handle the complex technical language used by the industry. Union workers must be trained in the special manufacturing techniques used in aircraft and spacecraft, including ceramics, fiber composites, and exotic metals. Many workers must obtain a security clearance, which includes extensive background checks.

Many companies hire more electrical engineers, mechanical engineers, and computer specialists than aerospace engineers. Also in high demand are materials scientists (to develop new alloys and composites), civil engineers (for site design and development), and chemical engineers (to study new fuels). Companies also hire safety engineers, manufacturing engineers (to help design efficient manufacturing processes), test and evaluation engineers, and quality control engineers.

A technical degree or advanced degree is not essential to work in the aerospace industry. Some jobs do not require a degree at all; however, they are becoming fewer in number each year. Engineers and scientists represent less than one-third of the workforce. The remaining two-thirds are nontechnical support personnel. In production companies that primarily manufacture hardware, the proportion of engineers and technicians may be as low as 10 to 15%.

The large portion of employees at a typical aerospace company includes 10 to 20% professional employees, such as managers, salespeople, and contract administrators. Mechanics, electricians, and drafters are another 5 to 10% of the employees. The remainder include human resource specialists, engineering records employees, secretaries, and assembly line workers.

Aerospace Program Management

The aerospace industry has managed some of the largest, most expensive, and complex projects ever undertaken by humans. Projects such as the Apollo missions, with the goal of landing humans on the Moon within a decade, and the ISS involved thousands of people, working all over the globe on different aspects of the project, who had to all come together at the right time and place. Learning to manage such huge projects requires excellent technical comprehension and outstanding management abilities.

Some critics have blamed NASA's "faster, better, cheaper" management approach, which was used on the series of "Discovery" class missions, for the spectacular failures of the *Mars Climate Orbiter* and *Mars Polar Lander*, in the 1998–1999 timeframe. Former NASA administrator Daniel Goldin has commented that in the 1990s NASA dramatically increased its number of missions and decreased the time for each, while at the same time reducing the size of its staff. This resulted in less experienced program managers who received insufficient training and mentoring.

The lack of qualified managers has led to the development of specialized training in program management. Scientists or engineers who have traditionally managed programs in space-related industries learned to manage programs while on the job. Former astronauts or others working in the aerospace industry have also managed such programs.

Some missions within the Discovery program, such as *NEAR Shoemaker*, *Mars Pathfinder*, *Lunar Prospector*, *Deep Impact*, and *Stardust*, have been much more successful. Teams from industry, government laboratories, universities, and elsewhere are selected by NASA to develop and implement their own proposals, with a principal investigator as its lead. Although this new approach has led to some spectacular successes in the space program, it has also led to some notable failures.

In response to criticism and failures of NASA, in particular, and the aerospace industry, in general, the National Academy of Sciences completed a study and published a white paper with a suggested new design for program management in the late 1990s. While the report specifically addresses human exploration of space and a potential Mars mission, its principles are applicable to any large-scale endeavor. The report grouped its recommendations into three broad areas.

The first recommendation made by the study group was that scientific study of specific solar system objects be integrated into an overall program of solar system exploration and science and not be treated as separate missions of exploration simply because of the interest in human exploration. All scientific solar system research would be grouped into a single office or agency.

The second recommendation made in the report was that a program of human spaceflight should have clearly stated program goals and clearly stated priorities. These would include political, engineering, scientific, and technological goals. The objectives of each individual part of a mission would have clearly stated priorities. These would be carefully integrated with the overall program goals.

The last recommendation made by the study group was that human spaceflight programs and scientific programs should work with a joint program office that would allow collaboration between the human exploration and scientific components. As a model, the study group suggested the successful Apollo, Skylab, and Apollo-Soyuz missions.

The National Academy of Sciences, through its Space Studies Board, continues to conduct reports annually on the aerospace industry. In addition, in 2009, NASA initiated an independent committee to study and analyze its planned human spaceflight activities. That committee issued its report in October 2009, outlining a number of options available for continued manned flights into space. It enunciated a number of general principles for future manned spaceflight programs, including the need to design programs that were consistent with economic reality, taking advantage of opportunities to use an international approach to programs and exploring a variety of space challenges ranging from low-Earth-orbit flights to the colonization of Mars and the Moon.

▶ *See also* **Career Astronauts (Volume 1)** • **Careers in Rocketry (Volume 1)** • **Careers in Space Law (Volume 1)** • **Careers in Space Medicine (Volume 1)** • **Careers in Writing, Photography, and Film (Volume 1)**

Resources

Books and Articles

Echaore-McDavid, Susan. *Career Opportunities in Aviation and the Aerospace Industry.* New York: Ferguson, 2005.

Longuski, Jim. *Advice to Rocket Scientists: A Career Survival Guide for Scientists and Engineers.* Reston, VA: American Institute of Aeronautics and Astronautics, 2004.

Maples, Wallace R. *Opportunities in Aerospace Careers.* Chicago: VGM Career Books, 2003.

National Research Council. *Space Studies Board Annual Report 2010.* (published annually) Washington, DC: National Academies Press, 2011.

Sacknoff, Scott, and Leonard David. *The Space Publications Guide to Space Careers.* Bethesda, MD: Space Publications LLC, 1998.

Senson, Ben, and Jasen Ritter. *Aerospace Engineering: From the Ground Up.* Sydney, Australia: Delmar Cengage Learning 2011.

Websites

Aerospace Careers. Federal Aviation Administration. <http://www.faa.gov/about/office_org/headquarters_offices/ast/careers/> (accessed October 12, 2011).

FAA Aerospace Forecast. Federal Aviation Administration. <http://www.faa.gov/about/office_org/headquarters_offices/apl/aviation_forecasts/aerospace_forecasts/2011-2031/media/2011%20Forecast%20Doc.pdf> (accessed October 12, 2011).

The Growing Role of Emerging Markets in Aerospace. <http://www.mckinsey.it/storage/first/uploadfile/attach/140142/file/grro08.pdf> (accessed October 12, 2011).

Industry Output and Employment Projections to 2018. U.S. Bureau of Labor Statistics (BLS). <http://www.bls.gov/opub/mlr/2009/11/art4full.pdf> (accessed October 12, 2011).

Seeking a Human Spaceflight Program Worthy of a Great Nation. National Aeronautics and Space Administration (NASA). <http://www.nasa.gov/pdf/396093main_HSF_Cmte_FinalReport.pdf> (accessed October 12, 2011).

Space Studies Board: The National Academies. <http://www7.nationalacademies.org/ssb/> (accessed October 12, 2011).

Careers in Rocketry

Three important developments during the first half of the twentieth century laid the foundation for both modern rocketry and careers within the field. The first was the inspired scientific and engineering work performed by Robert H. Goddard on solid propellant rockets, and subsequently on liquid propellant rockets, during the years 1915 through 1942. Remembered as the "Father of Modern Rocketry," Goddard was a physics professor at Clark University in Massachusetts. Working mostly alone, with limited funds, he built and launched experimental solid and liquid propellant rockets and established the physical principles that enabled future rocket development to proceed.

The second event took place in Germany in the early 1930s. Goddard's work generated little interest in the United States, but it excited a small

Robert Goddard, one of America's first rocket scientists, poses with one of his rockets at Roswell, New Mexico in 1938. © *AP Images.*

* **ballistic** the path of an object in unpowered flight; the path of a spacecraft after the engines have shut down

* **payload** any cargo launched aboard a rocket that is destined for space, including communications satellites or modules, supplies, equipment, and astronauts; does not include the vehicle used to move the cargo or the propellant that powers the vehicle

* **oxidizer** a substance mixed with fuel to provide the oxygen needed for combustion

organization of young German rocket enthusiasts called *Verein Zur Forderung Der Raumfahrt* (VFR). One of the leaders of that group was Wernher von Braun, who ultimately helped the United States in the space race against the Soviet Union to place men on the Moon. The VFR built successful experimental liquid propellant rockets and captured the interest of the German army. VFR members were then assigned to develop a long-range ballistic* missile that could deliver bombs to London, England. This huge effort ultimately resulted in the development of the V-2 rocket, which caused great devastation when it was used during World War II (1939–1945). In the space of a few months, over 1,300 V-2s were launched toward England. Technologically, the V-2 was an impressive development. It formed the prototype for most of the liquid-fueled rockets that were built over the next fifty years.

The third important development was the atomic bomb and the onset of the Cold War between the Soviet Union and the United States. This occurred immediately after the conclusion of World War II in 1945. Using V-2 technology, both nations embarked on enormous efforts to develop ballistic missiles that could deliver atomic bombs to any target. Coincidentally, that effort helped develop rockets capable of carrying payloads* into space.

Rocket Development

Professionals in the field of rocketry work on two general types of rockets: liquid propellant and solid propellant. Each type has applications where it is best suited. A third type, called a hybrid, combines a solid fuel with a liquid oxidizer*. At the beginning of the twenty-first century, hybrids were in early development by the National Aeronautics and Space Administration (NASA).

Liquid propellant rockets are generally preferred for space launches because of their flexibility of operation and better performance. For instance, the engines can be shut off and restarted, and the thrust can be throttled. On the other hand, solid propellant rockets have some tactical advantages. They do not require propellant loading on the launch pad, and they can be stored for long periods. Liquid propellant and solid propellant rockets have been used jointly to advantage in such cases as the Space Transportation System (STS), better known as the space shuttle. The shuttle was initially boosted by two solid rocket boosters working in tandem with the shuttle's liquid-fueled main engine. Other liquid rockets that employ solid propellant boosters are the Delta and Atlas space launchers.

The design and development of a rocket always begins with a requirement. That is, what is the nature of the mission that it is going to perform? The requirement could be established by the military, by NASA, or by a commercial enterprise concerned with exploiting opportunities in space. For the military, such a requirement might be putting a communications or spy satellite in Earth orbit. For NASA it might be launching a spacecraft to Jupiter or sending a lander to Mars. For commercial enterprises, the requirement usually centers on communications satellites or Earth-observation spacecraft.

As an example of the work involved in the field of rocketry, consider what happens when NASA comes up with a new requirement, and no existing rocket is capable of handling the mission. An entirely new rocket design is needed. Working to the requirements, a team of designers and systems engineers synthesizes several different concepts for the new rocket. Like all rockets, it basically consists of a propulsion system, propellant tanks to hold the propellant, guidance and electronics equipment to control and monitor the rocket in flight, structure to hold the parts together, and miscellaneous components, valves, and wiring needed to make the rocket function. An advanced rocket that could complete the mission in a single stage might be included in the investigations, as well as various arrangements of two or more stages. In this phase of work, coordination with rocket engine and electronics systems manufacturers begins. Working together, the designers, engineers, and manufacturing professionals determine what is available, or what would need to be developed to make a particular concept work. Then the various concepts are compared in what are called trade studies, to determine which one can be built with the least cost, with the least risk, and on time, with the additional consideration of operational costs. Eventually each team will submit its best technical proposal and business plan, and subsequent evaluations and negotiations with NASA will culminate in placing a development contract with the winner.

Professions in Rocketry

Looking back to the beginnings of rocketry, Goddard served the functions of inventor, scientist, engineer, machinist, and test engineer, all combined into one. Modern design, development, manufacture, and operation of rockets require a broad array of professionals, including: mechanical, chemical, electronics, and aerospace engineers; thermodynamicists; aerodynamics and structural designers and analysts; manufacturing and tooling engineers; systems engineers; project engineers; and test engineers. For instance, mechanical engineers can be further divided into more specialized areas such as structural mechanics, fluid mechanics, orbital mechanics, and flight dynamics. In addition, rocket scientists also include physicists and mathematicians, along with professionals involved with materials science, reliability and safety, noise control, and flight test and reliability. Rocketry is now heavily computer oriented, so persons preparing for a career in this field should become proficient in computer-aided analysis, design, and manufacturing.

Preparing for careers in rocketry is much more interesting than it used to be in the past. Today, rocket contests for students are carried out around the United States. In fact, the world largest rocket contest is called the Team America Rocketry Challenge (TARC). The Aerospace Industries

The design and development of a rocket is always guided by the nature of the mission that it is supposed to perform, and it is carefully tested to ensure its suitability. *NASA.*

Association (AIA) and the National Association of Rocketry (NAR) sponsor the contest, which has been held annually since 2003. In this event, teams of students from 7th through 12th grade compete to design, build, and fly model rockets for a specified altitude and duration. The best teams from around the country receive scholarships and other prizes.

The Future of Rocketry

As the twenty-first century unfolds, rocketry is still in its infancy, and there are many important areas in which rocketry professionals will be needed in the future. It is too expensive to travel to space as an everyday occurrence, so technology must be directed toward reducing space launch costs. One way this could occur is with the development of intercontinental ballistic travel. The development of a rocket engine that can operate on air and fuel would make this possible. With this innovation, hundreds of daily suborbital flights across continents can be envisioned. Travel to space in similar vehicles will be much more economical, too. For instance, Virgin Galactic is one such private enterprise developing rides into space, through its SpaceShipTwo (built by Scaled Composites), that will someday be affordable for the general public. In space travel, scientists can forecast nuclear propulsion as a way to travel past the Moon, particularly if it turns out that water under the lunar surface can be readily mined. Nuclear steam rockets will then become common. Pulse plasma rockets, huge butterfly-shaped rockets that collect solar energy from the Sun, may also be used. Ultimately, for travel to distant star systems, the tremendous energy available in particle annihilation systems (matter-antimatter rockets) could be applied in propulsion. Practical containers for antimatter* may be impossible to achieve. However, the secret may be to use antimatter as fast as it is generated—a challenge for rocketeers of the twenty-first century.

Other possible propulsion systems for future rocket engineers and scientists to study include ion propulsion (which uses an ion beam for thrust), along with solar and thermal propulsion (which uses light from the Sun or laser light to heat a hydrogen-based fluid). Nuclear fission and fusion also offer exciting possibilities for future research and development in the area of rockets.

 See also **Goddard, Robert Hutchings (Volume 1) • Launch Vehicles, Expendable (Volume 1) • Launch Vehicles, Reusable (Volume 1) • Nuclear Propulsion (Volume 4) • Reusable Launch Vehicles (Volume 4) • Rocket Engines (Volume 1) • Rockets (Volume 3) • Space Shuttle (Volume 3) • von Braun, Wernher (Volume 3)**

Resources

Books and Articles

Goddard, Robert H. *Rockets.* New York: American Rocket Society, 1946.

Hujsak, Edward J. *The Future of U.S. Rocketry.* La Jolla, CA: Mina Helwig Company, 1994.

———. *All about Rocket Engines.* La Jolla, CA: Mina Helwig Company, 2000.

Humphries, John. *Rockets and Guided Missiles.* New York: Macmillan Company, 1956.

Longuski, Jim. *Advice to Rocket Scientists: A Career Survival Guide for Scientists and Engineers.* Reston, VA: American Institute of Aeronautics and Astronautics, 2004.

Speth, Roland S. "Visiting the Mettelwerk, Past and Present." *Spaceflight* 42(2000):115–119.

Sutton, George P. *Rocket Propulsion Elements.* New York: Wiley, 1949 (and subsequent editions).

Taylor, Travis S. *Introduction to Rocket Science and Engineering.* Boca Raton, FL: CRC Press, 2009.

Websites

Aerospace Industries Association. <http://www.aia-aerospace.org/> (accessed October 26, 2011).

Career Corner. National Aeronautics and Space Administration. <http://www.nasa.gov/audience/foreducators/rocketry/careercorner/index.html> (accessed October 26, 2011).

* **antimatter** matter composed of antiparticles, such as positrons and antiprotons, as opposed to normal matter composed of particles, such as electrons and protons

National Association of Rocketry. <http://www.nar.org/> (accessed October 26, 2011).

Team America Rocketry Challenge. Aerospace Industries Association. <http://www.rocketcontest.org/> (accessed October 26, 2011).

Careers in Space Law

Attorneys have been involved in space law since the early 1960s, when the legal community started addressing many rules and regulations relating to outer space activities. Space law practice deals with the legally related behavior of governments and private individuals who have interacted in some manner with outer space.

Issues that Space Lawyers Address

Many situations requiring legal expertise crop up in the world of space. Space lawyers rely on already established space law but still enter into uncharted territory. An example of such an undefined area that affects what space attorneys do is the designation of where space begins. The Outer Space Treaty and most of the other international conventions do not define the boundary between Earth's atmosphere and outer space. However, many space lawyers use the Kármán line as the defining altitude above Earth for the boundary between Earth's atmosphere and outer space. That line is approximately 100 kilometers (62.1 miles) above Earth's surface, and is defined by the Fédération Aéronautique Internationale, an international standards-setting and record-keeping group for the aeronautics and astronautics community. Another dilemma confronting space lawyers is the many provisions of the treaties, such as the ban on claims of sovereignty and property rights in space as well as the prohibition against military operations in outer space.

Generally speaking, space law attorneys handle two areas of outer space law:

- International space law, which governs the actions of countries as they relate to other states.
- Domestic space law, which governs actions within the state.

The Five Core Space Treaties

Space attorneys conduct most of their legal activities in keeping with space treaties, which resulted from the establishment of the United Nations (UN) Committee on the Peaceful Uses of Outer Space (COPUOS), within its Office of Outer Space Affairs, in 1958. The COPUOS was established within the United Nations in order to coordinate the development of laws and principles applicable to the activities in outer space.

Many countries have ratified five major international treaties and conventions, which guide space law attorneys in international and domestic space law.

The first major space treaty was the 1967 Treaty on Principles Governing the Activities of States in the Exploration and Use of Outer Space, including the Moon and Other Celestial Bodies (known as the Outer Space Treaty). This treaty addresses many liability issues that attorneys would be involved in litigating. Countries that did not ratify the 1972 Liability Convention may still be legally obligated to abide by this treaty.

The 1968 Agreement on the Rescue of Astronauts, the Return of Astronauts, and the Return of Objects Launched into Outer Space (known as the Rescue Agreement) was the next major treaty. Attorneys play an important role with respect to this treaty by providing counsel to government and public organizations concerning rescue and recovery efforts.

This was followed by the 1972 Convention on International Liability for Damage Caused by Space Objects (known as the Liability Convention). One of the biggest concerns space attorneys deal with regarding these five treaties is the issue of liability. Therefore, space law practice is largely involved with such issues. Among the issues space law attorneys currently handle is damage caused by spacecraft and satellites, as well as indirect effects such as causing pollution in outer space that adversely affects Earth. In the future, as private tourism expands into space and private citizens go into outer space for pleasure, a very strong interest will arise in liability provisions and indemnification through the insurance industry.

Specifically, the Liability Convention requires payment of damages making restitution for "loss of life, personal injury or other impairment of health, or loss or damage to property of States or of persons, natural or juridical, or property of international governmental organizations" (Liability Convention, Article 1). A "launching state" is explicitly defined as a state that launches or procures the launching of a space object or a state from whose territory or facility a space object is launched, regardless of whether the launch was in fact successful.

The 1976 Convention on Registration of Objects Launched into Outer Space (known as the Registration Convention) requires adherence to regulations regarding the tracking of all spacecraft and satellites. Attorneys counsel organizations on how to comply with these requirements.

The final major space treaty was the 1979 Agreement Governing the Activities of States on the Moon and Other Celestial Bodies (known as the Moon Agreement). The United States has not ratified this treaty, but legal counsel still needs to be aware of its ramifications, especially when working with ratifying countries on joint projects.

In 1998, many countries entered into the Civil International Space Station Agreement Implementation Act, sometimes called the Space Station Agreement, when they became involved with activities involving the International Space Station (ISS). These countries include the United States, Japan, Russia, and Canada, along with the European Union. Among its statements, the Space Station Agreement acknowledges that the United States is the lead country responsible for the ISS, but that each participating country has responsibility over its own modules and equipment.

Now in the twenty-first century, more countries are becoming involved in space-based activities and private enterprises are beginning the first steps into space tourism and the commercialism of space. Thus, the expansion of space law is bound to happen as more disputes and disagreements occur in the interpretation of existing law and as humans expand into areas not even made yet into law. In the past, students studying law and wanting a career in space law did not have access to a specific legal degree for it. However, in preparation for more activities within space law, the first space lawyer was graduated, in May 2008, in the United States. U.S. space lawyer Michael Dodge earned his law degree at the University of Mississippi (Oxford), through its law school's National Center for Remote Sensing, Air, and Space Law.

 See also **Law (Volume 4)** • **Law of Space (Volume 1)** • **Legislative Environment (Volume 1)**

Resources

Books and Articles

Diederiks-Verschoor, Isabella Henrietta Philepina, and Vladimir Kopal. *An Introduction to Space Law.* 3rd ed. Alphen aan den Rijn, Netherlands: Kluwer Law International, 2008.

Gangale, Thomas. *The Development of Outer Space: Sovereignty and Property Rights in International Space Law.* Santa Barbara, CA: Praeger, 2009.

Hudgins, Edward L., ed. *Space, the Free-Market Frontier.* Washington, DC: Cato Institute, 2002.

Johnson-Freese, Joan, and Roger Handberg. *Space, the Dormant Frontier: Changing the Paradigm for the Twenty-first Century.* Westport, CT: Praeger, 1997.

Websites

Civil International Space Station Agreement Implementation Act. Department of Justice, Canada. <http://laws.justice.gc.ca/eng/acts/C-31.3/page-1.html> (accessed October 6, 2011).

First Space Lawyer Graduates. Space.com. <http://www.space.com/news/080508-first-space-lawyer.html> (accessed October 6, 2011).

Journal of Space Law.. University of Mississippi, National Center for Remote Sensing, Air, and Space Law. <http://www.spacelaw.olemiss.edu/jsl/index.html> (accessed October 6, 2011).

United Nations Committee on the Peaceful Uses of Outer Space. <http://www.unoosa.org/oosa/COPUOS/copuos.html> (accessed October 6, 2011).

Careers in Space Medicine

Outer space has a very different environment from that of Earth. The atmosphere, radiation, and gravity levels are so drastically varied that several major adjustments are made to protect astronauts from the deadly effects of the space environment on the human body. Thus, space medicine involves the practice of medicine on humans living and working in outer space. In particular, gravitational effects are not well controlled and the effects of long-term exposure to microgravity* are unknown.

Several experiments have already indicated that major biological changes begin in the human body within minutes of spaceflight. For example, when a person is exposed to microgravity, there is less blood volume in the legs and more in the upper body. This change makes the brain sense that there is too much fluid in the body and triggers an adaptive response such as increased urine production. This is similar to the sensation experienced by many people upon entering a swimming pool. Even the chemical composition of blood and urine is altered, which perhaps reflects that other body tissue and organ changes are taking place in response to the loss of gravity. Given all of the effects of spaceflight upon the human body, several career opportunities exist to study and treat future space travelers.

Medical Challenges

Several questions remain about the scope of space medicine because scientists do not understand all of the changes that take place in the human body either on Earth or in space. Can all of the problems be treated? Is a treatment really needed? Is the prevention of adaptive changes that occur when humans go to space better than treatment after a change has already taken place? How can medical problems in space be prevented? Will the body's cells, tissues, and organs return to normal at different rates upon landing back on Earth or on another planet with similar gravity?

Because scientists understand so little about how the human body works in normal gravity on Earth, few specific cures have been found for medical conditions. Doctors often can only treat symptoms and not the causes of diseases. For instance, allergy medication eliminates the symptoms of the malady*, but when a patient stops taking the medicine, the symptoms may return. A heart may no longer be able to beat properly, but a device such as a pacemaker can assist in doing the job. In neither

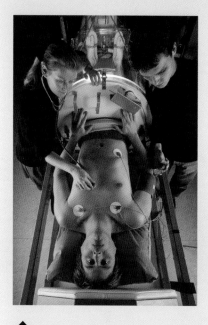

When in space, humans experience decreased aerobic capacity, endurance, strength, and muscle mass. Physicians in space medicine conduct tests to determine what types of exercise will most benefit astronauts who spend extended periods of time in space. © *Roger Ressmeyer/Corbis.*

* **microgravity** the condition experienced in freefall as a spacecraft orbits Earth or another body; commonly called weightlessness; only very small forces are perceived in freefall, on the order of one-millionth the force of gravity on Earth's surface

* **malady** a disorder or disease of the body

of these cases has the underlying disease been either treated or cured. If the main objective is to find a cure for a problem, how can this be done if what needs to be fixed is unknown? And how does one know what to fix if it is not known how the body normally works?

Scientists must therefore conduct experiments on Earth and in space to: (1) understand how the body normally functions in these two environments, (2) determine how diseases and other medical conditions develop, and (3) either find a specific cure or prevent the disease. Perhaps scientists will also discover that, in some instances, the adaptations the human body makes in space are not all necessarily bad or in need of medical attention.

Lessons from Experience

During the late 1950s to early 1960s, in the early years of spaceflight exploration, it was impossible to anticipate every change in bodily function that would occur to those venturing into space. It was unclear if even simple tasks such as swallowing would become a health hazard for the astronauts (i.e., is gravity needed to "pull down" food and water into the stomach?). The answers to this and other questions were found by successfully sending animals and then Russian cosmonauts and U.S. astronauts into short spaceflights.

By the early twenty-first century, scientists had gained a better understanding about how day-to-day, bodily activities are affected by microgravity. Though some negative side effects are indeed experienced (e.g., difficulty with eyesight focus, loss of balance, nausea), astronauts have by and large returned to live healthy lives on Earth.

Longer flights and numerous experiments later, scientists have a better idea of what may pose as a medical threat during long-term exposure to microgravity. These issues must be resolved to enable extended stays on the International Space Station (ISS) and flights to land humans on other planets such as Mars. The problems include a loss and reorganization of bone mass as well as loss of muscle strength and mass. If muscles are heavily used on Earth (such as with weight lifters), they become stronger and larger. Under the weightless conditions of space, the muscles no longer work as hard and become smaller (a condition known as atrophy). In space, the blood itself becomes weightless, and the heart will eventually atrophy because it has to work less to pump blood through the body. After a long trip in space, a sudden return to Earth might make an astronaut appear to have heart failure. Similarly, if bone does not sense the need to support the body against the effects of gravity, then spaceflight-induced bone loss might lead to osteoporosis-like problems upon a return to Earth.

Future Trends

To prevent microgravity-associated health problems and to ensure a safe return to normal bodily function, more studies are needed. Among the key areas for current and future research are diet, exercise, genetics, and whether or not hormones can produce their normal effects, both during and after spaceflight. Scientists and physicians trained to deal with these issues are needed both on the ground (e.g., in preparation for spaceflight) and as part of the flight team. Furthermore, understanding how the body changes in space will aid in the development of cures here on Earth, in addition to helping maintain the medical health of space travelers. For example, once it is known how bone is altered in space, would the discovery of a treatment also be useful to prevent or reverse osteoporosis?

The collective efforts of many biomedical-related fields are needed to fully understand and develop ways of coping with the effects of microgravity on the human body. The body is an integrated system in which different cells, tissues, and organs affect or interact with one another. When one system is altered, there are usually consequences to another system. Simply getting out of bed in the morning leads to many integrated changes—blood flows and pools in the legs, the blood vessels counteract this by contracting to "push" the blood upward, heart rate increases, and various hormones are released to prepare for the day's activities.

In 2008, a group of medical organizations signed a memorandum of understanding to work together for the betterment of the civilian spaceflight industry. With the intent to provide better space medicine resources, Wyle Laboratories, the University of Texas Medical Branch at Galveston, and Mayo Clinic (Arizona) joined forces for the research and development of a comprehensive space medicine resource. As humans venture more out into space, such agreements, from the private and public sectors, will help to assure safe journeys for these astronauts to distant worlds and around Earth.

Because of the integrative nature of human bodily functions, there are many career opportunities for basic research in such disciplines as pharmacology, biochemistry, biology, chemistry, physiology, and genetics. Physicians and other health-care professionals can then apply newly discovered biomedical information to ensure the continued improvement of human health on Earth and in space. When scientists work to make space a better place to work and live, they are also helping to make medical treatment on Earth better, especially in isolated rural areas, on the battle grounds of wars and conflicts, and at the scenes of natural and human-made disasters.

 See also **Aging Studies (Volume 1)** • **AIDS Research (Volume 1)** • **Cancer Research (Volume 1)** • **Medicine (Volume 3)**

Resources

Books and Articles

Lujan, Barbara F., and Ronald J. White, eds. *Human Physiology in Space.* Washington, DC: National Aeronautics and Space Administration Headquarters, 1995.

McPhee, Jancy C., and John B. Charles, eds. *Human Health and Performance Risks of Space Exploration Missions: Evidence Reviewed by the NASA Human Research Program.* Houston, TX: National Aeronautics and Space Administration, Lyndon B. Johnson Space Center, 2009.

Nicogossian, Arnauld E., Carolyn Leach Huntoon, and Sam L. Pool. *Space Physiology and Medicine.* 3rd ed. Philadelphia: Lea & Fibiger, 1994.

Sahal, Anil. "Neglected Obstacles to the Successful Exploration of Space." *Spaceflight* 42, no. 1 (2000):10.

Websites

"Leading Medical Institutes to Develop Collaborative Space Medicine Program." *Space Daily.* <http://www.spacedaily.com/reports/ Leading_Medical_Institutes_To_Develop_Collaborative_Space_ Medicine_Program_999.html> (accessed October 7, 2011).

Space Medicine. Japanese Aerospace Exploration Agency. <http://iss.jaxa. jp/med/index_e.html> (accessed October 7, 2011).

Space Medicine. National Aeronautics and Space Administration. <http://science.nasa.gov/headlines/y2002/30sept_spacemedicine. htm> (accessed October 7, 2011).

Careers in Writing, Photography, and Film

The ability to communicate with others is necessary in all avenues of life. However, in space science and technology it is critical to the successful furtherance of spaceflight objectives. One reason is the lesser importance that most individuals assign to the endeavor. Indeed, a central feature of communication efforts throughout the space age has been the coupling of the reality of spaceflight with the American imagination for exploring the region beyond Earth. Without it, humans might never have slipped the bonds of Earth, ventured to the Moon, and sent robots to the planets. In the process, the dreams of science fiction aficionados have been combined with developments in technology to create the reality of a spacefaring people.

An especially significant spaceflight "imagination" came to the fore after World War II (1939–1945) and urged the implementation of an aggressive spaceflight program. It was seen in science fiction books and film, but more important, serious scientists, engineers, and public intellectuals fostered it. The popular culture became imbued with the romance of spaceflight, and the practical developments in technology reinforced these perceptions that for the first time in human history space travel might actually be possible.

There are many ways in which the American public may have become aware that flight into space was both real and should be developed.

The communication of space exploration possibilities through the written, photographic, and electronic media began very early in the history of the space age. Since the 1940s, science writers such as Arthur C. Clarke and Willy Ley had been seeking to bring the possibilities of spaceflight to a larger audience. They had some success, but it was not until the early 1950s that spaceflight really burst into the public consciousness.

Wernher von Braun's Role in Promoting Spaceflight

In the early 1950s, the German émigré scientist Wernher von Braun, working for the army at Huntsville, Alabama, was a superbly effective promoter of spaceflight to the public. Through articles in a major weekly magazine, *Collier's,* von Braun urged support for an aggressive space program. The *Collier's* series, written in 1952, catapulted von Braun into the public spotlight as none of his previous activities had done. The magazine was one of the four highest-circulation periodicals in the United States during the early 1950s, with over three million copies produced each week. If the readership extended to four or five people per copy, as the magazine claimed, something on the order of 15 million people were exposed to these spaceflight ideas.

Von Braun next appeared on a variety of television programs, including a set of three highly rated Disney television specials between 1955 and 1957. These reached an estimated audience of 42 million each and immeasurably added to the public awareness of spaceflight as a possibility. As a result, von Braun became *the* public intellectual advocating space exploration, an individual recognized by all as an expert in the field and called upon to explain the significance of the effort to the general population.

The coming together of public perceptions of spaceflight as a near-term reality with rapidly developing technologies resulted in an environment more conducive to the establishment of an aggressive space program. Convincing the American public that spaceflight was possible was one of the most critical components of the space policy debate of the 1950s and 1960s. For realizable public policy to emerge in a democracy, people must both recognize the issue in real terms and develop confidence in the attainability of the goal. Without this, the creation of the National Aeronautics and Space Administration (NASA) and the aggressive piloted programs of the 1960s could never have taken place.

2001: A Space Odyssey

The powerful spaceflight concepts championed by von Braun found visual expression in a wide-screen Technicolor feature film released in 1968, *2001: A Space Odyssey.* Director Stanley Kubrick brought to millions a stunning science fiction story by Arthur C. Clarke about an artificially made monolith found on the Moon and a strange set of happenings at Jupiter. With exceptional attention to science fact, this film drew the contours of a future in which spaceflight was assisted by a wheel-like space station in orbit, a winged launch vehicle that traveled between Earth and

George Lucas, the creator of *Star Wars* saga, on the set of the second episode: *Attack of the Clones.* © LucasFilm/20th Century Fox/The Kobal Collection/ Tomasetti, Lisa.

the station, a Moon base, and aggressive exploration to the other planets. All of this was predicted to be accomplished by 2001, and the director and the technical advisors were concerned that their vision might be outdated within a few years by the reality of space exploration. It was not, and their vision still is far from becoming reality.

The Impact of Photography

The photographic record of spaceflight has also served to sustain interest in the endeavor. For example, the photographs taken of Earth from space sparked a powerful reaction among those who viewed them for the first time. Project Apollo forced the people of the world to view planet Earth in a new way. In December 1968, *Apollo 8* became critical to this change, for the image taken by the NASA crew, "Earthrise," showed a tiny, lovely, and fragile "blue marble" hanging in the blackness of space with the gray and desolate lunar surface in the foreground as a stark contrast to a world teeming with life. Poet Archibald MacLeish summed up the feelings of many people when he wrote, "To see the Earth as it truly is, small and blue and beautiful in that eternal silence where it floats, is to see ourselves as riders on the Earth together, brothers on that bright loveliness in the eternal cold—brothers who know now that they are truly brothers." (MacLeish, December 25, 1968) The modern environmental movement was galvanized in part by this new perception of the planet and the need to protect it and the life that it supports.

Carl Sagan as Public Intellectual

Astronomer Carl Sagan emerged as a public intellectual on behalf of space exploration in the 1970s in much the same way that von Braun had in the 1950s. An academic on the faculty at Cornell University (Ithaca,

New York), Sagan eschewed the scholarly trappings of the "ivory tower" to engage the broadest possible audience directly through writing, speaking, and television appearances. His brilliant thirteen-part *Cosmos* series on public television in 1980, like the *Collier's* series of von Braun, captured the imagination of a generation of Americans about the wonders of the universe and energized the public debate concerning space exploration in the 1980s. In part, the increases in federal budgets for space activities could be related to the excitement generated by Sagan's compelling arguments.

Sagan went on to fill von Braun's shoes as a public intellectual with verve until his death in 1996. He wrote best-selling nonfiction, such as *Cosmos* (1980) and *Pale Blue Dot: A Vision of the Human Future in Space* (1994), and a novel, *Contact* (1985). Always he drew a tight relationship among technical capabilities, philosophical questions, and human excitement and destiny. He rarely wrote for academic audiences, as was normal for other scholars, and published more articles in *Parade* magazine, reaching millions of readers, than in professional journals. He also took on the proponents of pseudoscience*, especially efforts to convince individuals of extraterrestrial visitations to Earth, publishing a major work on the subject, *The Demon-Haunted World* (1995), near the end of his life. Sagan appeared on popular talk shows such as *The Tonight Show* with Johnny Carson and, later, with Jay Leno to espouse his vision of a hopeful future in space. His belief in a universe filled with life, and humanity's place in that universe, came to the big screen in 1997 with the making of *Contact* into a feature film starring Academy Award winner Jodie Foster.

Opportunities in the Field

Opportunities to expound a compelling vision of a future in space exist in all arenas available for communication. Using written, photographic, and multimedia forms of communication, future writers and visual artists have the opportunity to become public intellectuals whose ideas expressed in these forms will shape the future of spacefaring in the United States. Indeed, it is largely up to such individuals to frame the debate on the future of spaceflight. Will the astronauts and their voyages to the Moon be remembered as being akin to Italian explorer Christopher Columbus and his voyages to the Americas—as vanguards of sustained human exploration and settlement? Or, will their endeavors prove to be more like Leif Eriksson's voyages from Scandinavia several hundred years earlier, stillborn in the European process of exploration to new lands? No one knows yet, but the public intellectuals of the future using all of the tools of communication available to them will be the ones to prompt both the policymakers and the public to make decisions about sustained exploration.

▶ *See also* **Clarke, Arthur C. (Volume 1)** • **Entertainment (Volume 1)** • **Literature (Volume 1)** • **Lucas, George (Volume 1)** • **Movies (Volume 4)** • **Roddenberry, Gene (Volume 1)** • **Sagan, Carl (Volume 2)** • **Science Fiction (Volume 4)** • *Star Trek* **(Volume 4)** • *Star Wars* **(Volume 4)** • **von Braun, Wernher (Volume 3)**

* **pseudoscience** a system of theories that assumes the form of science but fails to give reproducible results under conditions of controlled experiments

Resources

Books and Articles

Ashford, David. *Spaceflight Revolution.* London: Imperial College Press, 2002.

Bainbridge, William Sims. *The Spaceflight Revolution: A Sociological Study.* New York: Wiley, 1976.

Launius, Roger D. *Frontiers of Space Exploration.* 2nd ed. Westport, CT: Greenwood Press, 2004.

Ley, Willy. Paintings by Chesley Bonestell. *The Conquest of Space.* New York: Viking, 1949.

MacLeish, Archibald. "Riders on Earth Together, Brothers in Eternal Cold." *New York Times*, December 25, 1968.

McCurdy, Howard E. *Space and the American Imagination.* 2nd ed. Baltimore: Johns Hopkins University Press, 2011.

Ordway, Frederick I., III, and Randy L. Liebermann. *Blueprint for Space: Science Fiction to Science Fact.* Washington, DC: Smithsonian Institution Press, 1992.

Poundstone, William. *Carl Sagan: A Life in the Cosmos.* New York: Henry Holt, 1999.

Sagan, Carl. *Contact: A Novel.* New York: Simon and Schuster, 1985.

———. *Cosmos.* New York: Random House, 2002.

———. *The Demon-Haunted World: Science as a Candle in the Dark.* New York: Ballantine Books, 1997.

———. *Pale Blue Dot: A Vision of the Human Future in Space.* New York: Ballantine Books, 1997.

Websites

The Arthur C. Clarke Foundation. <http://www.clarkefoundation.org/> (accessed October 26, 2011).

Biography of Wernher Von Braun. National Aeronautics and Space Administration, Marshall Space Flight Center. <http://history.msfc.nasa.gov/vonbraun/bio.html> (accessed October 26, 2011).

The Carl Sagan Portal. Druyan–Sagan Associates. <http://www.carlsagan.com/> (accessed October 26, 2011).

Kubrick 2001: The Space Odyssey Explained. New Media Giants. <http://www.kubrick2001.com/> (accessed October 26, 2011).

Clarke, Arthur C.

British Science Fiction Writer and Visionary
1917–2008

Born at Minehead, Somerset, United Kingdom, on December 16, 1917, Arthur Charles Clarke, better known as Arthur C. Clarke, was fascinated by science fiction and astronomy at an early age. In the 1930s, he joined the British Interplanetary Society. After enlisting in the Royal Air Force in 1941, he became a radar* instructor and participated in the development of ground-controlled landings of aircraft under zero-visibility conditions. His only non-science-fiction novel was based on this experience: *Glide Path.*

In 1945, the technical journal *Wireless World* published Clarke's article "Extra-Terrestrial Relays," which proposed the use of three broadcast satellites in equatorial orbit* to provide worldwide communication. Clarke chose an orbital altitude of 35,786 kilometers (22,300 miles) because at that distance the angular velocity* of Earth's rotation would match that of the satellite. As a result, the satellite would remain fixed in the sky. Twenty years later, Early Bird was launched, the first of the commercial satellites that provide global communications networks for telephone, television, and high-speed digital communication, including the Internet.

Today the geostationary orbit that Clarke described is officially named The Clarke Orbit by the International Astronomical Union, an international organization responsible for the naming of such objects.

After World War II (1939–1945), Clarke obtained a bachelor's of science degree in physics and mathematics at King's College, London. In 1954, he became enchanted by underwater scuba diving, which simulated weightlessness in spaceflight. In 1969, Clarke moved to Ceylon, now called Sri Lanka, in order to pursue his interest in underwater exploration and scuba diving.

Clarke passed away in Sri Lanka on March 19, 2008. He had battled post-polio syndrome for many years. On the New Mexico Museum of Space History Web site Clarke, the visionary, is quoted: "Sometimes I think we're alone in the universe, and sometimes I think we're not. In either case the idea is quite staggering."

Clarke has written eighty books on science and technology, along with their sociological consequences. One of his best-known works is *The Sentinel,* which is a short story about humankind's contact with sentient (emotional and intelligent) life. He then collaborated with director Stanley Kubrick on the film *2001: A Space Odyssey* (1968), which was based on this short story. He wrote the novel *2001: A Space Odyssey* in 1964 to coincide with the movie of the same name, which premiered in 1968.

Clarke was the president of the British Interplanetary Society from 1947 to 1950, and in 1953. He has received many honors and awards,

* **radar** a technique for detecting distant objects by emitting a pulse of radio-wavelength radiation and then recording echoes of the pulse off the distant objects

* **equatorial orbit** an orbit parallel to a body's geographic equator

* **angular velocity** the rotational speed of an object, usually measured in radians per second

Arthur Clarke's published article "Extra-Terrestrial Relays" formed the basis for current global communications systems. © *AP Images/Lenin Kumarasiri.*

including knighthood (recognized by the Queen in 1998, and conferred by the Prince of Wales in 2000), the Franklin Institute Gold Medal, the UNESCO-Kalinga Prize, honorary fellow memberships and awards from major scientific and astronautical organizations, and a nomination for the Nobel Peace Prize in 1994. He was inducted into the International Space Hall of Fame, at the New Mexico Museum of Space History, in 1989. Asteroid 4923 Clarke, which was discovered in 1981, was named in his honor. His book *The Star* won the Hugo Award, while *Rendezvous with Rama* was honored with the Nebula Award and the Hugo Award.

Among Sir Arthur C. Clarke's works are the following books:

Nonfiction

- *Ascent to Orbit, a Scientific Autobiography: The Technical Writings of Arthur C. Clarke.* New York: John Wiley & Sons, 1984.
- *Astounding Days: A Science Fictional Autobiography.* New York: Bantam, 1989.
- *The Exploration of Space.* New York: Harper, 1951.
- *Greetings, Carbon-Based Bipeds!: Collected Essays, 1934–1998.* New York: St. Martin's Press, 1999.
- *How the World Was One: Beyond the Global Village.* New York: Bantam, 1992.
- *The Making of a Moon: The Story of the Earth Satellite Program.* New York: Harper, 1957.
- *Profiles of the Future: An Inquiry into the Limits of the Possible.* New York: Harper, 1962.
- *The Promise of Space.* New York: Harper, 1968.
- *Voices from the Sky: Previews of the Coming Space Age.* New York: Harper, 1965.

Fiction

- *Childhood's End.* New York: Ballantine, 1953.
- *The Fountains of Paradise.* New York: Harcourt, 1979.
- *The Hammer of God.* New York: Bantam, 1993.
- *Islands in the Sky.* Philadelphia: Winston, 1952.
- *Rendezvous with Rama.* New York: Harcourt, 1973.
- *The Sands of Mars.* London: Sidgwick & Jackson, 1951.
- *2001: A Space Odyssey.* New York: New American Library, 1968.
- *2010: Odyssey Two.* New York: Ballantine, 1982.
- *2061: Odyssey Three.* New York: Ballantine, 1988.
- *3001: Final Odyssey.* New York: Ballantine, 1997.
- *Time's Eye.* (with Stephen Baxter) New York: Ballantine, 2004.
- *Sunstorm.* (with Stephen Baxter) New York: Del Rey, 2005.

See also **Careers in Writing, Photography, and Film (Volume 1) • Communications Satellite Industry (Volume 1) • Entertainment (Volume 1) • Science Fiction (Volume 4)**

Resources

Books and Articles

Clute, John, and Peter Nicholls, eds. *Encyclopedia of Science Fiction.* New York: St. Martin's Press, 1995.

D'Ammassa, Don, ed. *Encyclopedia of Science Fiction.* New York: Facts on File, 2005.

McAleer, Neil. *Arthur C. Clarke: The Authorized Biography.* Chicago: Contemporary Books, 1992.

Websites

Arthur C. Clarke. New Mexico Museum of Space History, International Space Hall of Fame. <http://www.nmspacemuseum.org/halloffame/detail.php?id=98> (accessed October 13, 2011).

The Arthur C. Clarke Foundation. <http://www.clarkefoundation.org/> (accessed October 13, 2011).

Jonas, Gerald. "Arthur C. Clarke, Author Who Saw Science Fiction Become Real, Dies at 90." *New York Times.* <http://www.nytimes.com/2008/03/19/books/19clarke.html> (accessed October 13, 2011).

Commercialization

"Space commercialization" is a general term that distinguishes private activities from those public activities of the government in enabling the use of space from either an Earth-based operation or from space itself. Private-sector use of space involves activities that are expected to return a profit to investors, such as building, launching, and operating communications satellites or taking pictures of Earth from space to monitor crops.

In contrast to the private sector, government activities are performed to carry out specific missions for the public good. Examples range from national defense activities to scientific missions studying the planets, and also include satellites that monitor Earth's environment.

Because space research, development, and exploration are very expensive and risky, governments have funded most activities in the past and are continuing to do so today. However, during the 1990s, private companies began to expand beyond the already profitable communications satellite

* **fiber-optic cable** a thin strand of ultrapure glass that carries information in the form of light (radiation), with the light turned on and off rapidly to represent the information sent

* **payload** any cargo launched aboard a rocket that is destined for space, including communications satellites or modules, supplies, equipment, and astronauts; does not include the vehicle used to move the cargo or the propellant that powers the vehicle

* **expendable launch vehicles** launch vehicles, such as a rocket, not intended to be reused

services, and develop the use of the space environment for the introduction of new products. As such, the U.S. government requires a license for a U.S. firm to launch spacecraft and do business in space. Often when there is an overlap between a government mission and a private activity, the government will partner with the private company.

"Commercialization of space" is frequently confused with "privatization of space." Sometimes commercialization of space is used by the government to mean that a function previously performed by the government has been shifted to a private company, often with the government as a paying customer. "Privatization of space" involves the government reallocating authority, responsibility, and the risk of operations using government-owned assets and ultimately transferring asset ownership itself to the private sector. Because privatization is a process, there are many intermediate steps possible between total government management, control, and asset ownership and full privatization. And, because this process involves firms that are providing services for a profit, privatization and commercialization sometimes are used as synonyms even when they are not precisely the same.

Examples of Space Commercialization

The largest commercial use of space is by satellite communications and associated services. Long-distance communications are dependent on two major transmission modes: satellites and fiber-optic cables*. Satellites are the cheapest and best providers of point-to-multipoint communications while fiber-optic cables provide efficient high-capacity point-to-point services. In 2011, the Satellite Industry Association reported that total revenues worldwide for the industry amounted to $168.1 billion. This total represented an increase of five percent over the previous year, and an overall increase of 89.3 percent between 2005 and 2010. The annual growth rate during that period averaged 11.2 percent.

Other uses of space for commercial purposes generate relatively small revenues but hold growth potential. Remote sensing (taking digital pictures of Earth from satellites) is used to monitor Earth and for mapping and discovering new sources of natural resources. Google, one of the commercial leaders in providing Earth mapping images to the public, released Google Earth in 2005, which provides images obtained from satellites, aerial photography, and three-dimensional geographic information system (3D GIS). Revenue from remote sensing satellites in 2010 amounted to $1 billion, double the amount realized from that source in 2005.

Launch vehicles that boost payloads* into space also provide business opportunities for firms. Since the late 1980s, expendable launch vehicles* (ELVs) have been privately manufactured and operated in the United States. Of course, the need for launch vehicles is determined by the need to place payloads in space. In 1996, the U.S. government, primarily through the Department of Defense, began a program to develop evolved expendable launch vehicles (EELVs) in order to reduce the costs of such vehicles and to assure reliable, safe, and cost-effective vehicles would continue to be built in the future. In 2006, the government awarded the EELV contract to Lockheed Martin and The Boeing Company after they formed the organization United Launch Alliance (ULA). In 2010, ULA conducted seven Atlas and Delta launches for the government; it followed with three additional such launches in the first half of 2011.

In 2010, the launch vehicle component of the satellite industry reported revenues of $4.3 billion, a small decrease from $4.5 billion in the preceding year. Launch revenues grew consistently in the period from 2005 ($3.0 billion) to 2010, at an average rate of about 8.6 percent per year.

Several firms are designing and developing commercial reusable launch vehicles (RLVs). One major goal of this effort is to produce vehicles capable of launching people into space for recreational purposes. The first success in this area was achieved in 2004 when the Tier One project of the Scaled Composites corporation won the $10 million Ansari X Prize for launching two sub-orbitals flights containing humans within a two week period. The space ship used for these flights, SpaceShipOne, was immediately retired, and work began on its next version, ShapeShipTwo. SpaceShipOne was carried into space by a jet-powered aircraft called White Knight One and then released. At that point, the space ship fired its own engines and traveled through space before returning to Earth. As with SpaceShipOne, White Knight One was then retired and development began on its successor, White Knight Two. Scaled Composites is now working in collaboration with Virgin Galactic to provide the first sub-orbital commercial flights using SpaceShipTwo. The first such flights are expected to occur sometime after 2012, with tickets currently available at a cost of $200,000 through Virgin Galactic.

The European Space Agency announced in January 2009 that it would begin work on a commercial spacecraft similar in concept to SpaceShipOne called the Future High-Altitude High-Speed Transport (FAST) 20XX project. The immediate goal of that project is to develop the technology required for sub-orbital human flight, with the production of an actual spacecraft scheduled for some years down the road.

Finally, the International Space Station (ISS) is the result of an international partnership among governments (primarily the United States, Russia, Japan, Canada, and the European Union), which is promoting a wide variety of commercial opportunities. Completion of the ISS was originally planned for the mid-2000s, but has been delayed to May 2012. At that point, the last remaining component, the Russia Nauka module, is expected to be delivered and attached to the ISS. Companies are being encouraged to perform research and development onboard the ISS. In addition, there are proposals to have private firms provide power and other "utilities" for the ISS, but so far (as of 2011) such proposals have not been completed. There will also be a market for

One area of growth potential in space commercialization is remote sensing—taking digital pictures of Earth from satellites. This orbital image is of Gezira Scheme in Sudan, Africa, one of the world's largest irrigation systems. *NASA.*

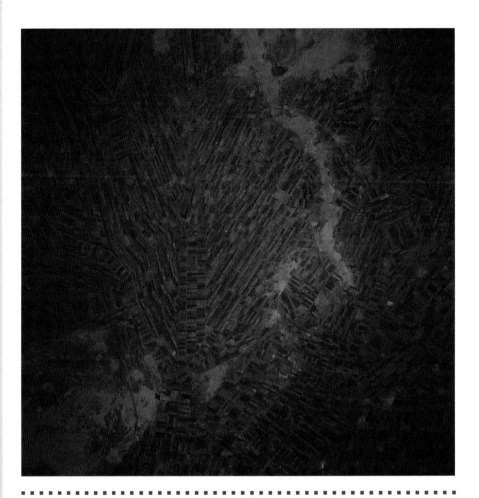

boosting cargo and human beings to and from the ISS, perhaps creating modest business opportunities. The American firm, Space Exploration Technologies Corporation (SpaceX), is developing a system for the National Aeronautics and Space Administration (NASA) consisting of Falcon 9 launch vehicles and Dragon cargo capsules. The company is hoping to provide around 9,900 kilograms (22,000 pounds) of cargo per flight to the International Space Station or, eventually, a human crew of seven astronauts to low-Earth orbit.

The Value of Technology Spinoffs from Space

When technologies developed for the space program are used for other purposes, they are termed "spinoffs." Since the beginning of the space program, the cutting-edge research and development required for the unique environment of space has generated inventions and innovations. Many of the technologies have their largest applications within the aerospace industry, but many also find their way into industrial applications and retail stores.

Examples of space spinoffs fall into several categories. First are the new products and services that consumers can purchase. Beyond satellite-based voice, television, video, and paging communication services, there arc many other spinoff products. Materials such as lightweight carbon-fiber composites* used in tennis racquets, boats, and other products were developed for the doors of the space shuttle. Insulating fabrics and thermal protection equipment used in space suits and onboard space equipment are now available for household uses as well as for firefighters and industrial safety equipment.

Second, the need for precision instruments to remotely monitor astronauts' health and to conduct other space activities has generated a vast new array of scientific and medical applications that permit better research and more accurate and less invasive medical procedures.

Many less obvious procedures and equipment developed for space have resulted in manufacturing improvements. For example, advanced clean room procedures needed for assembling satellites have been used to manufacture high-technology electronics. Research into new lubrication techniques has made industrial equipment last longer. Cheaper and more efficient water purification devices aid people in remote areas.

It is difficult to precisely measure the economic impact of space spinoffs. However, various studies clearly illustrate that the income and jobs created from these space technologies have contributed greatly to the long-run productivity of the economy and to improving the quality of everyday life.

▶ See also **Emerging Space Businesses (Volume 1) • International Space Station (Volumes 1 and 3) • Made with Space Technology (Volume 1) • United Space Alliance (Volume 1)**

▲

Cordless power drills are a spinoff of technology first used by NASA's Apollo program. © *warren0909/ShutterStock.com.*

* **carbon-fiber composites** combinations of carbon fibers with other materials such as resins or ceramics; carbon fiber composites are strong and lightweight

Resources

Books and Articles

Handberg, Roger. *International Space Commerce: Building from Scratch.* Gainesville: University Press of Florida, 2006.

Morris, Langdon, and Kenneth J. Cox, eds. *Space Commerce: The Inside Story by the People Who Are Making It Happen.* [n.p.]: Aerospace Technology Working Group, 2010.

Olla, Phillip. *Commerce in Space: Infrastructures, Technologies, and Applications.* Hershey, PA: Information Science Reference, 2008.

Payne, Silvano, and Hartley Lesser. *2010 International Satellite Directory: The Complete Guide to the Satellite Communications Industry.* Sonoma, CA: SatNews Publishers, 2010.

Websites

"Europe Aims for 2015 SpaceShipTwo Competitor." *Flight Global.* <http://www.flightglobal.com/articles/2009/01/28/321749/europe-aims-for-2015-spaceshiptwo-competitor.html> (accessed October 24, 2011).

International Space Station. National Aeronautics and Space Administration. <http://www.nasa.gov/mission_pages/station/main/index.html> (accessed October 24, 2011).

NASA Spinoff. Office of Technology Transfer, National Aeronautics and Space Administration. <http://www.sti.nasa.gov/tto/> (accessed October 24, 2011).

Scaled Composites Web Site. <http://www.scaled.com/> (accessed October 24, 2011).

State of the Satellite Industry Report 2011. <http://www.sia.org/PDF/2011%20State%20of%20Satellite%20Industry%20Report%20(June%202011).pdf> (accessed October 24, 2011).

What We Offer. Scientific and Technical Information, National Aeronautics and Space Administration. <http://technology.jsc.nasa.gov/> (accessed October 24, 2011).

Communications Satellite Industry

The beginning of the satellite communications era is often considered to have begun with the publication of a paper written by British science fiction writer/futurist Arthur C. Clarke in 1945. The paper described human-tended space stations designed to facilitate communications links for points on Earth. The key to this concept was the placement of space stations in geostationary* Earth orbit (GEO), a location 35,800 kilometers (22,300 miles) above Earth. Objects in this orbit will revolve about Earth along its equatorial plane at the same rate as the planet rotates. Thus, a satellite or space station in GEO will seem fixed in the sky and will be directly above an observer at the equator when both are at the same longitude. A communications satellite in GEO can "see" about one-third of Earth's surface, so to make global communications possible, three satellites need to be placed in this unique orbit.

Clarke envisioned a space station, rather than a satellite, as a communications outpost because he felt that astronauts would, for example, be needed to change vacuum tubes for the receivers and transmitters. However, the concept became extraordinarily complex and expensive when life support, food, and living quarters were factored into the mix. For this reason, and because telephone and television services were perceived at the time as adequate, Clarke's idea was not given much attention. In 1948, however, longer-lived solid-state transistors, which marked the dawn of microelectronics, replaced the vacuum tube. Humans, it seemed, might not be required to tend space-based communications systems after all. Nonetheless, questions remained: Would

*geostationary a specific altitude of an equatorial orbit where the time required to circle the planet matches the time it takes the planet to rotate on its axis. An object in geostationary orbit will always remain over the same geographic location on the equator of the planet it orbits

there be a demand for communications satellites, and, if so, how would they be placed in orbit?

During the mid–twentieth century, people were generally satisfied with telephone and television service, both of which were transmitted by way of cable and radio towers (where radio is a type of radiation [light] with frequencies just below that of visible light). However, telephone service overseas was exceptionally bad and live television could not be received nor transmitted over great distances. Properly positioned satellites could provide unobstructed communications for nearly all points on Earth as long as there was a method to put them in orbit.

Shortly after World War II (1939–1945), the United States acquired the expertise of German rocket engineers through a secret mission called Operation Paper-clip. The German rocket program, which produced the

A Juno 1 rocket is launched from Cape Canaveral, Florida, in 1958, carrying *Explorer 1,* the first American artificial satellite, into orbit. © *Bettmann/Corbis.*

* **low Earth orbit** an orbit between 300 and 800 kilometers (185 and 500 miles) above Earth's surface

world's first true rocket, the V-2, was highly valuable to the United States. These engineers were sent to New Mexico to work for the U.S. Army using hundreds of acquired German V-2 missiles. Within a decade, the German engineers produced powerful missiles called Jupiter, Juno, and Redstone. At the same time, the U.S Air Force was interested in fielding intercontinental ballistic missiles (ICBMs) and was separately developing the Atlas, Thor, and Titan rockets to meet this mission. The U.S. Navy also had a rocket program and was working on a medium-range missile called Vanguard.

On October 4, 1957, the Soviet Union launched *Sputnik 1,* a satellite whose purpose was to demonstrate Soviet technology. Americans were afraid and alarmed and politicians from both political parties demanded that the federal government establish a space program to regain prestige. President Dwight Eisenhower, they felt, did not do enough to prevent the United States from lagging behind the Soviets technologically. In truth, Eisenhower had directed the U.S. Navy to launch a satellite on Vanguard, but the rocket was encountering setbacks. The mission to launch the first American satellite, instead, fell to the U.S. Army, whose Juno 1 was doing remarkably well. The satellite *Explorer 1* finally went up onboard a modified Jupiter-C rocket (now called Juno 1), on January 31, 1958. With this first artificial satellite successfully launched into space, the United States found that launching satellites was possible and, now, communications satellite concepts were now seriously being considered.

The First Communications Satellites

On December 18, 1958, the military's Satellite Communication by Orbit Relay Equipment (SCORE) was launched into low Earth orbit* (LEO) by a U.S. Air Force Atlas rocket. SCORE was designed to receive a transmission, record it on tape, and then relay the transmission to another point on Earth within hours. President Eisenhower used the opportunity to demonstrate American technology by transmitting a recorded Christmas greeting to the world, the first time in history a satellite was used for communications.

Recognizing the potential of satellite communications, John Pierce, director of AT&T's Bell Telephone Laboratories (where AT&T stands for American Telephone & Telegraph), developed projects designed to test various communications satellite concepts. The National Aeronautics and Space Administration (NASA), only two years old, planned to send an inflatable sphere into space for scientific research. Pierce wanted to use the opportunity to reflect signals off the balloon's metallic surface. On August 12, 1960, the sphere, called *Echo 1a* (commonly referred to as *Echo 1*) was successfully launched, and Pierce was encouraged by the reflective signal tests. Because *Echo 1* had no electronic hardware, the satellite was described as passive. For communications to be effective, Pierce felt that active satellites were required.

Meanwhile, the military was continuing with the tape-recorded communications concept, developing new satellites called Courier.

The first one was destroyed when the rocket exploded. *Courier 2* was successfully launched on October 4, 1960, but it failed after seventeen days of operation. During this time, significant military resources were being allocated to Atlas, Titan, and intelligence satellites, which took priority.

Two years after the Echo experiments, Bell Laboratories created the Telstar program, what would turn into the world's first active communications satellites. Consisting of two almost identical satellites (*Telstar 1* and *Telstar 2*), they were designed to operate in medium Earth orbit (MEO), about 5,000 kilometers (3,000 miles) above Earth's surface. During this time, NASA selected a satellite design from RCA (Radio Corporation of America, now RCA Corporation) called Relay to test MEO communications but agreed to launch the Telstar satellite as soon as it was ready. *Telstar 1* was launched on July 10, 1962, *Relay 1* was sent up on December 13 of the same year, *Telstar 2* was launched on May 7, 1963, and *Relay 2* was launched on January 21, 1964. All four were successful, and despite the greater sophistication of *Relay 1* and *Relay 2,* people remembered the live television broadcasts of *Telstar 1* from the United States to locations in Europe.

Advantages and Disadvantages Soon the advantages and disadvantages regarding LEO and MEO communications satellites were being studied. One problem with communications satellites in orbits lower than geosynchronous is the number of satellites required to sustain uninterrupted transmissions. Whereas a single GEO satellite can cover approximately 34% of Earth's surface, individual LEO and MEO satellites cover only between two and 20%. This means that a fleet of satellites, called a "constellation," is required for a communications network.

The major advantage in using LEO and MEO communications satellites is a "minimization of latency," or the time delay between a transmitted signal and a response, often called the "echo effect." Even though transmissions travel at the speed of light, an average time delay of 0.24 seconds for a round-trip signal through a GEO satellite can make phone calls problematic. Despite this drawback, sending three communications satellites to GEO would save money, and people would not need to wait years for an LEO or MEO constellation to be complete.

COMSAT

Shortly after the Soviet Union launched the first human into space, President John Kennedy wanted a national plan for space exploration and settled on a series of programs that included the famous Apollo* missions to the Moon. Less familiar but perhaps more significant for the long term, Congress, with the support of President Kennedy, authorized the establishment of an organization designed to integrate the nation's space-based communications network.

* **Apollo** lunar program to land men on the Moon; *Apollo 11, Apollo 12, Apollo 14, Apollo 15, Apollo 16,* and *Apollo 17* delivered twelve men to the lunar surface (two per mission) between 1969 and 1972 and returned them safely back to Earth

Formed in February 1963 by the Communications Satellite Act of 1962, the Communications Satellite Corporation, or COMSAT, was given the task of creating a national communications satellite system in the earliest possible time. Half of COMSAT would be publicly traded, while satellite manufacturers would purchase the other half. COMSAT's first major hurdle was deciding what kind of satellite system it would pursue: LEO, MEO, or GEO. Because the Telstar and Relay programs were successful, these MEO systems seemed the default choice. For uninterrupted communications service, however, about twenty satellites such as Telstar or Relay were needed, costing an estimated $200 million. The president of COMSAT, Joseph Charyk, a veteran of satellite engineering programs, was not sure that this was the right way to proceed.

Meanwhile, Hughes Aircraft Company was developing the Syncom (where Syncom stood for synchronous communication satellite) series of satellites, each designed to test communications technologies in GEO. The first two satellites were not entirely successful, but *Syncom 3,* launched on August 19, 1964, achieved a stationary GEO with a Delta launch vehicle. Charyk was aware of the Syncom project early on and followed its progress closely. COMSAT was beginning to realize that a GEO communications satellite network was the most practical in terms of cost. Nevertheless, COMSAT asked a variety of companies to study the feasibility of LEO communications constellations in the event that a GEO system was unsuccessful. AT&T and RCA researched the merits of a random system, in which satellites drifted freely without any particular relationship to one another. At the same time, STL (Standard Telecommunications Laboratory) and ITT (International Telephone & Telegraph) studied the phased approach, where strings of satellites orbiting at LEO were spaced

President Kennedy speaks at Rice University, Houston, Texas, on September 12, 1962 on the space effort: "We choose to go to the moon in this decade and do other things, not because they are easy, but because they are hard..." © *Corbis.*

in such a way to allow for continuous, uninterrupted communications. COMSAT finally decided on a GEO system, and on April 6, 1965, it launched *Early Bird* (Intelsat 1), the world's first commercial communications satellite in geosynchronous orbit. This satellite also became a test bed for the latency problem, and methods to suppress the echo effect were successfully employed.

Bandwidth Capacity

During this time, NASA continued to fund research in communications satellite technology, contributing to programs such as Applications Technology Satellites (ATS). Six ATS units were developed and launched, and each was designed to test various technologies related to bandwidth capacity and new components. Of particular importance was bandwidth capacity, the range of frequencies* used in a satellite.

Satellite communications providers were particularly interested in boosting the capacity of transponders used for telephone conversations and television broadcasts. A telephone call, for example, uses about five kilohertz of bandwidth. A satellite with 50 kilohertz of bandwidth can handle ten calls simultaneously. Early satellites could handle only about thirty calls at one time and were easily overwhelmed. Research continued to improve the capacity problems, and digital technologies have significantly increased the number of simultaneous calls. Satellite engineers also designed antennas that did not interfere with systems orbiting nearby and recommended adequate separation between satellites to prevent signals from interfering.

* **frequencies** the number of oscillations or vibrations per second of an electromagnetic wave or any wave

---

WHAT IS BANDWIDTH?

Bandwidth capacity refers to the range of frequencies. All frequencies are classified according to the electromagnetic spectrum and are measured in hertz (Hz). At one end, where frequencies are low, the spectrum includes radio and microwaves. In the middle, the spectrum is characterized by infrared radiation (IR), visible light, and ultraviolet (UV) radiation. High-frequency energy, such as x rays, gamma rays, and cosmic rays, occupies the other end of the spectrum. The radio spectrum, which is divided into eight segments ranging from extremely high frequency (EHF) to very low frequency (VLF), provides the communications spectrum.

This spectrum, which is important to communications satellites, is divided into ten parts. These are—from the highest frequency to the lowest—millimeter (in the EHF range), W, V, Ka, K, Ku, X, C, S, and L (in the ultra-high-frequency, or UHF, range). A typical communications satellite in orbit today will have a series of bandwidth-specific transmitter-receiver units, called transponders, classified as C-band or Ku-band.

---

* **payload** any cargo launched aboard a rocket that is destined for space, including communications satellites or modules, supplies, equipment, and astronauts; does not include the vehicle used to move the cargo or the propellant that powers the vehicle

Becoming Global

After the establishment of COMSAT, efforts were under way to approach the international community about setting up a global communications satellite network. COMSAT dispatched several key people, along with U.S. State Department officials, to a dozen nations interested in the communications satellite market. In 1964, International Telecommunications Satellite Organization (INTELSAT) was formed, and it started operations using part of the new *Early Bird* satellite launched in 1965. The company grew rapidly over the next three decades and by 2001, it had 146 members, almost half of whom (70 nations) relied entirely on the system for its international communications. In that year, the company became a private corporation and changed its name to Intelsat, Inc. In January 2005, the company was purchased by a consortium of four private equity firms, Madison Dearborn Partners, Apax Partners, Permira, and Apollo Management. Four years later, the new corporation also purchased PanAmSat, a satellite service provider based in Greenwich, Connecticut. The deal made Intelsat the world's largest fixed satellite provider, with 52 satellites in orbit around the Earth. In June 2007, the London-based private equity firm of BC Partners purchased a controlling interest in Intelsat and moved its corporate headquarters to Luxembourg, where it remains today.

Other international communications satellite organizations have since formed, such as Eutelsat, a cooperative formed in 1977 providing communications services for all of Europe, along with Africa, the Middle East, India, and parts of Asia and the Americas. France, England, and Germany established the European Space Research Organization (ESRO) and the European Launch Development Organization (ELDO) shortly after the launch of an experimental communications satellite called *Symphonie* in 1967. ESRO was responsible for research, development, construction, and operation of payloads* and ELDO handled launch activities. Because of management and system integration concerns, ESRO and ELDO merged to form the European Space Agency (ESA) in 1974. Three years later, the Conference of European Posts and Telecommunications (CEPT) approved the formation of Eutelsat S.A., which by 2000 had nearly fifty members. In 2005, the organization formed Eutelsat Communications, the holding company for Eutelsat S.A. As of 2011, the company was the third largest satellite operator in the world, with a global share of 14 percent of all satellite communications.

COMSAT was also asked to assist in the development of a regional communications satellite organization for southwestern Asia, northern Africa, and areas of southern Asia. COMSAT agreed and was contracted to develop and build what later became known as Arabsat, an acronym for Arab Satellite Communications Organization (or ASCO). Arabsat is headquartered in Riyadh, Saudi Arabia, and it the leading communications satellite organization for Arabic-speaking countries. Inmarsat, founded in 1982 as the International Maritime Satellite

* **elliptical** having an oval shape

Organization, is another international organization providing global communications services to seagoing vessels and oil platforms. When it began to provide services to aircraft and portable users its name was changed to International Mobile Satellite Organization, but the name Inmarsat was retained.

The Soviet Union, recognizing the benefits of a global communications satellite network, was not interested in a GEO system because of the country's northern location. A GEO system comprised of three satellites would miss parts of the Soviet Union. The Soviets developed an ingenious solution by launching communications satellites into highly elliptical* orbits. The orbit consisted of a very close and fast approach over the Southern Hemisphere while tracing a slow and lengthy arc over the Soviet mainland. In 1965, the Soviet Union launched its first communications satellite as part of an ongoing system called Molniya, a name also assigned to the unusual orbit it occupies.

The Soviet Union, despite being approached by representatives of COMSAT and the U.S. State Department to join Intelsat, declined membership and initiated a regional network, in 1971, called Intersputnik International Organization of Space Communications, commonly shortened to Intersputnik. It was successful during the following decades with its Gorizont, Express, and Gals satellites but experienced funding difficulties after the collapse of the Soviet Union in 1991. In the 1990s, however, Intersputnik was revitalized with a membership of twenty-three nations and the introduction of its series of satellites called Express-A. As of 2011, twenty six nations partipcate in Intersputnik. In addition to most of the former members of the Soviet Union, Intersputnik members include a number of nations around the world, including Afghanistan, Cuba, India, Laos, Mongolia, Nicaragua, Syria, and Yemen.

Back to LEO?

In the early 1990s, LEO communications satellite constellations were revisited. Microelectronics was allowing for smaller satellites with greater capacities, and the launch industry was stronger than it was thirty years earlier. Two companies that pursued this concept were Iridium SSC and Teledesic.

Iridium's plan was to loft about 100 satellites into several LEOs to provide uninterrupted cell phone and pager services anywhere on Earth. Iridium became the first company to provide these services on November 1, 1998. Sixty-six Iridium satellites, all built by Motorola, were launched in the late 1990s. Unfortunately, Iridium filed for bankruptcy in 1999. The company was reorganized in 2001 by a group of private investors at a fraction of its worth, and is now known as Iridium Communications, Inc. As of 2011, the company had nearly 400,000 subscribers worldwide. Iridium announced that it would begin replacing the original fleet of Motorola satellites in 2015.

Teledisc planned to provide computer networking, wireless Internet access, interactive media, and voice and video services for the world. It also planned to use LEO satellites developed and built by Motorola. Founded by Craig McCaw and Microsoft founder Bill Gates with $9 billion in 1990, Teledesic also experienced financial troubles. It ceased operations in 2002. Globalstar began operations in 1991 as a joint project of Loral Corporation and Qualcomm. Its plan to provide a LEO satellite constellation for satellite phone and low-speed data communications failed when it filed for bankruptcy in 2002. It reorganized as Globalstar LLC in 2003 and was converted to Globalstar, Inc. in 2006. As of 2011, Globalstar operates mobile satellite voice and data services to over 120 countries around the world with a total of just over 400,000 subscribers.

ORBCOMM began LEO satellite operations in 1993 with the help of the two companies Orbital Sciences Corporation and Teleglobe (now called VSNL International Canada). Hoping to develop a constellation of LEO communications satellites, it also claimed bankruptcy protection in 2000. Since then, it sold millions of common shares of stock in 2006 in order to bolster its position. As of 2011, ORBCOMM provides machine-to-machine (M2M) monitoring and messaging services from its nearly 30 LEO communications satellites and terrestrial gateways.

By 1998, satellite communications services included telephone, television, radio, and data processing, and totaled about $65.9 billion in revenues, or almost 7% of the total telecommunications industry. During that year, about 215 communications satellites were in GEO

Satellite phones enable people to communicate from out-of-the-way spots like mountaintops and other locations where traditional communication networks are unavailable. © *Philip and Karen Smith/ Iconica/Getty Images.*

and 187 in LEO. The Satellite Industry Association (SIA) reported that total world revenues for satellite communications services totaled $168.1 billion for the year 2010. The SIA also state that the annual growth rate from 2005 to 2010 averaged 11.2%. As of June 2011, 986 satellites were in orbit, with 37% of them being commercial communications satellites.

 See also **Clarke, Arthur C. (Volume 1) • Communications, Future Needs in (Volume 4) • Ground Infrastructure (Volume 1) • Satellite Industry (Volume 1)**

Resources

Books and Articles

Alper, Joel, and Joseph N. Pelton, eds. *The Intelsat Global Satellite System.* New York: American Institute of Aeronautics and Astronautics, 1984.

Clarke, Arthur C. "Extraterrestrial Relays: Can Rocket Stations Give World-wide Radio Coverage?" *Wireless World,* October (1945):305–308.

Elbert, Bruce R. *Introduction to Satellite Communication.* Boston: Artech House, 2008.

Ford, Steve. *The ARRL Satellite Handbook.* Newington, CT: American Radio Relay League, 2008.

Goodman, John M. *Space Weather and Telecommunications.* New York: Springer, 2005.

Gorn, Michael H. *Superstructures in Space: From Satellites to Space Stations.* New York: Merrell, 2008.

Johnson, Rebecca L. *Satellites.* Minneapolis, MN: Lerner Publications, 2006.

Payne, Silvano, and Hartley Lesser. *2010 International Satellite Directory: The Complete Guide to the Satellite communications Industry*, 25th ed. Sonoma, CA: SatNews Publishers, 2010.

Websites

2011 State of the Satellite Industry Report Shows Continued Growth in 2010. Satellite Industry Association. <http://www.sia.org/PDF/FINAL%20Press_Release_State%20of%20the%20Satellite%20Report%202011%20JUNE%202011.pdf> (accessed October 7, 2011).

Eutelsat Communications. <http://www.eutelsat.com/home/index.html> (accessed October 7, 2011).

▲

Robert Crippen on April 29, 1979, in a photo taken prior to his first spaceflight in 1981.

© *Bettmann/Corbis.*

Globalstar. <http://www.globalstar.com/en/> (accessed October 7, 2011).

Iridium Everywhere. <http://www.iridium.com/default.aspx> (accessed October 7, 2011).

Whalen, David J. *Communications Satellites: Making the Global Village Possible.* National Aeronautics and Space Administration. <http://www.hq.nasa.gov/office/pao/History/satcomhistory.html> (accessed October 7, 2011).

What's In Space? Union of Concerned Scientists. <http://www.ucsusa.org/nuclear_weapons_and_global_security/space_weapons/technical_issues/satellites-types-orbits.html> (accessed October 7, 2011).

Crippen, Robert

American Astronaut
1937–

Robert Laurel Crippen has been a major contributor to America's space exploration efforts. From making the first historic flight of the space shuttle, to directing the Kennedy Space Center, to exploring opportunities in the private sector, Crippen has provided experience and leadership for both piloted and unpiloted spaceflight.

Crippen was born in Beaumont, Texas, on September 11, 1937. He graduated from New Caney High School in Caney, Texas, and received a bachelor's of science degree in aerospace engineering from the University of Texas in 1960.

Crippen received his commission through the U.S. Navy's Aviation Officer Program at Pensacola, Florida. He continued his flight training at Whiting Field, Florida, and went from there to Chase Field in Beeville, Texas, where he received his "wings," becoming a qualified pilot. From June 1962 to November 1964, he was assigned to Fleet Squadron VA-72, where he completed two and a half years of duty as an attack pilot aboard the aircraft carrier USS *Independence.* He later attended the U.S. Air Force (USAF) Aerospace Research Pilot School at Edwards Air Force Base, California, and remained there as an instructor after his graduation. In October 1966, Crippen was among the second group of aerospace research pilots to be selected to the USAF Manned Orbiting Laboratory Program.

Crippen joined the National Aeronautics and Space Administration (NASA) as an astronaut in September 1969 following the cancellation of the Manned Orbiting Laboratory program. He was a crewmember of the Skylab Medical Experiments Altitude Test, a fifty-six-day simulation of the Skylab mission. He was also a member of the astronaut support crew for the Skylab 2, 3, and 4 missions and the Apollo-Soyuz Test Project mission. Crippen's first spaceflight was during April 12–14, 1981, as the pilot of the space shuttle *Columbia* and STS-1 (Space Transportation System 1), the first space shuttle mission. In 1983, Crippen was spacecraft commander of STS-7 (on June 18–24, 1983). He completed two more space shuttle flights as commander in 1984: STS-41C (April 6–13) and STS-41G (October 6–13).

In 1987, Crippen was stationed at NASA's John F. Kennedy Space Center (KSC) serving as the deputy director of shuttle operations for NASA Headquarters. He was responsible for final shuttle preparation, mission execution, and the return of the orbiter* to KSC following landings at Edwards Air Force Base. From 1990 to 1992, he was responsible for the overall shuttle program at NASA Headquarters in Washington, D.C. From 1992 to 1995, during his tenure as director of KSC, Crippen presided over the launch and recovery of twenty-two space shuttle missions, establishing and developing new quality management techniques while ensuring the highest safety standards in an extremely hazardous environment.

Crippen left NASA in 1995 and joined the Lockheed Martin Information Systems Company as their vice president of automation systems. The following year he became their vice president of simulation and training systems. In October of that year, he was named to the newly created position of president of the Thiokol Aerospace Group. He remained at Thiolol Aerospace until April 2001.

Crippen's accomplishments have earned him many awards. Among them are the NASA Exceptional Service Medal, the Department of Defense Distinguished Service Award, the American Astronautical Society of Flight Achievement Award, and four NASA Space Flight Medals. The highest award for spaceflight activities was awarded to Crippen on April 6, 2006: the Congressional Space Medal of Honor. He is also a fellow of the American Institute of Aeronautics and Astronautics (and serving as its president for its 1999–2000 term), Society of Experimental Test pilots, and American Astronautical Society.

 See also **History of Humans in Space (Volume 3)** • **Skylab (Volume 3)** • **Space Shuttle (Volume 3)**

> * **orbiter** spacecraft that uses engines and/or aerobraking, whose main function is to orbit about a planet or natural satellite

Resources

Books and Articles

Reichhardt, Tony, ed. *Space Shuttle: The First 20 Years—The Astronauts' Experiences in Their Own Words.* New York: DK Publishing, 2002.

Websites

Astronaut Bio: Robert L. Crippen. National Aeronautics and Space Administration, Johnson Space Center. <http://www.jsc.nasa.gov/Bios/htmlbios/crippen-rl.html> (accessed October 13, 2011).

Robert L. Crippen. New Mexico Museum of Space History, International Space Hall of Fame. <http://www.nmspacemuseum.org/halloffame/detail.php?id=107> (accessed October 13, 2011).

Crippen Awarded Space Medal of Honor. National Aeronautics and Space Administration. <http://www.nasa.gov/mission_pages/shuttle/sts1/index.html> (accessed October 13, 2011).

E

Education

In the 1960s, many young people in the United States were inspired to pursue aerospace-related careers because of the commitment by the United States to send humans to the Moon. Universities saw an influx of enthusiastic students ready to take on the challenges of the Apollo program. Six Apollo Moon landings brought twelve astronauts (two per mission) to explore the lunar surface. However, moonwalkers are part of history lessons to students in the twenty-first century. Consequently, universities today put forth the challenge of a human mission to Mars to attract students.

Rapid advances in technology and computers have influenced more students to pursue courses of study in the sciences and space-related engineering and technology programs. Many computer experts who lost their jobs in the crash of the "dot-com" industry, from 1998 to 2001, subsequently explored the field of aerospace engineering. Even if students do not decide on a space-related career, an aerospace engineering degree provides them with a wide variety of employment choices.

What are these students looking for in a college or university? They not only want a good selection of courses in the fields of their interests, but students also want exposure to innovative research in the field. Colleges and universities are addressing these needs largely by building valuable relationships with space-related organizations, aerospace companies, governmental agencies like the National Aeronautics and Space Administration (NASA), and other colleges and universities. Internships are frequently beneficial experiences for students, and often lead to employment opportunities at the sponsoring facility.

How Universities Attract New Students to Space Sciences

The public affairs departments at some universities have realized the potential of promoting their students' and professors' accomplishments. An example of this approach is the University of Arizona in Tucson, whose Department of Aerospace and Mechanical Engineering posts a quarterly online newsletter, Arizonaengineer_online, four times a year describing the department's programs and achievements.

A university whose graduates become astronauts or known in a field of space science or aerospace engineering is also a pull for students. This is not only true for the University of Arizona at Tucson, but also the

The Discovery Laboratory at NASA's Marshall Space Flight Center (Huntsville, Alabama) provides hands-on educational workshops for teachers and students. *NASA.* ▶

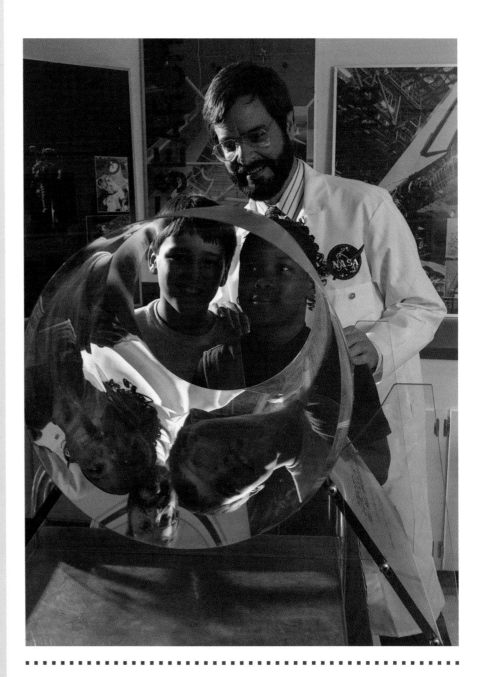

Massachusetts Institute of Technology (MIT, Cambridge), the California Institute of Technology (Pasadena), and Purdue University (West Lafayette, Indiana), among others.

One of the opportunities Purdue University affords both its graduate and undergraduate students is the chance to be a part of a tight-knit academic community with top professors in the aerospace field. This personal attention makes their program a popular one with students. Purdue claims to have produced more astronauts than any other public university, explaining its nickname of *Cradle of Astronauts.* For private universities, MIT claims the top spot. Of course, the U.S. Air Force Academy, the U.S. Naval Academy, and the U.S. Naval Postgraduate School also tout a large number of their graduates that have been selected for the NASA astronaut program.

A New Array of Space Courses

Many colleges and universities have expanded their degree programs and course offerings in the fields of space sciences, astronomy, and Earth sciences to attract more students, as well as professors and research grants. The future holds a vast array of space-related careers. For example, space tourism in the decades to come will require a wide range of careers, and students at Rochester Institute of Technology (Henrietta, New York) are getting ready. In the departments of hotel management, food management, and travel programs, students are enrolled in what is likely the world's first college course on space tourism.

Promoting Space in Universities

National Space Grant Consortium One of the most effective programs for bringing more space research and related projects, as well as funding, to universities is NASA's National Space Grant College and Fellowship Program. This program funds space research, education, and public service projects through a network of consortia in each of the fifty states, Puerto Rico, and the District of Columbia.

Each state's space grant consortium provides the students with information about local aerospace research and financial assistance. They also develop space education projects in their states. Some space grant projects, such as the one at the University of Colorado at Boulder (CU Boulder), involve students in specially designed space projects. Students at CU Boulder can participate in the university's RocketSat program, in which they spend ten months during their freshman and sophomore years studying every step involved in the launch of a rocket payload, from design and review to the final launch on a sounding rocket out of Wallops Flight Facility. In 2011, Adams State College in Alamosa and the Colorado Space Grant Consortium announced the fifth annual competition for the construction of an autonomous robot capable of navigating the Great Sand Dunes National Park. The actual competition was scheduled to be held on April 7, 2012. The competition was open to all students enrolled at Colorado colleges and universities.

Universities Space Research Association The Universities Space Research Association (USRA) is a private nonprofit corporation formed under the auspices of the National Academy of Sciences. All member institutions have graduate programs in space sciences or aerospace engineering. Besides ninety-five member institutions in the United States, there are two member institutions in Canada, four in Europe, two in Israel, one in Hong Kong, and one in Australia.

USRA provides a mechanism through which universities can cooperate effectively with one another, with the government, and with other organizations to further space science and technology and to promote education in these areas. A unique feature of USRA is its system of science councils, which are standing panels of scientific experts who provide

* **microgravity** the condition experienced in freefall as a spacecraft orbits Earth or another body; commonly called weightlessness; only very small forces are perceived in freefall, on the order of one-millionth the force of gravity on Earth's surface

program guidance in specific areas of research. Most of USRA's activities are funded by grants and contracts from NASA.

Universities Worldwide

The International Space University (ISU), through both its summer courses and its permanent campus near Strasbourg, France, has made major contributions to establishing new curricula. It draws the top students worldwide, because their professors are leading figures from space-related industries, government and international organizations, and universities around the world. ISU students come to the university with their specialist backgrounds and broaden their perspectives through increased knowledge in other relevant fields. Another example of international efforts to attract students is found at Saint Louis University, in St. Louis, Missouri, which offers a degree in aerospace engineering jointly at its Madrid, Spain, and Parks College, St. Louis, campuses.

Student Space Competitions

Universities are also involved in efforts to reach out to younger students and expose them to space sciences. Space-related projects and competitions for kindergarten through twelfth-grade students sponsored by a university member of the National Space Grant—or in collaboration with other organizations such as the National Space Society, the Challenger Center for Space Science Education, the Space Foundation, and the Planetary Society—can make an impression on students that will influence their career decisions much later.

The experience of being involved in science fair projects also provides students with a sense of ownership and interest that lasts throughout their careers. Many university scientists and engineers, as well as experts from aerospace companies, are involved in helping and judging science fairs.

Through space-related professional organizations like the American Institute of Aeronautics and Astronautics, the aerospace division of American Society of Civil Engineers, and the Institute of Electrical and Electronics Engineers, universities are providing opportunities for students to submit papers and projects to be judged by experts in the field. These competitions, which are held at the organizations' conferences, provide an avenue for building relationships with aerospace professionals, as well as other students. These relationships can form an essential network of colleagues as students launch into their careers.

NASA and other organizations sponsor an array of design projects for students of all ages. Projects can include flying their experiment on a McDonnell Douglas C-9B Skytrain II airplane, nicknamed the "Weightless Wonder," that provides twenty to thirty seconds of microgravity* at a time. Other competitions involve designing space settlements, along with Moon and Mars bases.

On the Internet, NASA maintains many Web sites that challenge the minds of schoolchildren with regards to space-based activities.

Its NASA Quest Web site (http://quest.arc.nasa.gov/) provides interactive explorations designed to simulate real-life scientific and engineering processes. And, its NASA Education Web site (http://www.nasa.gov/offices/education/about/index.html) supports educators across the country in order to help prepare, encourage, and nurture the young minds that will soon provide the manpower to pursue space missions in the future.

NASA's Commercial Space Centers

NASA's commercial space centers are consortia of academia, government, and industry who partner to develop new or improved products and services, usually through collaborative research conducted in space. The NASA Innovative Partnerships Program office manages the organizational structure for acquiring, developing, and maturing technologies and capabilities for NASA's Mission Directorates through partnerships and investments with academia, government agencies, industry, and national laboratories. Topics of interest include space power, satellite communication networks, remote sensing*, mapping, microgravity materials processing, medical and biological research and development, crystallography*, space automation and robotics, engineering, space technology, and combustion in space.

▶ *See also* **Career Astronauts (Volume 1) • Careers in Astronomy (Volume 2) • Careers in Business and Management (Volume 1) • Careers in Rocketry (Volume 1) • Careers in Space (Volume 4) • Careers in Space Law (Volume 1) • Careers in Space Medicine (Volume 1) • Careers in Space Science (Volume 2) • Careers in Spaceflight (Volume 3) • Careers in Writing, Photography, and Film (Volume 1) • International Space University (Volume 1)**

Resources

Books and Articles

Flowers, Lawrence O, ed. *Science Careers: Personal Accounts from the Experts.* Lanham, MD: Scarecrow Press, 2003.

Sachnoff, Scott, and Leonard David. *The Space Publication's Guide to Space Careers.* BethSpace Publications, 1998.

Websites

Innovative Partnership Program. National Aeronautics and Space Administration. <http://www.nasa.gov/offices/oct/partnehttp://www.nasa.gov/offices/oct/partnership/index.htmlrship/index.html> (accessed November 3, 2009).

Look to the Future: Careers in Space. NASA Goddard Space Flight Center. <http://mgs-mager.gsfc.nasa.gov/Kids/careers.html> (accessed November 3, 2009).

* **remote sensing** the act of observing from orbit what may be seen or sensed below on Earth or another planetary body

* **crystallography** the study of the internal structure of crystals

MIT Has Produced More Astronauts than Any Other Private University. Massachusetts Institute of Technology News Office. <http://web.mit.edu/newsoffice/2001/commnasanuts-0606.html> (accessed November 3, 2011).

NASA Education. National Aeronautics and Space Administration. <http://www.nasa.gov/offices/education/about/index.html> (accessed November 3, 2011).

NASA Quest. National Aeronautics and Space Administration. <http://quest.arc.nasa.gov/> (accessed November 3, 2011).

National Space Grant Foundation. Web Site. <http://www.spacegrant.org/> (accessed November 3, 2011).

Opportunities in Our Nation's Space Program.. National Aeronautics and Space Administration. <http://nasajobs.nasa.gov/> (accessed November 3, 2011).

Space Grant. National Aeronautics and Space Administration. <http://www.nasa.gov/offices/education/programs/national/spacegrant/home/index.html> (accessed November 3, 2011).

Space, Tourism's Final Frontier, Beckons Visionaries. <http://www.onepaper.com/deals/?v=d&i=&s=Caribbean:Business&p=20783> (accessed November 3, 2011).

Welcome to the NASA Career Expo. National Aeronautics and Space Administration. <http://kids.earth.nasa.gov/archive/career/index.html> (accessed Novembe 3, 2011).

ELV *See Launch Vehicles, Expendable (Volume 1)*

Emerging Space Businesses

The first human to go into space was Soviet cosmonaut Yuri Gagarin (1934–1968). His *Vostok 1* spacecraft completed just over one orbit around the Earth in April 12, 1961. For the next forty years only government sponsored ("professional") astronauts and cosmonauts, mostly from the United States and the Soviet Union (now Russia), ventured into space.

However, on April 28, 2001, Italian-American engineer and multimillionaire Dennis Tito became the first "space tourist" to go into space. His over US$20 million space adventure in the Russian *Soyuz TM-32* space capsule opened the door to private space companies hoping to make money from tourists going on suborbital flights into space and on the equipment, materials, and spacecraft needed to send both private and government travelers into space.

In the early part of the 2010s, many private companies are entering the fledgling private space services industry, with the hopes of making huge profits and accomplishing what few private companies have done before: go into space. Many companies have already failed in their attempts to launch their vehicles into space. Others have continued to pursue their space-related goals, often times with the financial backing of either wealthy benefactors or lucrative contracts from the U.S. federal government.

Armadillo Aerospace

Armadillo Aerospace is based in Mesquite, Texas, and was founded in 2001, by John Carmack, the co-founder of id Software (the maker of computer games such as the series *Doom* and *Quake*). In 2008, Armadillo won the top prize at the Level 1 Lunar Lander Challenge and, in 2009, won the second prize at the Level 2 Lunar Lander Challenge. The company is working with Space Adventures (as of May 2010) on a suborbital commercial rocket that will take tourists into space for about $102,000 per customer. The spacecraft, tentatively called Black Armadillo, will be a vertical takeoff, vertical landing (VTVL) suborbital vehicle. No date has been set for the maiden voyage.

Bigelow Aerospace

Bigelow Aerospace, headquartered in North Las Vegas, Nevada, bought the commercial rights to the National Aeronautics and Space Administration (NASA) patents involving "TransHab" (for transit habitat). The TransHab module was a NASA-developed inflatable living space made of lightweight, flexible materials stronger than steel. It was designed to fold inside the cargo bay of the space shuttle for transportation into space, then be inflated to its full size when in orbit about Earth.

Bigelow Aerospace was founded by American hotel and aerospace entrepreneur Robert Bigelow (1945–) in 1998. Bigelow owns the hotel chain Budget Suites of America. As of 2009, Bigelow is pursuing plans for a privately funded and expandable space station based on the TransHab design. In July 2006, Bigelow launched its *Genesis I* module, which was followed up about eleven months later with its *Genesis II* module. As of October 1, 2011, both modules were in orbit about Earth, and were being evaluated by their ground-control personnel back on Earth with respect to pressure, temperature, radiation in and around their shells, along with other measurements.

The company is expecting to launch its *BA 330* module (formerly called Nautilus space complex module) in 2014 or 2015. The space habitation module, expandable to 12,000 cubic feet (330 cubic meters), will directly support zero-gravity scientific and manufacturing research, along with indirect activities involving space tourism.

Bigelow is also designing a *BA 2100* module that would require a heavy-lift rocket to send it to low-Earth orbit. If manufactured, the module would be six times larger than the *BA 330*—at 77,690 cubic feet (2,200 cubic meters). In addition, Bigelow Aerospace is also developing a series of inflatable modules, under the name of Bigelow Next-Generation Commercial Space Station. The complex will include spacecraft modules, along with a central docking node, solar arrays, propulsion, and crew capsules. The company hopes to launch the components in 2014, with leasing options starting in 2015. So far, Bigelow has contracts with seven countries interested in leasing its space station modules.

Blue Origin

Blue Origin, which is based in Kent, Washington, but with its main flight facility in Culberson County, Texas, was founded by Jeff Bezos (the founder of Amazon.com) in 2000. The company is working on its VTVL *New Shepard* spacecraft, which is based on the technology for the McDonnell Douglas DC-X. Unmanned test flights took place in 2011 with more flights planned for 2012 and beyond.

Excalibur Almaz

Excalibur Almaz is based in Douglas, Isle of Man, and founded in 2005, by Arthur M. Dula and Buckner Hightower. The company is designing a spacecraft based on the Soviet TKS space capsules, which were originally designed to go to the Soviet *Almaz* space station, which was never developed. According to its website, testing of its capsule is expected to begin in 2012, with operational flights no earlier than 2013.

Masten Space Systems

Masten Space Systems, headquartered in Mojave, California, is developing a VTVL spacecraft for unmanned suborbital flights. It currently is developing its Xaero reusable VTVL launch vehicle as part of NASA's Flight Opportunities Program.

Scaled Composites and Virgin Galactic

U.S. private enterprise Scaled Composites, founded by Burt Rutan, built *SpaceShipOne* and *WhiteKnightOne*. The *SpaceShipOne* was a suborbital air-launched spaceplane that won the $10 million Ansari X Prize in 2004, with the help of its carrier aircraft *WhiteKnightOne*. At that time, the spaceplane became the first privately-funded manned spacecraft to go into space. The venture was paid for by American businessman Paul Allen, who is the co-founder of Microsoft, in an amount of approximately US$25 million, through Mojave Aerospace Ventures, a joint effort between Allen and Rutan.

Since then, Scaled Composites has teamed up with Richard Branson's Virgin Galactic. Scaled Composites is building a reusable spacecraft that Virgin Galactic intends to use to make suborbital flights available to the private sector. The reusable spacecraft, a series of spacecraft called *SpaceShipTwo* (actually about 10% of the craft, excluding the

SpaceShipTwo is suspended
under the wing of its mothership,
WhiteKnightTwo, is debuted at the
Mojave Air and Space Port on December
7, 2009. © *Robyn Beck/AFP/Getty Images.*

fuel, is not reusable), will be launched with the help of one of a series
of *WhiteKnightTwo* launching vehicles, which take the spacecraft from
launch to about 18 kilometers (11 miles) above Earth. The spacecraft
then completes the trip to space, about 100 kilometers (62 miles) above
Earth. Test flights were conducted in 2011, with more scheduled in 2012.
The first commercial suborbital flight is planned for sometime in 2013.

Sierra Nevada Corporation

Sierra Nevada Corporation (SNC) is a private company specializing in
commercial orbital transportation services and microsatellites, along with
energy, telemedicine, and nanotechnology. Headquartered in Sparks,
Nevada, SNC does most of its work through contracts secured through
private spacecraft companies and such government entities as the military and NASA. For instance, it supplied the rocket engine for Scaled
Composites' RocketMotorTwo for *SpaceShipTwo*. In December 2008, the
company acquired SpaceDev, which was developing the DreamChaser
orbital spacecraft. Just over a year later, SNC was awarded a NASA contract for the development of the DreamChaser craft for its Commercial
Crew Development program.

One of SpaceX's Dragon capsules parachutes into the Pacific Ocean on December 8, 2010. This was the first time a private business launched a capsule into orbit and recovered it on their own. © *AP Images.*

Space Exploration Technologies Corporation

Space Exploration Technologies Corporation (SpaceX) was founded in 2002 by engineer and entrepreneur Elon Musk, the co-founder of the Internet payment company PayPal and the electric car company Tesla Motors. SpaceX is developing the Dragon spacecraft and the Falcon 9 launch vehicle, along with the Falcon Heavy, a super-heavy lift launch vehicle. On December 23, 2008, NASA announced that SpaceX, which is headquartered in Hawthorne, California, had been awarded a contract under a NASA Commercial Resupply Services (CRS) program. The contract requires the delivery of a minimum of 12 unmanned missions to the International Space Station between 2009 and 2016, for US$1.6 billion.

In December 2010, SpaceX became the first private company to successfully launch, orbit, and deorbit a spacecraft, its unmanned Dragon space capsule, which made two orbits around Earth. On July 13, 2011, SpaceX began construction of a launch site at Vandenberg Air Force Base, in California. The site will launch its Falcon Heavy rocket. SpaceX is proceeding with the goal of beginning cargo missions to the ISS by 2012. SpaceX has also announced plans to develop a manned version of the Dragon capsule, capable of carrying people into orbit.

XCOR Aerospace

In 1999, XCOR Aerospace, based at the Mojave Air and Space Port (Mojave, California) was founded by members of Rotary Rocket, Inc., headed by Gary Hudson, who were developing a single-stage-to-orbit (SSTO) vehicle. Jeff Greason, who now heads the company, is developing the Lynx spaceplane, which is expected to carry one pilot and one passenger on suborbital flights. In 2011, Space Expedition Curacao (SXC) signed a lease contract for XCOR's Lynx suborbital spaceplane to fly from its spaceport, possibly as early as 2012.

Others

Other private space companies include Interorbital Systems Corporation (IOS, http://www.interorbital.com/), based in Mojave, California, which was founded in 1996 by Roderick and Randa Milliron. According to its website, IOS "develops and manufactures low-cost, state-of-the-art orbital launch vehicles and satellites for private, commercial, governmental, academic, and military applications."

Based in Chicago, Illinois, the PlanetSpace Corporation (http://www.planetspace.org) was founded in 2005, by Geoff Sheerin and Dr. Chirinjeev Kathuria. The company is working on the Canadian Arrow rocket (a two-stage rocket based on the German V-2 rocket) and the Silver Dart orbital spaceplane (based on the U.S. Air Force's Flight Dynamics Laboratory-7 [FDL-7] program). The company is also working with Lockheed Martin, the Boeing Company, and Alliant Techsystems on the Athena III booster rocket, based on the solid rocket boosters from the space shuttle.

Space Tours

Several companies in the United States are offering suborbital flights into space for future space tourists. One has successfully sent seven spaceflight participants (space tourists) to the International Space Station via the Russian space agency.

Space Adventures Space Adventures Ltd., headquartered in Vienna, Virginia, is a space tourism company co-founded in 1998, by American entrepreneur and aerospace engineer Eric C. Anderson. Space Adventures brokers trips to space for tourists. As of October 2011 it offers trips involving zero-gravity atmospheric flights, cosmonaut training, spacewalk training, launch tours, suborbital spaceflights, orbital spaceflights (with or without a spacewalk), and lunar spaceflights. So far, seven tourists have paid millions of U.S. dollars to participate in spaceflights to the International Space Station through contracts with the Russian space agency Russian Federal Space Agency (Roscosmos, or RKA).

American businessperson Dennis Tito became the world's first space tourist. South African Mark Shuttleworth followed Tito to the ISS in April 2002. The third space tourist was American Gregory Olsen in 2005. Iranian-American Anousheh Ansari (2006), Hungarian-American Charles Simonyi (2007), and American Richard Garriott (2008), followed these wealthy tourists to space. Simonyi became the first spaceflight participant to fly twice into space when he returned to space in March 26, 2009. Canadian businessperson Guy Laliberte was launched into space on September 30, 2009, for a 12-day visit to the International Space Station.

In 2010, the Russians temporarily halted the program for allowing space tourists to fly along with its cosmonauts in their Soyuz capsules. This was due to the size of the ISS Expedition crews increasing from three to six beginning in 2011, along with the fact that the U.S. space shuttle program was ending in that same year. However, with the temporary ending of space tourism with the Russians, Space Adventures, as of April 2010, is now offering suborbital commercial flights with Armadillo Aerospace. The team is contemplating a variety of suborbital flight options in the near future. Many customers have made reservations for these upcoming suborbital spaceflights.

As an option once suborbital flights become established, orbital spaceflights are also on the agenda for Space Adventures. Recently, Space Adventures contracted with the Boeing Company to market orbital flights for its customers aboard its *CST-100* spacecraft. Boeing's Crew Space Transportation (CST) crew capsule is being developed in collaboration with Bigelow Aerospace to transport astronauts to and from the ISS and to proposed private space stations. The spacecraft is expected to be operational in 2015.

Space Adventures announced on July 21, 2006, that they will begin offering 1.5-hour spacewalks for about $15 million as part of its future orbital spaceflights. According to its website, "The first spacewalk was performed by Alexei Leonov in March of 1965. Since his historic mission, spacewalks have been conducted to complete the construction, and facilitate routine maintenance and repair of the space station, and to conduct scientific experimentation that cannot be performed from within the ISS. Today, for the first time, Space Adventures offers spaceflight participants the unique opportunity to walk in space."

Incredible Adventures Incredible Adventures, a space tourism company based in Sarasota, Florida, offers atmospheric flights in the Russian MiG-29 Fulcrum aircraft. The company calls these trips the "Russian Edge of Space" flights. In the future, the company is offering "RocketShip Adventures" inside a XCOR Aerospace Lynx rocket plane for a suborbital flight into space. Incredible Adventures is also offering future "Orbital Space Flights" with Roscosmos. Some training will take place at the Gromov Flight Research Institute, an aircraft test base and scientific research center, outside Moscow, Russia, while other training will take place at Star City, the area where Russian cosmonauts train. If selected as a tourist within the Russian astronaut program, the full package, according to the company, "can range from $15 million to $25 million, depending on the mission profile."

RocketShip Tours RocketShip Tours, based in Phoenix, Arizona, is also teaming up with XCOR Aerospace for suborbital flights into space with its Lynx rocket plane. Founded by American travel entrepreneur Jules H. Klar, RocketShip Tours states that the "mission of RocketShip Tours is to make space travel accessible and affordable to all those who aspire to experience this great new adventure."

▶ *See also* **Aerospace Corporations (Volume 1)** • **Barriers to Space Commerce (Volume 1)** • **Getting to Space Cheaply (Volume 1)** • **Hotels (Volume 4)** • **Launch Industry (Volume 1)** • **Launch Services (Volume 1)** • **Market Share (Volume 1)** • **Musk, Elon (Volume 1)** • **Space Tourism, Evolution of (Volume 4)** • **Tourism (Volume 1)** • **TransHab (Volume 4)** • **United Space Alliance (Volume 1)**

Resources

Books and Articles

McCurdy, Howard E. *Space and the American Imagination.* Baltimore: Johns Hopkins University Press, 2011.

Parker, Martin, and David Bell, editors. *Space Travel and Culture: From Apollo to Space Tourism.* Malden, MA: Wiley-Blackwell/Sociological Review, 2009.

Pelt, Michael. *Space Tourism: Adventures in Earth's Orbit and Beyond.* New York: Springer, 2005.

Websites

Armadillo Aerospace. <http://www.armadilloaerospace.com/n.x/Armadillo/Home> (accessed October 3, 2011).

Bigelow Aerospace. <http://www.bigelowaerospace.com/> (accessed October 7, 2011).

Blue Origin. <http://www.blueorigin.com/> (accessed October 3, 2011).

Excalibur Almaz. <http://www.excaliburalmaz.com/> (accessed October 7, 2011).

Incredible Adventures. <http://www.incredible-adventures.com/> (accessed October 7, 2011).

Masten Space Systems. <http://masten-space.com/> (accessed October 7, 2011).

Rocketship Tours. <http://www.rocketshiptours.com/> (accessed October 7, 2011).

Scaled Composites. <http://www.scaled.com/> (accessed October 7, 2011).

Sierra Nevada Corporation. <http://www.sncorp.com/> (accessed October 7, 2011).

Space Adventures. <http://www.spaceadventures.com> (accessed October 3, 2011).

Virgin Galactic. <http://www.virgingalactic.com/> (accessed September 28, 2011).

XCOR Aerospace. <http://www.xcor.com/> (accessed October 3, 2011).

* **base-load** the minimum amount of energy needed for a power grid

Energy from Space

Forecasts indicate that worldwide demand for new base-load* electrical power generation capacity will continue to grow throughout the twenty-first century. However, evidence is mounting to show that the use of fossil fuels—coal, oil, and natural gas—and the resulting increase in greenhouse gas emissions may be leading to measurable global climate change. New energy technologies are needed to offset the future growth in fossil fuel use.

One concept that would provide power on a significant scale for global markets is energy from space. Earth is about 150 million kilometers (93 million miles) from the Sun. Sunlight constantly delivers, on average, 1,358 watts of energy per square meter of area to the part of Earth facing the Sun. However, by the time the sunlight reaches the surface of Earth,

* **solar arrays** groups of solar cells or other solar power collectors arranged to capture energy from the Sun and use it to generate electrical power

* **multi-bandgap photovoltaic cells** photovoltaic cells designed to respond to several different wavelengths of electromagnetic radiation

* **frequencies** the number of oscillations or vibrations per second of an electromagnetic wave or any wave

* **high-power klystron tubes** a type of electron tube used to generate high frequency electromagnetic waves

* **phased array** a radar antenna design that allows rapid scanning of an area without the need to move the antenna; a computer controls the phase of each dipole in the antenna array

atmospheric filtering has removed about 30% of the initial energy even on a clear day. Moreover, the effects of the seasons and the day-night cycle further reduce the average energy received by an additional 90%. Often, the remaining 100 to 200 watts per square meter can be blocked completely by weather for days at a time.

A solar power satellite, by contrast, could collect sunlight in space and convert the energy into electrical current to drive a wireless power transmission system, which would in turn beam the power down to receiving antennas on Earth. Earth-bound receiving antennas could capture the transmitted energy—almost twenty-four hours a day, seven days a week—and deliver it to local electrical grids as base-load power. This approach could eliminate the need for extremely large solar arrays* on the ground and dramatically expensive energy storage systems, but would require a number of new technological advances.

Details of the Concept of Energy from Space

Power Generation in Space Typically, photovoltaic arrays are used to generate power in space. These solid-state devices exploit the characteristic of semiconductors such as silicon to allow incoming photons to readily dislodge electrons, a process that produces voltage (or, electrical potential difference). Key measures of the effectiveness of these technologies are the specific power (e.g., watts produced per kilogram of solar array mass) and the efficiency (e.g., watts produced per square meter of solar array area). By the early twenty-first century, space solar power technology research programs were using concentrating lenses and multi-bandgap photovoltaic cells* to achieve specific power levels approaching 300 watts per kilogram and efficiencies of over 40%. These figures represent advances of more than a factor of four over the state-of-the-art technology of the late 1990s.

Wireless Power Transmission Power can be transmitted using radio frequencies*, microwaves or visible light. In the case of radio frequencies, a wide range of devices could be used to generate the power beam, including high-power klystron tubes*, low-power, highly efficient, solid-state devices, and magnetrons, which are a type of vacuum tube providing in-between levels of power (and are also used in household microwave ovens). In all of these cases, a number of the devices would be arranged and operated in a lockstep phased array* to create a coherent, collimated (parallel) beam of energy that would be transmitted from space to the ground. The efficiency of the transmitter can be as high as 80% or more. On the ground, a radio frequency power beam would be converted back to voltage by a rectifying antenna (also known as a "rectenna") operating at about 90% efficiency. Taking into account losses of the collimated beam at the edges of the rectenna and small levels of interference from the atmosphere, a power beam might generate over 200 watts per square meter on the ground on average.

Energy collection in space from other planets or satellites could revolutionize the method of power delivery on Earth. *NASA.*

* **payload** any cargo launched aboard a rocket that is destined for space, including communications satellites or modules, supplies, equipment, and astronauts; it does not include the vehicle used to move the cargo nor the propellant that powers the vehicle

The Challenge of Large Systems in Space A central challenge of space solar power is that of launching and building these exceptionally large systems in space. As of the early twenty-first century, the cost of space transportation ranged from $6,600 to $11,000 per kilogram of payload*, launched to Earth orbit. In order to be economically viable, space solar power systems must be launched at costs of no more than $400 per kilogram. Such a dramatic improvement requires the development of a range of new technologies and new types of space transportation systems.

Base-load Solar Power Systems: Ground and Space The projected costs of base-load solar power using ground-based solar arrays are clearly dominated by the cost of the energy storage system needed to allow energy received during a clear day to be delivered to customers at night—or during several consecutive days of cloudy weather. These costs can be greater than $15,000 per kilowatt-hour of energy stored. In other words, to power a house using two kilowatts of power over five days of cloudy weather would require about $4 million to build the energy storage system alone. Conversely, the installation cost for a space-based solar power system (providing power for hundreds of thousands of homes) might be expected to range between $100,000 and $300,000 per home, for a comparable power-using home. This is still much greater than the cost of installing new fossil-fuel power-generating capacity, but the cost of a space-based solar power system could be as little as 1% of the cost of a comparable ground-based system.

History and Future Directions

In the 1970s, the U.S. Department of Energy (DOE) and the National Aeronautics and Space Administration (NASA) extensively examined large solar power satellite systems that might provide base-load power into terrestrial markets. From 1995 to 1997 and in 1998, NASA reexamined space solar power (SSP), with both encouraging technical results and cautionary findings concerning the economics of introducing the technology during the first two decades of the twenty-first century. As a result, from 1999 to 2000, NASA conducted the SSP Exploratory Research and Technology program, which refined and modeled SSP systems concepts, conducted research and development to yield "proof-of-concept" validation of key technological concepts, and laid the foundation for the creation of partnerships, both national and international. A number of innovative concepts and technology advances resulted from these efforts, including a new solar power satellite concept, the Integrated Symmetrical Concentrator, and new technologies such as lightweight, high-efficiency photovoltaic arrays, inflatable heat radiators, and new robots for space assembly.

Recognizing that SSP research is being conducted by many different government and private entities, the U.S. National Security Space (NSS) office was formed in 2004 to facilitate and integrate space activities. In October 2008, the NSS published a report, entitled "Space Based Solar Power (SBSP) As An Opportunity For Strategic Security," which resulted from a global collaboration of over one hundred government, academic, technical and legal entities. While the report addressed many issues concerning SBSP, the major recommendation put forth was for the U.S. government to establish a federal agency to act as a spearhead for the development and encouragement of public and private initiatives to further promote all phases of SBSP technology.

Advances in space solar power generation took a giant leap in April 2009 when Pacific Gas and Electric (PGE), an electric utility company based in California, announced plans to purchase base-load space solar power starting in 2016. The Solaren Corporation is planning to build the first SSP plant to market space based solar energy to PGE. Using solid-state power amplifiers, the solar electricity will be collected by solar arrays aboard a satellite, converted into radio frequency (RF) energy, and beamed down to receiving stations on Earth.

Other means of transporting energy being researched are microwave and laser transmission. While economics continues to be the largest obstacle in the path of achieving a feasible SBSP program, non-technical issues will also play an important role. Impact on communications, air travel, and astronomy in addition to environmental impact, safety, and security concerns are some of the issues that will have to be addressed. Future technology efforts will focus on providing the basis for better-informed decisions regarding solar energy in space and related research and development.

See also **Power, Methods of Generating (Volume 4) • Solar Power Systems (Volume 4)**

Resources

Books and Articles

Glaser, Peter E., et al. "First Steps to the Solar Power Satellite." *Institute of Electrical and Electronic Engineers (IEEE) Spectrum* 16, no. 5 (1979):52–58.

Iles, Peter A. "From Vanguard to Pathfinder: Forty Years of Solar Cells in Space." *Proceedings of Second World Conference and Exhibition on Photovoltaic Solar Energy Conversion, Vienna, Austria* (1998):LXVII–LXXVI.

———. "Evolution of Space Solar Cells." *Solar Energy Materials and Solar Cells* 68(2001):1–13.

Nansen, Ralph. *Energy Crisis: Solution from Space.* Burlington, ON: Apogee Books, 2009.

U.S. House of Representatives. *Solar Power Satellite.* Honolulu, HI: University Press of the Pacific, 2004.

Websites

Center for Space Power. Texas A&M University. <http://engineering.tamu.edu/tees/csp/> (accessed October 10, 2011).

Landis, Geoffrey A. *Reinventing the Solar Power Satellite.* National Aeronautics and Space Administration, Glenn Research Center. <http://gltrs.grc.nasa.gov/reports/2004/TM-2004-212743.pdf> (accessed October 10, 2011).

Logan, James. *Safety of Space-Based Solar Power.* Space Energy AG. <http://www.spaceenergy.com/i/pdf/safety_paper.pdf> (accessed October 10, 2011).

Pacific Gas and Electric Company. *Contract for Procurement for Renewable Energy Resources Resulting from PGE's Power Purchase Agreement with Solaren Corporation.* <http://www.pge.com/nots/rates/tariffs/tm2/pdf/ELEC_3449-E.pdf> (accessed October 10, 2011).

Space-Based Solar Power as an Opportunity for Strategic Security. National Security Space Office. <http://www.nss.org/settlement/ssp/library/final-sbsp-interim-assessment-release-01.pdf> (accessed October 10, 2011).

URSI White Paper on Solar Power Satellite (SPS) Systems. International Union of Radio Science. <http://ursi.ca/SPS-2006sept.pdf> (accessed October 10, 2011).

Entertainment

Outer space is big business for the entertainment world. The earliest record of an authored work of science fiction, written to fuel the imagination and entertain the public, was the Greek satirist Lucian's *Vera historia* (True history), penned around AD 170. In Lucian's tale, a sailing vessel is caught up by a whirlwind and after a journey of eight days arrives at the Moon. Lucian's description of this imaginary lunar voyage set the scene for many stories, films, and even computer games that have followed.

Science fiction novels sell in phenomenal numbers, appealing to the reader's wish to escape the everyday and stimulating the imagination with the possibilities of tomorrow. Many novelists such as Ben Bova (1932–) and Ray Bradbury (1920–) have made careers in science fiction writing. Others, such as James Michener (1907–1997), author of *Space,* Isaac Asimov (1920–1992), and Arthur C. Clarke (1917–2008) have been lured by the theme of space, as one compass of a much broader writing career. Science-fiction conventions celebrate this genre and allow fans an opportunity to meet with famous authors, hear how they develop their themes related to the future of space exploration, and dissect the plots. These conventions are also major business enterprises.

In modern times, the most notable entertainment of the first half of the twentieth century was Orson Welles' broadcast of *The War of the Worlds.* English novelist and historian H. G. Wells (1866–1946) wrote this tale of a Martian invasion as a magazine serial in 1897. However, Welles' eerie radio rendition of the tale in 1938 sent shock waves through the United States as listeners tuned in to what they thought was a serious report of alien invasion.

▶ Hollywood's depictions of alien life forms have ranged from the gentle extra-terrestrial E.T. to the terrifying alien of *Aliens* (pictured). *The Kobal Collection.*

Television: From *Star Trek* to *Nova*

Some of the most successful and longest-running series on television have had themes of space exploration. *Star Trek,* the brainchild of the legendary Gene Roddenberry (1921–1991), through its various generational formats has made the careers of several actors and actresses and met with so much enthusiasm that it has spawned Star Trek conventions. The British invention, *Doctor Who,* also met with universal, long-term success and was assimilated as one of the "cult" shows of the twentieth century. *Babylon Five* also developed a very significant following, and the Jim Henson-backed series *Farscape,* featuring a lost astronaut thrown into the distant regions of space, was a long-running Sci-Fi Channel original series from March 1999 to March 2003.

Space themes are not confined to futuristic fictional series on television, although these are by far the best known and the greatest revenue generators. Aliens are a common theme both as a dramatic effect in a storyline and as the subject matter of serious newsmagazine programs about scientific exploration and pseudoscience*. Educational programs about space exploration have great popular appeal and, by extension, attract significant advertising dollars to television stations. One of the great television successes of the 1990s was Tom Hanks' HBO series *From the Earth to the Moon,* the story of the Apollo missions that landed twelve humans on the Moon in six separate missions. In 2000, the Discovery Channel's in-depth study of the International Space Station represented a significant programming investment.

From July 1997 to March 2007, *Stargate SG-1* was a military science-fiction series on the Showtime channel (first five years) and, later, on the Sci-Fi Channel (last five years). In a similar vein, *Battlestar Galactica* was a military science-fiction series, based on the 1978 television series, that ran from December 2003 to March 2009. Finally, news magazine programs such as *Nova* frequently return to stories of space exploration because the human fascination with the unknown and the "great beyond" of the universe draws a large audience.

Films: Special Effects and Special Stories

Outer space can be daunting, fascinating, and mysterious—a gift to moviemakers. Facing the challenges of working in microgravity* calls for fearless heroes and feats of courage. And re-creating outer space for the motion picture audience offers numerous possibilities for the special effects department.

People may snicker at the title of the 1956 film *Invasion of the Body Snatchers,* but this movie is an influential classic and still very scary. It tells the story of residents of a small town who are replaced by inert duplicates, which are hatched from alien "pods."

2001: A Space Odyssey, Stanley Kubrick's influential 1968 masterpiece (with the screenplay written by Arthur C. Clarke), opened the imagination to the possibility of other intelligent entities developing in time frames different from the evolution of humans on Earth, while also

* **pseudoscience** a system of theories that assumes the form of science but fails to give reproducible results under conditions of controlled experiments

* **microgravity** the condition experienced in freefall as a spacecraft orbits Earth or another body; commonly called weightlessness; only very small forces are perceived in freefall, on the order of one-millionth the force of gravity on Earth's surface

featuring alien encounters and a computer with an attitude called HAL (short for Heuristically programmed ALgorithmic computer). In *2001,* possibly the most influential space movie to date, Kubrick enticed the audience with the vastness and timelessness of space in comparison to the current human condition.

From the days of the earliest space-themed movies, directors have been awed by the subject matter and have worked studiously to be as authentic as possible in the representation of spaceflight and off-world locations. This is how space historian Fred Ordway and space artist Robert McCall (who painted the lunar mural in the National Air and Space Museum in Washington, D.C.) found themselves in London, consulting on the making of *2001,* and how countless astronauts have been called upon to advise actors on how to realistically simulate behavior in microgravity.

The 1977 blockbuster *Star Wars,* now also called *Episode IV: A New Hope,* and the two subsequent episodes in the trilogy, *Episode V: The Empire Strikes Back* (1980) and *Episode VI: Return of the Jedi* (1983), opened a new era in opening the imagination of moviegoers to space. In its next series of films, starting with *Episode I: The Phantom Menace* (1999), legendary producer George Lucas introduced audiences to tales of life and conflict in a vast universe populated by creatures of mind-boggling diversity and cunning. Since then, the pre-sequel series continued with *Episode II: Attack of the Clones* (2002) and *Episode III: The Revenge of the Sith* (2005). In 2008, the animated film *Star Wars: The Clone Wars* was released. It introduced the 2008 weekly animated television series of the same name.

The 1977 film *Close Encounters of the Third Kind,* directed by Steven Spielberg, describes a first contact with alien beings. Impressive cinematography won an Oscar award for Vilmos Zsigmond. Spielberg's *E. T. The Extra-Terrestrial,* released in 1982, cemented Spielberg's reputation as a director and won John Williams an Academy Award for his score, with additional Oscars going to the sound and visual effects teams. *E. T.* is a classic of the sympathetic alien genre of movies, which developed along with the growing understanding of the unique nature of human life in the solar system and with the increasing knowledge about the origins of life. Its enduring influence is demonstrated by its rerelease to the big screens in 2002.

The blockbuster of the 1990s was *Apollo 13,* based on the book *Lost Moon* by *Apollo 13* commander Jim Lovell and Jeffrey Kluger. This exhilarating story of the ill-fated NASA (National Aeronautics and Space Administration) *Apollo 13* Moon mission was directed by Ron Howard and starred Tom Hanks, Bill Paxton, Kevin Bacon, Gary Sinise, and Kathleen Quinlan. Sticking painstakingly close to the true story of *Apollo 13,* this 1995 movie told the tale of the human ingenuity, fast thinking, and enormous courage that brought the crew of *Apollo 13* (commander James Lovell, command module pilot Jack Swigert, and lunar module pilot Fred Haise) back safely to Earth from its path toward the Moon after a catastrophic explosion deprived them of the majority of their air supply. The film provided a marked contrast with the media headlines of failure (because the crew failed to land on the Moon) that had formed public opinion about the mission twenty-five years earlier.

In the late 1990s, as scientific understanding of asteroids grew as a result of better telescopes and the detailed images from robotic missions such as *Galileo* and *Near Earth Asteroid Rendezvous–Shoemaker,* a crop of movies about the threat of asteroid or comet collisions with Earth were released. Both *Deep Impact* and *Armageddon* did well at the box office and served to broaden the public debate on the threat of asteroid impacts.

Space Cowboys, released in 2000 and starring Clint Eastwood (also as director), Tommy Lee Jones, Donald Sutherland, and James Garner, reflected growing concern with the amount of space debris circling the planet and, on occasion, falling uncontrolled to Earth. And movies telling of the human exploration of Mars—one of the great space challenges for the human race in the twenty-first century—were on the rise. Such movies include *Mission to Mars* (2000), directed by Brian de Palma, which tells of a rescue mission to the planet Mars after the first manned mission to the planet has taken place. Other movies about Mars include the American-made *Red Planet* (2000) and the Spanish film *Stranded: N´ufragos* (2002).

Other notable science fiction movies of the 2000s and early 2010s, include *Star Trek* (2008), which chronicles the early lives of the men and women that will comprise the *U.S.S. Enterprise,* *The Time Traveler's Wife* (2009), which describes a man with a genetic disorder who is able to travel in time to confront special moments in his life, while developing a romantic relationship; *Avatar* (2009), which tells the mid-22nd century story of humans mining precious minerals on the planet Pandora within the Alpha Centauri star system; and *Tron: Legacy* (2010), which is a sequel to the 1982 film *Tron* that tells the story of a software engineer who finds a hidden computer laboratory that transports him to a virtual-reality world inside a video arcade game.

It is impossible to discuss films about space without mentioning the large-scale IMAX films on a range of space topics that are screened at numerous science museums around the world. These films trace the history of the human exploration of space with awe-inspiring visual effects provided by Mother Nature. Such movies brought to IMAX include *Magnificent Desolation: Walking on the Moon 3D* (2005), which tells the stories of the twelve astronauts who walked on the Moon, *Roving Mars* (2006), which entails the development, launch, and operation of the NASA Mars Exploration Rover (MER) mission consisting of *Spirit* and *Opportunity,* and *Hubble 3D* (2010), the story of the *Hubble Space Telescope.*

Exhibits and Theme Parks

The best-known visitor attractions with space themes include the most visited museum in the world, the National Air and Space Museum in Washington, D.C.; the Epcot Center at Disney World in Orlando, Florida; and Tomorrowland in California. Space theme parks, which allow visitors to sample the technologies of the future or simulate a spaceship ride or a walk on the surface of the Moon or Mars, have been developed by visionaries who foresee hundreds of thousands of people routinely traveling in space in the future.

Video and Computer Games

Computers play a major role in simulating complex rendezvous, docking, and landing maneuvers for space missions. They also provide exciting games that test a player's skill in retrieving a satellite, docking or maneuvering a spacecraft in zero gravity, and much more. If the majority of people cannot experience space travel themselves, some of the computer games available are the next best thing. Some recent space simulation games include *Lunar Pilot* (2004, by Things-To-Come), which simulates piloting a spacecraft to the Moon's surface; *Claimstake* (2007, by Claimstake Team), which allows users to land on an asteroid and mine it for valuable minerals; *Space Shuttle Mission 2007* (2007, by Simsquared Ltd.), which simulates a flight on the NASA space shuttle; and *Infinite Space* (2010), which simulates the design and operation of a spaceship within two galaxies.

Space Tourism

The business of ferrying space tourists back and forth from space is being actively pursued by several private companies in the 2010s. Some of the companies are developing a single-stage system in which the entire vehicle travels from the launch pad all the way to a suborbital altitude (and later to orbit) and returns to the launch site. Others propose two-stage vehicles in which a booster/orbiter combination leave the launch pad together and return separately. One of them is Richard Branson's Virgin Group (Virgin Galactic), which is proposing to provide sub-orbital spaceflight missions to paying customers. U.S. enterprise Scaled Composites, founded by Burt Rutan, has teamed up with Richard Branson's Virgin Galactic. The design and manufacturing company Scaled Composites is building reusable spacecraft and launch vehicles for Virgin Galactic so that suborbital flights are available for the private sector. The reusable spacecraft, a series of spacecraft called *SpaceShipTwo* (actually about 10% of the craft, excluding the fuel, is not reusable), will be launched with the help of one of a series of *WhiteKnightTwo* launching vehicles, which take the spacecraft from launch to about 18 kilometers (11 miles) above Earth. The spacecraft then completes the trip to space, about 100 kilometers (62 miles) above Earth. Test flights have taken place in 2011 with more planned for 2012. The first commercial flight is scheduled for sometime in 2013.

Other companies are attempting to develop private manned suborbital flights into space for space tourists. Armadillo Aerospace is based in Mesquite, Texas, and founded in 2001, by John Carmack, the co-founder of id Software (maker of computer games such as the series *Doom* and *Quake*). The company is working with Space Adventures (as of May 2010) on a suborbital commercial rocket that will take tourists into space for about $102,000 per customer (priced as of 2010). The spacecraft, tentatively called Black Armadillo, will be a vertical takeoff, vertical landing (VTVL) suborbital vehicle. No date has been set for the maiden voyage.

Blue Origin, which is based in Kent, Washington, but with its main flight facility in Culberson County, Texas, was founded by Jeff Bezos (the founder of Amazon.com) in 2000. The company is working on its VTVL *New Shepard* spacecraft, which is based on the technology for the McDonnell Douglas DC-X. Unmanned test flights took place in 2011 with more flights planned for 2012 and beyond.

Excalibur Almaz is based in Douglas, Isle of Man, and founded in 2005, by Arthur M. Dula and Buckner Hightower. The company is designing a spacecraft based on the Soviet TKS space capsules, which were originally designed to go to the Soviet *Almaz* space station, which was never developed. According to its website, testing of its capsule is expected to begin in 2012, with operational flights no earlier than 2013.

In 2011, XCOR Aerospace and Space Expedition Curacao (SXC) signed a lease contract for XCOR's Lynx suborbital spaceplane. The Caribbean island of Curacao now expects to begin construction of its spaceport with an expected beginning date of operations in 2014. As of September 19, 2011, SXC had signed 35 space tourists for the first suborbital flights of XCOR, and was expecting to sign up an additional 50 space tourists by the end of the year. Each space experience will involve only one paying customer and a pilot.

Space hotels are also being planned for the future. In the United States, Bigelow Aerospace is developing various inflatable habitat modules. In July 2006, Bigelow launched its *Genesis I* module, which was followed up about eleven months later with its *Genesis II* module. As of November 1, 2011, both modules are in orbit about Earth. The company is expecting to launch its *BA 330* (formerly called Nautilus space complex module) in 2014 or 2015. The space habitation module, expandable to 330 cubic meters (12,000 cubic feet), will directly support zero-gravity scientific and manufacturing research, along with indirect activities involving space tourism.

Bigelow is also designing a *BA 2100* module that would require a heavy-lift rocket to send it to low-Earth orbit. If manufactured, the module would be six times larger than the *BA 330*—at 2,200 cubic meters (77,690 cubic feet). In addition, Bigelow Aerospace is also developing a series of inflatable modules, under the name of Bigelow Next-Generation Commercial Space Station (also called *Space Complex Alpha*). The complex will include spacecraft modules, along with a central docking node, solar arrays, propulsion, and crew capsules. The company hopes to launch the components in 2014, with leasing options starting in 2015. So far, Bigelow has contracts with seven countries interested in leasing its space station modules. The company also has plans for a second orbital space station called *Space Complex Bravo*, with launches scheduled to begin in 2016 and possible commercial operations commencing in 2017.

 See also **Apollo (Volume 3)** • **Careers in Writing, Photography, and Film (Volume 1)** • **Clarke, Arthur C. (Volume 1)** • **Impacts (Volume 4)** • *Star Trek* **(Volume 4)** • *Star Wars* **(Volume 4)**

Resources

Books and Articles

Clarke, Arthur C. *The Coming of the Space Age.* New York: Meredith Press, 1967.

McCurdy, Howard E. *Space and the American Imagination.* Baltimore: Johns Hopkins University Press, 2011.

von Braun, Wernher, Frederick I. Ordway III, and Dave Dooling. *Space Travel: A History.* New York: Harper & Row, 1985.

Websites

Armadillo Aerospace. <http://www.armadilloaerospace.com/n.x/Arma-dillo/Home> (accessed October 3, 2011).

Blue Origin. <http://www.blueorigin.com/> (accessed October 3, 2011).

Future of Space Tourism: Who's Offering What. Space.com. <http://www.space.com/11477-space-tourism-options-private-spaceships.html> (accessed October 3, 2011).

Knapp, George. *I-Team: Bigelow Aerospace Begins Big Expansion.* 8News-Now.com. <http://www.8newsnow.com/story/13967660/i-team-big-elow-aerospace-begins-big-expansion> (accessed September 29, 2011).

Kubrick2001: the space odyssey explained. New Media Giants. <http://www.kubrick2001.com/> (accessed October 31, 2011).

RocketShip Adventures: XCOR Lynx Overview. Incredible Adventures. <http://www.incredible-adventures.com/xcor-lynx.html> (accessed September 30, 2011).

Star Trek. <http://www.startrek.com/> (accessed October 31, 2011).

Star Wars. <http://www.starwars.com/> (accessed October 31, 2011).

Study Guide for H. G. Wells: The War of the Worlds (1898). Paul Brians, Washington State University. <http://www.wsu.edu/~brians/science_fiction/warofworlds.html> (accessed October 31, 2011).

Virgin Galactic. <http://www.virgingalactic.com/> (accessed September 28, 2011).

XCOR Aerospace. <http://www.xcor.com/> (accessed October 3, 2011).

Expendable Launch Vehicles   *See Launch Vehicles, Expendable (Volume 1)*

F

Financial Markets, Impacts on

The space industry in 2011 is in the midst of a revolution driven to a large extent by the expansion of private industry into an arena that has long been dominated by governmental programs. According to the Space Foundation's *Space Report 2011*, worldwide space industry revenues reached a total of $276.52 billion in 2010, an increase of 7.7 percent over the previous year. That increase was substantially greater than the five percent improvement seen in two previous years. It was especially impressive given the general financial sluggishness in financial markets worldwide between 2008 and 2011. Of the total space revenues, the largest share came from commercial space products and services ($102.00 billion; 37 percent of total revenue), closely followed by commercial infrastructure and support industries ($87.39 billion; 32 percent). Lagging behind these two sources were U.S. government space expenditures ($64.63 billion; 23 percent), space budgets of non-U.S. governmental agencies ($22.49; eight percent), and commercial space transportation services ($0.01 billion; less than one percent). These numbers reflected a significant increase in commercial infrastructure and support revenue (13 percent) and commercial products and services (nine percent) compared to only miniscule changes in governmental programs (about one percent and 0.3 percent in non-U.S. agencies). The success of the commerical space industry was also reflected in dramatic improvements in the Space Foundation market index, which grew by nearly 20 percent during 2010.

Changes in the governmental-private mix in the space industry has had a varied impact on employment patterns around the world. In the United States, the discontinuation of the U.S. space shuttle program and the cancellation of the proposed Constellation space program resulted in a loss of 2,700 space jobs between March 2008 and March 2010. Additional losses of 3,000 jobs are anticipated following the announced reductions at other space facilities, such as the Kennedy Space Center, in the near future. Still, business improved sufficiently in the private sector to result in a total employment figure of 260,000 workers in 2009, a number that has remained relatively constant for the past few years. The situation was about the same in Europe, where employment figures rose by three percent in 2009. But the scene was quite different in countries just entering the commerical space race, such as Japan, where employment in the space industry grew by 22 percent in 2009. The growth of the space industry in countries like Japan, India, and China was perhaps most obvious in the last of these countries, where the number of space and engineering graduates more than doubled (from 325,000 to 770,000) between 2002 and 2006.

Space industry analysts have pointed to a number of trends that are likely to affect future patterns in the field. Perhaps most important of these is the greater emphasis on the role of private industry, rather than governmental agencies, in the design and implementation of new space products and infrastructure. The cancellation of the Constellation program to return humans to the Moon, for example, has forced the U.S. National Aeronautics and Space Administration to shift its budget from direct federal programs to the support of private space industries. Changing financing modalities have accompanied this shift, with greater impetus to the space industry company coming from less traditional sources of funding, such as prize awards (for example, the Ansari X Prize, a $10 million prize awarded by the X Prize Foundation for the first non-governmental organization to fly a reusable manned spacecraft twice) and the use of export credits for the support of space projects. Accompanying these changes has also been a sheer increase in the number of countries successfully involved in space projects. As of 2011, for example, more than 50 nations had active satellite programs which have already launched or are expected soon to launch their own satellite.

Projections for the future of the worldwide space market are difficult to make, given the current state of flux in most countries. In the United States, the shift from governmental to private initiatives has already begun, and a number of start-up, young, and established companies are working on the development of a variety of satellite and commercial space transportation programs. Many analysts believe that the United States is likely to lose market share in the space industry because of increasingly tight export restrictions on U.S.-produced components and the increased sophistication of space technology in other countries. In some cases, a particular company or market appears to have the field "tied up" for the foreseeable future because of its monopoly on technology. France's use of the Russia Soyuz rocket for medium-weight cargo and its own Ariane-5 for heavier cargo virtually assures its dominance in those fields for some years to come. Some observers suggest that the fate of the U.S. space industry may depend, to some extent, on its ability to partner with other nations of the world, especially those in the developing world, in the development of their own space programs.

Satellites: A Driving Force

Of all the activities associated with commercial space, satellites are likely to drive the space industry's growth for the foreseeable future. Satellites have the advantage of speed, mobility, and costs, independent of their Earth orbit. A single satellite system can reach every potential user across an entire continent. For many applications, satellite technology provides the most cost-effective way of providing service over a wide area. As a result, satellites will be instrumental in helping to raise the standard of living in many underdeveloped countries where there is little or no communications infrastructure*. In that role, the economic impact of commercial space may be incalculable.

Satellites provide a broad menu of services. These range from mobile telephony to direct-to-home broadcast of television, cable, and video programming. The Satellite Industry Association (SIA) reported that total world revenues for satellite communications services totaled $168.1 billion for the year 2010, which was a 4.5% increase over 2009.

The wave of the future is broadband, which refers to a frequency on the electromagnetic spectrum* that will allow satellites to provide high-speed Internet access, interactive video, and video on demand.

In 1998, a combination of failed attempts to launch satellites, in-orbit satellite failures, and business plans gone awry sent investors scrambling as they pulled their support from ventures. This could happen again. On the other hand, commercial space is still in an early stage of development, and if recent history is any guide, investors' understanding of this unique business will continue to increase, and sufficient venture capital will remain available.

In 2011, the U.S. Federal Aviation Administration (FAA) presented its forecast for the next decade of satellite launches in the United States. FAA projected a relatively constant number of geosynchronous orbit (GSO) satellites until 2020, at about 20 satellites per year. In contrast, they predicted that the number of satellites launched into non-geosynchronous orbit (NGSO) would decrease over the decade, from a maximum of 45 in 2015 to a minimum of 15 at the end of the decade. The "hump" in the prediction curve around 2015 reflects the peak time during which the five dozen satellites in the Iridium program are being replaced. Without that program, the number of satellites expected to be launched annually is relatively constant at about 15 per year.

* **communications infrastructure** the physical structures that support a network of telephone, Internet, mobile phones, and other communication systems

* **electromagnetic spectrum** the entire range of wavelengths of electromagnetic radiation

Satellites are likely the force that will drive the space industry, as they have advantages that are independent of their Earth orbit, such as speed and mobility. © *AP Images/Alexander Merkushev.*

Commercial Space Transportation

A growing component of the space industry has been the development of space transportation programs, designed to provide a regular flow of space tourists coming from and going into space. A number of aerospace companies are currently involved in one or another aspect of this endeavor. One such company is Virgin Galactic, a subsidiary of British entrepneur Richard Branson's Virgin Group. Virgin Galactic plans on using a spacecraft called the SpaceShip to carry six passengers and two pilots 70,000 feet (21,000 meters) into space on a voyage that will include a six-minute period of weightlessness. After successfully testing its SpaceShipOne spacecraft in 2004, it began testing its updated model of the ship, SpaceShipTwo. In anticipation of that ship's first commercial flight at some unannounced date in the future, more than 450 individuals (as of late 2011) had made a 10 percent downpayment on the $200,000 cost of the trip. The company is also planning to build a fleet of SpaceShipThree spacecraft to take people into orbit about Earth. If space tourism is successful, experts are predicting a $100 billion industry by 2017. Orbital hotels are also being designed by various companies, such as Bigelow Aerospace, which was founded in 1999 by American hotel and aerospace entrepreneur Robert Bigelow.

 See also **Commercialization (Volume 1) • Communications Satellite Industry (Volume 1) • Market Share (Volume 1)**

Resources

Books and Articles

Belfiore, Michael P. *Rocketeers: How a Visionary Band of Business Leaders, Engineers, and Pilots Is Boldly Privatizing Space.* New York: Smithsonian Books, 2007.

Brennan, Louis, and Alessandra Vecchi. *The Business of Space: the Next Frontier of International Competition.* Basingstoke: Palgrave Macmillan, 2011.

"Iridium's Slow Ramp-Up Has Investors on Edge." *Aviation Week & Space Technology* April 5, 1999, S18.

National Research Council. Committee on the Rationale and Goals of the U.S. Civil Space Program, et al. *America's Future in Space: Aligning the Civil Space Program with National Needs.* Washington, DC: National Academies Press, 2009.

Seedhouse, Erik. *Tourists in Space: A Practical Guide.* Berlin; New York: Springer, 2008.

The Space Report 2011. Washington, DC: The Space Foundation, 2011.

Websites

2011 Commercial Space Transportation Forecasts. <http://www.faa. gov/about/office_org/headquarters_offices/ast/media/2011%20 Forecast%20Report.pdf> (accessed October 20, 2011).

American Asronautics. <http://www.americanastronautics.com/> (accessed October 20, 2011).

Book Your Place in Space. Virgin Galactic. <http://www.virgingalactic. com/> (accessed October 20, 2011).

Forecast 2011: Space—Changes Ahead. Flightglobal. <http://www. flightglobal.com/news/articles/forecasts-2011-space-changes-ahead-351294/> (accessed October 20, 2011).

Future of U.S. Manned Spacecraft Looks Bleak. Physorg.com <http:// www.physorg.com/news/2011-07-future-spaceflight-bleak.html> (accessed October 20, 2011).

Space Investment Summit. <http://spaceinvestmentsummit.com/> (accessed October 20, 2011).

SpaceX. <http://www.spacex.com/index.php> (accessed October 20, 2011).

G

Getting to Space Cheaply

Getting to space is not cheap. It costs quite a bit of money to send a payload into low Earth orbit (LEO) and even more for higher orbits. In 2010, the cost of sending a payload into low-Earth orbit* was about $11,000 per kilogram ($5,000 per pound of weight on Earth), with a range of anywhere from $3,000 to $30,000 per kilogram ($3,600 to $13,600 per pound). The wide range is brought about by the variability on the size and type of rocket, type and amount of fuel, altitude and inclination or orbit, and other such pertinent factors.

For instance, to get a payload into LEO with the European Space Agency's (ESA) Ariane 5G (which has the capability of sending about 18,000 kilograms [39,648 pounds] to LEO), it costs approximately $10,476 per kilogram ($4,762 per pound), according to the Federal Aviation Administration (in 2009). For Russia, the Proton rocket, which can carry up to 19,760 kilograms (43,524 pounds) into LEO, is approximately $4,302 per kilogram ($1,953 per pound), also according to the FAA. The U.S. space shuttle, which could take about 28,800 kilograms (63,400 pounds) into LEO, totaled at about $10,416 per kilogram ($4,729 per pound).

Anyone who wants to perform experiments, launch communications satellites or telescopes, or other activities in space has a large cost hurdle to overcome that is not encountered in any other area of human endeavor. In other words, launching payloads into space is expensive. Some organizations and companies are trying to make it less expensive to get into space, especially now that in the 2010s commercial organizations are getting more interested in space-related activities. However, for them, the perceived high cost of space transportation is one of the biggest hurdles for the growth of space commercialization and exploration. Some estimates show that to make it cost effective for commercial space companies, the price for launching payloads into LEO must come down to about $2,200 per kilogram ($1,000 per pound).

What Makes It Expensive

Low Earth orbit (LEO) is a few hundred miles up in space. To get to LEO from the surface of Earth, it takes 30,000 feet per second (9.14 kilometers [5.86 miles] per second) of velocity change; the total energy needed to get to the Moon requires a speed of about 45,000 feet per second (13.72 kilometers [8.52 miles] per second). LEO is therefore two-thirds of the way to the Moon; that is, energy-wise.

* **low Earth orbit** an orbit between 300 and 800 kilometers (185 to 500 miles) above Earth's surface

* **drag** a force that opposes the motion of an aircraft or spacecraft through the atmosphere

* **aerodynamic heating** heating of the exterior skin of a spacecraft, aircraft, or other object moving at high speed through the atmosphere

* **payload** any cargo launched aboard a rocket that is destined for space, including communications satellites or modules, supplies, equipment, and astronauts; does not include the vehicle used to move the cargo or the propellant that powers the vehicle

Two factors make the step from Earth to LEO difficult. The first factor is Earth's atmosphere, which causes drag* and aerodynamic heating*. The second factor is the gravity gradient, or the change in the force of gravity as one moves away from Earth. The force of gravity declines inversely as the square of the distance from Earth, meaning that the farther away one gets from Earth, the easier it is to overcome the force of gravity. As a result, it is more difficult (with respect to the amount of fuel) to get from Earth to LEO than from LEO to almost anywhere else in the solar system.

Expendable Launch Vehicles

Throughout the twentieth century, getting to space was accomplished almost exclusively with expendable launch vehicles (ELVs)—rockets that are used once and then discarded in the process of putting their payload* into orbit.

As of October 2011, every rocket used to place payloads into orbit has used multiple parts, or stages. Each stage is itself a working rocket. One or more stages are discarded and dropped off as the vehicle ascends, with each discard eliminating mass, enabling what is left over to make orbit.

ELVs available at the beginning of the twenty-first century include the two EELV ("Evolved ELV") families paid for by the U.S. Air Force: the Atlas V and Delta IV. Still available for purchase are the Delta II and IV, as well as the Atlas V Heavy, and the Boeing Sea Launch ELV, a converted Zenit rocket launched from a ship at sea. However, in June 2009, Sea Launch Company LLC, the company that provides the Sea Launch service, filed for bankruptcy protection. Over a year later, the company emerged from bankruptcy, with the Russian company Energia Overseas Limited acquiring a majority stake in the operation. Sea Launch expects to resume land-based launches no earlier than late 2011, with its ocean-based launches already back in operation.

There is also the market leader, the French Ariane 5 ELV (which is expected to launch at least through 2015). The Russian Proton and Rokot ELVs are also available, as are the Chinese Long March families of throwaway boosters. ELVs are inherently incapable of providing cheap access to space for the same rationale that throwing away an automobile after each use is also not economical. Nevertheless, ELVs are here for the foreseeable future; at least through the early part of the twenty-first century.

Reusable Launch Vehicles

The closest type of rocket that launched payloads into space, which was not expendable, was the solid rocket boosters (SRBs) on the space shuttle. These two SRBs were ejected from the orbiter after they used up their fuel, and they fell back to Earth (into the water) to be used over again. In other words, they were reusable. In a modified form, it may also be used in the future to lift the new Orion space capsules (also called Multi-Purpose Crew Vehicles [MPCVs]) into space, as part of the National Aeronautics and Space Administration's (NASA's) new program to go to asteroids and the planet Mars. This program replaces NASA's space shuttle program (or, formally called the Space Transportation System program) that was terminated on August 31, 2011. NASA's new rocket is called the Space Launch System (SLS). When built, it is expected to exceed the power generated by the gigantic Saturn 4 rocket that sent U.S. astronauts to the Moon in the late 1960s and early 1970s. NASA expects to begin unmanned test flights in 2017 of its new SLS system, with manned missions beginning later in the 2010s.

As part of the Commercial Orbital Transportation Services (COTS) contract by NASA, two companies—Orbital Sciences Corporation (Orbital) and Space Exploration Technologies Corporation (SpaceX)—were awarded, on February 2008, contracts for a minimum of eight Orbital missions and twelve SpaceX missions. Orbital will launch its Cygnus spacecraft from the Taurus II rocket at a launch pad on Wallops Island (Virginia), while SpaceX will launch its Dragon spacecraft from its Falcon 9 rocket at Cape Canaveral Air Force Station (Florida). The first test flight for Orbital's Cygnus is tentatively scheduled for late 2011, while SpaceX's Dragon began its first test flight in December 2010, with a second plan scheduled for November 2011.

Further, in 2011, SpaceX announced that it would attempt to develop the world's first completely reusable spacecraft and rocket system. The founder of SpaceX, Elon Musk, stated that its fully reusable rocket would greatly decrease the cost of lifting cargo and astronauts into space. The reusable system, based on its Falcon 9 rocket and Dragon space capsule, would consist of a two-stage rocket, where each would land back to Earth under its own power, while the space capsule, after performing its mission

◄

The X-34 Technology Testbed Demonstrator is delivered here to NASA on April 16, 1999. Results gleaned from this technology will contribute to the development of lower-cost reusable launch vehicles. *NASA.*

in space, would do likewise, descending vertically through the atmosphere and landing under its own power on four legs. The system would then be refueled, reintegrated, and relaunched, ready for another mission into space.

There are also other small company start-ups competing in the market to provide low-cost RLVs. The Silver Dart is an orbital spaceplane being developed by PlanetSpace. Based on the U.S. Air Force's Flight Dynamics Laboratory 7 (FDL-7) program, the Silver Dart will launch on a NOVA rocket for its ride into space and then glide back to Earth after completing its mission.

The Sierra Nevada Dream Chaser is a vertical-takeoff, horizontal-landing (VTHL) spaceplane that is being developed to carry seven people back and forth from LEO. The vehicle would launch from a traditional rocket, most likely an Atlas V, and land horizontally on traditional runways. On April 18, 2011, NASA awarded Sierra Nevada, which is headquartered in Sparks, Nevada, funding for its Dream Chaser spacecraft as part of NASA's Commercial Crew Development Phase 2 (CCDev-2) program.

Single-Stage-to-Orbit Reusable Rockets

A single-stage-to-orbit (SSTO) reusable rocket would probably be the best technical solution to inexpensively get to LEO. Even when all parts of a multistage rocket are reused, the rocket still needs to be put back together again. An SSTO rocket would not have to be reassembled, reducing the number of people required for operations.

Unfortunately, it is difficult to get to LEO without staging. To get to LEO with a single stage, a rocket has to be 90% fuel, leaving only 10% for everything else. Such a rocket has proven easy to build but difficult in practice to get into orbit. Several spacecraft have been designed and constructed, such as the McDonnell Douglas Delta Clipper Experimental (DC-X), tested in the early 1990s, the Lockheed Martin X-33 (VentureStar), in the 1995 to 2001 timeframe, and the Rotary Rocket Roton SSTO, between 1999 and 2001; however, none have achieved orbit.

For instance, in 1994, the National Aeronautics and Space Administration (NASA) decided to develop technologies to lead to a reusable launch vehicle (RLV) with a single-stage rocket. Its major step toward this goal was the $1.4 billion X-33 program, which aimed to fly a vehicle to demonstrate some of these technologies. Like many X-vehicle programs before it, the X-33 encountered severe technical difficulties as well as budget overruns and schedule delays. As a result, NASA terminated the X-33 program in early 2001. Consequently, the use of multistage rockets has continued.

NASA is also investigating technical paths to SSTO other than with conventional rockets. These primarily involve air-assisted propulsion, such as ramjets (which uses the forward motion of an air-breathing jet engine to compress incoming [subsonic] air without a compressor), supersonic combustion ramjets ("scramjets"; a type of ramjet that compresses incoming air at supersonic speeds), or liquid air cycle rockets (also called liquid air cycle engines, which propel rockets by using cryogenic hydrogen fuel to liquefy the air; thus, part of its oxidizer comes directly from the atmosphere). These rockets all lie in the future.

Other Possibilities

Another possible concept to help to make access to space less expensive is beamed thermal propulsion. This involves directing a rocket into space with the use of beamed energy in the form of focused microwave or laser beams positioned on the Earth's surface. A heat exchanger, positioned on the rocket, would transfer the energy from the energy beam to its store of liquid propellant, such as hydrogen. The heated propellant would be converted to a hot gas that is pushed out of the rocket's nozzle. Experts state that this technology could send two to five times more payload weight into space than conventional chemical rockets. A reusable rocket using beamed thermal propulsion could make, in the future, for a low-cost way to go into LEO.

Still another option is an experimental pressurized light-gas-powered cannon that totally removes the need for rockets. Such a device could make an inexpensive way to get to space in the future. Dr. John Hunter, from Lawrence Livermore National Laboratory, is working on the concept. His "space gun" would insert satellites and cargo into space for under $2,200 per kilogram ($1,000 per pound). At one end of the space cannon is a long tube containing highly pressurized gas, probably hydrogen, helium, or methane. When the pressure is released, a payload inside the cannon is shot out at speeds over 21,000 kilometers (13,000 miles) per hour.

The space elevator is another concept that has been discussed for many years. It consists of a space launch structure that does not use rockets. Instead, it uses a super-strong cable—one that extends at least 36,000 kilometers (22,370 miles) in length—which reaches from the Earth's surface (on the equator) to geostationary orbit. A counterweight is attached onto the space elevator beyond geostationary orbit. Human passengers and/or cargo are transported to and from space inside an "elevator" attached to the cable. To provide low-cost, safe, and reliable transportation to space, the cable would have to resist a tremendous amount of stress. Although materials do not exist to make this cable, carbon nanotube technology may eventually become a material needed to make such an Earth/space structure.

The Marketplace

There are three currently viable commercial space payload markets: geostationary communications satellites, a market that has had dependable growth for decades, and three newcomers to the market, low Earth orbit communications satellites, remote sensing, and space tourism. As of June 2007, the commercial payload market was valued at under $3 billion per year. Over half of the market is dominated by Arianespace (with its Ariane rocket family), with Russian launch vehicles taking the second position.

Communications Satellites Communications satellites are satellites sent to space for the express purpose of telecommunications for people on Earth. These satellites, sometimes abbreviated with the term COMSATs, are launched in many different types of orbits, such as geostationary and low-Earth orbits. These can be further divided into polar and non-polar orbits, elliptical and circular, and others.

GEOSTATIONARY COMMUNICATIONS SATELLITES A satellite in a geostationary orbit appears to an observer at the equator on the Earth's surface to be fixed at a specific position in the sky. In reality, it is revolving around the Earth at the same angular velocity as the Earth itself. Therefore, in each 24-hour Earth day, it revolves around the Earth once. To accomplish this action, the satellite is positioned approximately 35,786 kilometers (22,236 miles) above the Earth. A series of communications satellites placed in such orbits can then be used to transmit information back and forth to people on Earth.

The major television networks, such as ABC, NBC, CBS, Fox, and others, use these satellites to transmit their television shows to their subscribers. In addition, DirecTV and Dish Network, along with other direct-broadcast satellite-service providers, also use communications satellites for their operations. They use expendable rockets to launch their satellites. For instance, Dish launched its EchoStar XIV satellite with a Russian Proton rocket on March 20, 2010, using the company International Launch Services. Some of the U.S. companies that build communications satellites are Boeing Satellite Development Center, Loral Space Systems, Orbital Sciences Corporation, Lockheed Martin, and Northrop Grumman. Companies outside of the United States that build communications satellites include Thales Alenia Space and Astrium.

From the first communications satellite placed in geostationary orbit on August 19, 1964—often considered to be *Syncom 3*, launched by NASA from a Delta rocket—to early in the 2010s, these companies all use expendable launch vehicles to place their satellites into space because of the proven record of expendables and the lack of other alternatives.

LOW EARTH ORBIT COMMUNICATIONS SATELLITES Another possible new market, LEO communications satellite constellations, was halted temporarily by the business failure of Iridium, the first such system. However, in late 2000, the newly formed Iridium Satellite LLC, headquartered in Bethesda, Maryland, purchased the failed Iridium satellite system and associated ground systems for $25 million. The company soon began selling Iridium phone service at much lower prices than those of its predecessor. As of October 2011, Iridium Satellite LLC operates its 66 satellites for worldwide voice and data communications. Iridium is in the process of updating its satellites, with launches planned from 2015 to 2017, using the Falcon 9 rocket from Space Exploration Technologies Corporation (SpaceX). The launches will take place from Space Launch Complex 3 at Vandenberg Air Force Base in California.

The X-33, an unpiloted vehicle, was designed to launch vertically like a rocket and land horizontally like an airplane. *NASA.*

Beginning in 2003, Globalstar LLC also provides LEO satellites for satellite phone services and low-speed data communications. In 2007, it launched eight spare satellites to accommodate lack of coverage from its other satellites until its second-generation series of satellites are developed and launched into space. The first second-generation satellite, manufactured by Thales Alenia Space, was launched October 19, 2010, from the Russian Baikonur Cosmodrome, in Kazakhstan, using a Soyuz rocket. It is the first of 48 satellites for Globalstar. Six more satellites were launched on July 13, 2011, with a Soyuz launcher at the same Russian launch site.

However, each company uses ELVs to launch its communications satellites. Without new markets, few businesses are in need for anything but ELVs. Few incentives exist to develop new launch systems to get payloads to space cheaply.

Remote Sensing (Earth Observation) Satellites
Remote sensing is the act of observing from orbit what may be seen or sensed below on Earth. Various techniques are used to observe Earth from orbit. Synthetic

Space Sciences, 2ⁿᵈ Edition

aperture radar (SAR) is used to make precise elevation models of rough terrain features. RADARSAT-2, which was launched in 2007, uses SAR to observe the Canadian coastline and the Arctic down to a resolution of about 100 meters (330 feet). Laser and radar altimeters work by measuring altitude changes on the surface of the Earth and under Earth's oceans. Remote sensing can also use light detection and ranging (LIDAR), which detects concentrations of chemicals in the atmosphere. Other remote sensing devices include radiometers, photometers, aerial photographs, and hyperspectral imagers and devices. However, remote sensing equipment and satellites only require a reliable launch system to send such payloads into orbit.

Space Tourism Some companies are proposing two-stage vehicles in which a booster/orbiter combination leave the launch pad together and return separately. One private company has succeeded in building a RLV vehicle. U.S. enterprise Scaled Composites, founded by Burt Rutan, built *SpaceShipOne* and *WhiteKnightOne* The *SpaceShipOne* was a suborbital air-launched spaceplane that won the $10 million Ansari X Prize in 2004, with the help of its carrier aircraft the *WhiteKnightOne*. At that time, the spaceplane became the first privately funded manned spacecraft to go into space. The venture was paid for by Paul Allen, an American businessman and co-founder of Microsoft, in an amount of approximately $25 million, through Mojave Aerospace Ventures, a joint effort between Allen and Burt Rutan's Scaled Composites.

Since then, Scaled Composites has teamed up with Richard Branson's Virgin Galactic. The design and manufacturing company Scaled Composites is building the RLVs for Virgin Galactic so that suborbital flights are available for the private sector. The reusable spacecraft, a series of spacecraft called *SpaceShipTwo* (actually about 10% of the craft, excluding the fuel, is not reusable), will be launched with the help of one of a series of *WhiteKnightTwo* launching vehicles, which take the spacecraft from launch to about 18 kilometers (11 miles) above Earth. The spacecraft then completes the trip to space, about 100 kilometers (62 miles) above Earth. Test flights are scheduled for 2011 and 2012 with the first commercial suborbital flight scheduled sometime in 2013.

Other companies are attempting to develop private manned suborbital flights into space for space tourists. John Carmack, the co-founder of id Software, founded Armadillo Aerospace, which is based in Mesquite, Texas, in 2001. The company is working on a suborbital commercial rocket that will take tourists into space. The spacecraft, tentatively called Black Armadillo, will be a vertical takeoff, vertical landing (VTVL) suborbital vehicle.

Blue Origin, which is based in Kent, Washington, but with its main flight facility in Culberson County, Texas, was founded by Jeff Bezos (the founder of Amazon.com) in 2000. The company is working on its VTVL *New Shepard* spacecraft, which is based on the technology for the McDonnell Douglas DC-X. Unmanned test flights took place in 2011 with more flights planned for 2012 and beyond.

In 2011, XCOR Aerospace is developing its Lynx suborbital spaceplane. The suborbital horizontal-takeoff, horizontal-landing (HTHL) rocket-powered spaceplane will have one pilot that will take one passenger or a payload over the 100-kilometer (62-mile) altitude that is the boundary into outer space. If successful, with its first test flights beginning in 2012, XCOR plans to fly its second-generation Mark II spaceplane next.

 See also **Emerging Space Businesses (Volume 1) • Launch Industry (Volume 1) • Launch Vehicles, Expendable (Volume 1) • Launch Vehicles, Reusable (Volume 1) • Reusable Launch Vehicles (Volume 4) • Spaceports (Volume 1)**

Resources

Books and Articles

McCurdy, Howard E. *Space and the American Imagination.* Baltimore: Johns Hopkins University Press, 2011.

Parker, Martin, and David Bell, editors. *Space Travel and Culture: From Apollo to Space Tourism.* Malden, MA: Wiley-Blackwell/Sociological Review, 2009.

Pelt, Michael. *Space Tourism: Adventures in Earth's Orbit and Beyond.* New York: Springer, 2005.

Websites

Armadillo Aerospace. <http://www.armadilloaerospace.com/n.x/ Armadillo/Home> (accessed October 3, 2011).

Beaming Rockets into Space. AstroBiology Magazine. <http://www. astrobio.net/exclusive/3747/beaming-rockets-into-space> (accessed October 19, 2011).

Bergin, Chris. *SpaceDev announce Dream Chaser agreement with ULA Atlas V."* NASASpaceflight.com. <http://www.nasaspaceflight. com/2007/04/spacedev-announce-dream-chaser-agreement-with- ula-atlas-v/> (accessed October 19, 2011).

Blue Origin. <http://www.blueorigin.com/> (accessed October 3, 2011).

Globalstar. <http://www.globalstar.com/> (accessed October 19, 2011).

Hydrogen Gas Cannons Could Launch Payloads to Orbit (w/video). UniverseToday.com. <http://www.universetoday.com/51532/ hydrogen-gas-cannons-could-launch-payloads-to-orbit-wvideo/> (accessed October 19, 2011).

Iridium Satellite. <http://iridium.com/default.aspx> (accessed October 19, 2011).

NASA Announces Design for New Deep Space Exploration System. Federal Aviation Administration. <http://www.nasa.gov/exploration/systems/sls/sls1.html> (accessed October 19, 2011).

NASA Exploring Laser Beams To Zap Rockets Into Outer Space. Fox News. <http://www.foxnews.com/scitech/2011/01/25/nasa-exploring-lasers-beams-zap-rockets-outer-space/> (accessed October 19, 2011).

Semi-Annual Space Launch Report: Second Half of 2009. Federal Aviation Administration. <http://www.faa.gov/about/office_org/headquarters_offices/ast/media/10998.pdf> (accessed October 19, 2011).

Sierra Nevada Corporation Space Systems. <http://www.spacedev.com/> (accessed October 19, 2011).

SpaceX Undercut Competition To Clinch Head-turning Iridium Deal. SpaceNews.com. <http://www.spacenews.com/launch/100617-spacex-undercut-competition-clinch-492m-iridium-deal.html> (accessed October 19, 2011).

SpaceX Unveils Plan for World's First Fully Reusable Rocket. Space.com. <http://www.space.com/13140-spacex-private-reusable-rocket-elon-musk.html> (accessed October 19, 2011).

Soyuz Rocket Launch Beefs Up Globalstar Satellite Fleet. Space.com. <http://www.space.com/12314-soyuz-rocket-launch-globalstar-satellites.html> (accessed October 19, 2011).

Virgin Galactic. <http://www.virgingalactic.com/> (accessed September 28, 2011).

XCOR Aerospace. <http://www.xcor.com/> (accessed October 3, 2011).

Global Industry

From the earliest forays into commercial space, competition has been fierce. The United States has an established lead in the design, construction, and marketing of satellites. At the end of the twentieth century, U.S. satellites were being launched routinely for a significant number of nations by launch vehicles provided by eight different nations from launch sites all around the world. The transfer of export license processing from the U.S. Department of Commerce to the U.S. Department of State in 1999 resulted in a conspicuous slowdown in satellite exports, but even this impediment did not compromise U.S. leadership in this world market. The decision to change licensing authority arrangements in the United States was a response to instances of transfer of technological information to China by U.S. companies that the American government deemed inappropriate. This article will explain how space business enterprises had become internationally interwoven to a deep degree in the first decade of the twenty-first century.

Changes Following the *Challenger* Disaster

Space business underwent some major changes between the mid–1980s and the end of the twentieth century. Following the explosion of the space shuttle *Challenger* in 1986 because of a technical malfunction and poor management decisions, the United States decided to no longer use the space shuttle to carry commercial payloads* into orbit. It quickly became clear that the United States had put too much reliance on one launch vehicle, the shuttle. Lack of a good alternative launch system sent aerospace companies scurrying to develop suitable rockets to fill the gap. More importantly, it meant that the United States had lost its dominant position in the launch business to the growing competition from other nations.

Initially the main competition came from the European Ariane launchers. Over time, more and more nations became involved in the commercial launch business, most notably Russia (after the end of the Cold War), Japan, and China. By 2011, ten countries had independently launched their own satellites into space: Russia, United States, France, Japan, China, United Kingdom, India, Israel, Ukraine, and Iran. France and the United Kingdom now operate under the auspices of the European Space Agency. Other countries, such as Argentina, Australia, Canada, Egypt, Germany, Italy, South Africa, and Spain, have developed their own launch capabilities but, as of late 2011, had not had a successful launch. Several other countries are at various stages of development for launch capabilities. In addition to governmental agencies, a number of private commercial firms have been working on satellite-launch programs. Private companies that have thus far successfully launched one or more satellites include Orbital Sciences (United States), RocketLab (New Zealand), Scorpius Space Launch (United States), and SpaceX (United States). Other companies are in the planning or development phase of producing rocket launchers. They include Air Launch and Interorbital Systems (both of the United States) and ArcaSpace (Romania).

Mergers and Acquisitions

In the largest American mergers of the 1990s, Lockheed Corporation merged with Martin Marietta Corporation to become Lockheed Martin Corporation, and the Boeing Company absorbed McDonnell Douglas Corporation as well as elements of Rockwell International Corporation (the corporation that had built the space shuttles) and Hughes Electronics Corporation. A decade later, another merger brought together the space programs in Boeing and Lockheed Martin in the formation of United Launch Alliance (ULA). This climate of mergers and acquisitions started in response to events in the United States but continued as the industry reacted to events worldwide that impact this global market. The trend of mergers and acquisitions was

* **payload** any cargo launched aboard a rocket that is destined for space, including communications satellites or modules, supplies, equipment, and astronauts; does not include the vehicle used to move the cargo or the propellant that powers the vehicle

A rocket blasts off from Sea Launch's floating *Odyssey* launch platform, on the equator in the midst of the Pacific Ocean, on July 16, 2008. It carried Dish Network's EchoStar 11 satellite into orbit. © *AP Images/Sea Launch Company* ▶

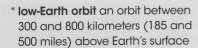

* **low-Earth orbit** an orbit between 300 and 800 kilometers (185 and 500 miles) above Earth's surface

not confined to the United States. For example, in 1999, Aérospatiale Matra, Daimler Chrysler AG, and Construcciones Aeronauticas S.A. merged their space capabilities to create the transnational firm EADS, the European Aeronautic Defence and Space Company. In 2000, Matra, BAE Systems, and DaimlerChrysler formed Astrium, covering the whole spectrum of space business. In 2003, EADS N.V. acquired Astrium, making it one of its divisions.

The focus on low-Earth orbit* (LEO) communications networks and on the design of launch vehicles limited in capability to delivery to near-Earth space was another shaping factor in the development of the global industry. Companies such as the unsuccessful Iridium planned to provide worldwide communications capability primarily for an international business clientele. This market initiative stalled because of a combination of technological developments in fiber-optic networks (greatly undercutting the cost of space-based communication systems), and a poor assessment of the market niche. However, in late 2000, the newly formed Iridium Satellite LLC (Maryland) purchased the failed Iridium satellite system and related ground systems for about $25 million. The company soon began selling Iridium telephone service at much lower prices than those of its predecessor. As of late 2011, Iridium Satellite operates its 66 satellites for worldwide voice and data communications. It expects a rosy future with the launch of its second generation Iridium NEXT series of satellites beginning in about 2015. Iridium's success, along with that of companies such as Globalstar LLC, has significantly increased emphasis on the global marketplace.

Globalization—the worldwide expansion of corporate business activity—also became a factor in the space industry towards the beginning of the twenty-first century. The creation of companies such as U.S. based International Launch Services (ILS), which combined American business savvy with access to customers and merchandizing from Lockheed Martin and with the technical reliability of Russian rocketry from Khrunichev (Khrunichev State Research and Production Space Center) and RKK Energia (S.P. Korolev Rocket and Space Corporation Energia), was typical of the trend towards international leveraging of assets for market success. ILS marketed the U.S. Atlas V rocket and the Russian Proton rocket. In 2006, Lockheed Martin withdrew from the company, selling its share to Space Transport. Consequently, ILS solely deals with the Russian Proton rocket. In 2008, Russian company Khrunichev bought out Space Transport, and became the majority stockholder in ILS. The company remains headquartered in McLean, Virginia.

Another company, Sea Launch, brought even more partners together, teaming U.S. corporate business leadership, operations and management from Boeing; a Norwegian-built oceangoing launch platform; a two-stage Zenit rocket made in the Ukraine; and initially, a Cayman Islands registry. Sea Launch, which is owned partially by Boeing, filed for Chapter 11 bankruptcy protection in June 2009, but continued operating as usual while filing for reorganization. The company emerged from bankruptcy on October 27, 2010, with a new majority owner, the Russian firm of Energia Overseas Limited.

The Global Nature of the Commercial Space Business

Even before the trend for mergers and acquisitions and the development of corporations geared to worldwide operation in a global market, the commercial aerospace industry was intensely internationally interconnected. For example, while a launch vehicle might be built in the United States or Japan, some component parts would come from elsewhere and payloads—the satellites that were lofted into space by the launch vehicles—frequently brought together instruments and components from several additional nations. Furthermore, each launch and its payload has to be insured against catastrophic failure of the launch vehicle, failure of the launch vehicle to place the satellite in the correct orbit, and malfunction of the satellite itself. The United States develops the majority of the world's satellites and it has regained a significant share of the launch market. However, the majority of launch insurance comes from Europe. Australian-based insurance companies have also been a significant provider of insurance and reinsurance for the space business. Insurers require detailed knowledge of the vehicle and payload to be insured, resulting in the necessary transfer of detailed information about a planned launch to individuals outside the originating nation. The insurance element provides one more illustration of the global nature of the space industry with its many component parts.

Growth Trends in the Global Space Industry

The first year in which commercial space revenues exceeded revenues from government space contracts was 1998. In subsequent years, the growth of commercial space business has widened the gap between commercial revenues and government revenues from space commerce. As of mid-2011, 986 satellites were in orbit around the Earth, the majority of which (49 percent) were in LEO, with slightly fewer (41 percent) in geosynchronous orbit. By far the greatest fraction of those satellites (365 satellites, or 37 percent) were being used for telecommunications purposes. Other satellite functions included civil communications (108; 11 percent), remote sensing (92; 9 percent), military communications (84; 9 percent), other military purposes (89; 9 percent), navigation (75; 8 percent), and space science (59; 6 percent). Total revenue from all satellite industry activities amounted to $168.1 billion in 2010, an increase of 5 percent over 2009, and an increase of 135.8 percent over 2002 revenues.

Over the past decade, direct broadcast satellite television has been the fastest-growing consumer electronics product in history. At the beginning of 2011, there were more than 147 million subscribers worldwide to satellite pay TV services, an increase of about five million from the previous year. About a quarter of all those subscribers lived in the United States, which experienced a generally flat market in the year because of general economic concerns. The two largest satellite pay TV providers in the United States are Dish Network and DirecTV, which had, at the end of 2010, about 14 million and 19 million subscribers, respectively. In 2007, the two largest U.S. satellite radio companies, Sirius Satellite Radio and XM Satellite Radio merged to form Sirius XM Radio, Inc. At the end of 2010, the company had about 20.6 million subscribers.

As of late 2011, the space industry employed about 260,000 people in the United States and an almost equal number (250,000) in Russia. The European industry was significantly smaller in size (about 30,000 workers), and even smaller in other nations, such as Japan (6,300 workers). As space programs in countries outside the United States, Russia, and Europe continue to grow, these numbers will almost certainly increase substantially.

 See also **Aerospace Corporations (Volume 1)** • **Business Failures (Volume 1)** • **Insurance (Volume 1)**

Resources

Books and Articles

Andrew, Steven. *Newspace: How NASA and Private Enterprise Will Re-launch America's Space Industry.*. Sausalito, CA: Polipoint Press, 2011.

Brennan, Louis, and Alessandra Vecchi. *The Business of Space: the Next Frontier of International Competition.* Basingstoke: Palgrave Macmillan, 2011.

The Space Report 2011. Washington, DC: The Space Foundation, 2011.

Websites

2011 Commercial Space Transportation Forecasts. <http://www.faa.gov/about/office_org/headquarters_offices/ast/media/2011%20Forecast%20Report.pdf> (accessed October 20, 2011).

Aerospace Product and Parts Manufacturing. <http://www.bls.gov/oco/cg/cgs006.htm> (accessed October 21, 2011).

Satellite Markets and Research. <http://www.satellitemarkets.com/> (accessed October 21, 2011).

Space Launch Report. <http://www.spacelaunchreport.com/log2010.html> (accessed October 21, 2011).

Global Positioning System

One hazard of human existence is being geographically lost, which can sometimes mean the difference between life and death. The ability to know one's position has been considerably enhanced by the U.S. Global Positioning System (GPS), officially called NAVSTAR GPS but often simply referred to as GPS. This global navigation satellite system (GNSS) enables users anywhere on Earth to determine their location to a high degree of accuracy. The Global Positioning System became fully operational on April 27, 1995.

Components of the System

GPS is a satellite-based navigation system consisting of three segments: space, ground, and user. The space and ground segments are run by a military organization called the United States Space Command, which is located in Colorado Springs, Colorado. This command, composed of components of the United States Air Force (USAF), the U.S. Army, and the U.S. Navy, launches the NAVSTAR satellites and is responsible for space and ground operations. The user segment includes any organization, ship, person, or airplane that uses GPS.

The space segment consists of a constellation of twenty-four to thirty-two satellites based in six different orbital planes at an altitude of 20,000 kilometers (12,400 miles). In this orbit, each satellite circles the planet twice in twenty-four hours and travels at the speed of 3.89 kilometers per second (8,640 miles per hour). Each satellite has an inclination of 55 degrees with respect to the equator, which means that it flies to a maximum of 55° north latitude and 55° south latitude during its orbits. As of 2011, the GPS system consists of thirty-one active satellites, with several "retired" satellites in orbit that can be reactivated if the primary GPS satellites malfunction. The number of GPS satellites has stayed relatively consistent over the past several years. The ground segment consists of the

* **radar** a technique for detecting distant objects by emitting a pulse of radio-wavelength radiation and then recording echoes of the pulse off the distant objects

* **solar arrays** groups of solar cells or other solar power collectors arranged to capture energy from the Sun and use it to generate electrical power

* **oscillation** energy that varies between alternate extremes with a definable period

radar* stations that monitor the satellites to determine the position and clock accuracy of each satellite. The locations of these ground stations are: Hawaii; Ascension Island, located in the southern Atlantic; Diego Garcia, an island in the Indian Ocean; Kwajalein, part of the Marshall Islands of the western Pacific; and Schriever Air Force Base, Colorado.

The stations are staffed continuously to ensure that GPS broadcasts the most accurate data possible.

Each NAVSTAR satellite possesses a mass of about 1,000 kilograms (2,200 pounds, equivalent weight on Earth) and is 5.25 meters (17 feet) long with its solar arrays* extended. The spacecraft transmits its timing information to Earth with the power of 50 watts, obtained from the solar panels and augmented battery power. Using its 50 watts, the satellite transmits two signals called "Links," L_1 and L_2, shorthand for Link1 and Link2. L_1 and L_2 are "downlinks" because their signals go to Earth. Two cesium and two rubidium atomic clocks provide signal timing. Atomic clocks are not powered atomically; they measure the precise oscillations* of cesium and rubidium atoms. These oscillation measurements are so accurate that an atomic clock, if left unadjusted, would gain or lose one second every 160,000 years. But how does accurate timing from a satellite at an altitude of 20,000 kilometers translate into a position within meters on Earth?

How Positions Are Determined

Distance to the satellite—the range—is the key for determining positions on Earth. Time is related to range by a very simple formula: Range = Velocity × Time. For GPS, the range is the distance from the receiver to the satellite; the velocity equals the speed of light (300,000 kilometers per second [186,300 miles per second]); and the time is the time it takes to synchronize the satellite signal with the receiver. Because the speed of light is so fast, the key to measuring range is the accurate timing provided by the atomic clocks.

What is meant by synchronizing the satellite signal with the receiver? First, imagine that a GPS satellite begins to play the song "Twinkle, Twinkle, Little Star." Simultaneously, a GPS receiver starts playing the same song. The satellite's signal has to travel 20,000 kilometers to the receiver, and by the time it does, the words are so late that when the receiver says "Star" the satellite's signal starts its first "Twinkle." If two versions of the song were played simultaneously, they would interfere with one another. Consequently, the receiver determines the delay time when it receives the satellite's first "Twinkle" and then starts to play the receiver's tune with a delay time calculated, thereby synchronizing with the satellite's signal. The amount of delay time is the signal travel time. This signal travel time is multiplied by the speed of light to determine the range.

Obviously, the GPS does not use "Twinkle, Twinkle, Little Star," but rather it generates an electronic signal. This signal is similar to the interference heard on the radio when one cannot tune in the correct station or the "snow" one sees on one's television when the set is not on an operational channel. This electronic signal from the GPS satellite is called the Pseudo Random Code (PRC).

A PRC is a very complex electronic signal that repeats its pattern. The pattern of zeros and ones in the digital readouts ensures that the user segment receivers synchronize only on a NAVSTAR satellite downlink and not on some other electronic signal. Because each satellite has its own unique PRC, the twenty-four satellites do not jam each other's signals. This allows all the satellites to use the same GPS frequencies*. Each satellite transmits two PRCs, over L_1 and L_2. The L_1 PRC is known as Coarse Acquisition (CA), and it allows civilian receivers to determine position within 100 meters (330 feet). The second PRC is called the precise code, or "P," and is transmitted on L_2. The P combines with the CA for orientation and then encrypts the signal to permit only personnel with the correct decoding mechanism, called a key, to use it. When L_2 is encrypted, it is called the Y code and has an accuracy of 10 meters (33 feet).

Besides clock accuracy and PRC reception, the receiver needs to know the satellite's location. A typical receiver anywhere on Earth will see about five satellites in its field of view at any given instant. The USAF uses the GPS Master Plan for satellites to ensure that a minimum number are always in view anywhere on Earth. Additionally, all GPS receivers produce an almanac (technically referred to as an ephemeris) that is used to locate each GPS satellite in its orbital slot. The USAF, under the control of U.S. Space Command, monitors each satellite to check its altitude, position, and velocity* at least twice a day. A position message, a clock correction, and an ephemeris (giving the satellite's predicted position) are also updated and uplinked to the GPS satellite daily.

A receiver needs ranges and satellite location information from three satellites to make a position determination. To obtain this, the receiver determines the range while synchronizing its internal clock on the first satellite's correct Universal Time* (UT), which is based on the time in Greenwich, England. Once the clocks have been synchronized and the range to the first satellite has been determined, the receiver also determines the ranges to two other satellites. Each satellite's range can be assumed to be a sphere with the receiver at the center. The intersection of the three spheres yields two possible positions for the receiver. One of these positions must be invalid because it will place the user in outer space or deep inside Earth, so the receiver has to be at the second position. Then the receiver compares the satellite's ephemeris and current almanac location to obtain the receiver's latitude and longitude. A fourth GPS satellite's range synchronizes the receiver's clock with all the atomic clocks aboard the spacecraft, narrows the accuracy of the receiver's position to only one intersecting point, and determines the receiver's altitude.

* **frequencies** the number of oscillations or vibrations per second of an electromagnetic wave or any wave

* **velocity** speed and direction of a moving object; a vector quantity

* **Universal Time** current time in Greenwich, England, which is recognized as the standard time that Earth's time zones are based

A commerical GPS locator device, being used to find a geocache. Geocaching is a sport centered on the use of GPS devices. First, someone caches a "treasure," then they share the GPS coordinates for the cache for others to try and find. © *Fort Worth Star-Telegram/Contributor/ McClatchy-Tribune/Getty Images.*

Selective Availability and Differential GPS

There are several errors in timing, ephemeris, and the speed of light for which the system must correct. However, the crews of U.S. Space Command occasionally must induce errors to keep the accuracy of the GPS system from falling into the hands of a hostile force. This error inducement is called "selective availability." To accomplish this, the crew inserts intentional clock or ephemeris errors. On May 1, 2000, President Bill Clinton ordered the removal of selective availability, greatly enhancing the public use of GPS.

However, the probability that access to navigation data would be blocked in times of hostilities has led to the initial construction of an independent GPS-style system called Galileo, funded by the countries of the European Union (EU). The Galileo navigation system is expected to be operational by 2014, at which time it will consist of 30 satellites providing worldwide navigation coverage. As in the U.S. GPS system, the optimal number of satellites in Galileo will be 24, with the additional satellites used as spares in case of equipment failures. The first two Galileo navigation satellites were launched atop a Russian Soyuz rocket at the spaceport located in Kourou, French Guiana on October 21, 2011. Galileo is designed to be compatible with the U.S. GPS navigation system. However, there are differences between the EU Galileo system and the U.S. GPS system. Some of the satellites in Galileo's constellation will be in orbits having a greater inclination to the equatorial plane than the GPS satellites. This will give northern Europe better coverage than that provided by GPS today. Another difference between the two systems is that Galileo will have a higher accuracy than the U.S. system—within a meter of the actual position.

When selective availability was introduced, a number of people wanted more accurate GPS readings, leading to the invention of Differential GPS. This system uses a known surveyed position, such as an airport tower, upon which is placed a GPS receiver. The GPS receiver determines its position constantly, compares the GPS position to the surveyed position, and develops a "correction" factor that can be applied to make the accuracy of the GPS in the range of inches. Applications of Differential GPS include precision landings with aircraft and precision farming, which allows a farmer to know exactly where to apply fertilizer or pesticide, or both, within a field. Differential GPS is so accurate that it also permits scientists to accurately measure the movement of Earth's tectonic plates, which move at the speed of fingernail growth.

GPS receivers can be found on ships, trains, planes, cars, elephant collars, and even whales. This system is changing the way humans live, and satellite-based navigation has become a multibillion-dollar industry.

Commercial Enterprises Involved in GPS

There are a number of commercial companies involved in the GPS industry. The largest are the companies that make the satellite itself, Lockheed Martin, Hughes (taken over by Boeing), Rockwell (taken over by Boeing), and Boeing Space. The satellites are built in generations, or "Blocks", with each Block of satellites usually representing an improvement in technology of some sort. Lockheed Martin and Boeing have both produced Blocks of GPS satellites—the latest Block of GPS satellites (as of 2011) are the Block III satellites, and are under development by Lockheed Martin. The first Block III GPS satellite is scheduled to be placed into orbit in 2014.

Upgraded GPS satellites are launched on an ongoing basis. Starting in 2010, the GPS Block IIF (sometimes simply referred to as GPS IIF) was the next generation of GPS satellite to be launched into orbit. These Block IIF satellites will keep the GPS system functional until the first Block III satellites are launched, which is scheduled to begin in 2014.

* **payload** any cargo launched aboard a rocket that is destined for space, including communications satellites or modules, supplies, equipment, and astronauts; does not include the vehicle used to move the cargo or the propellant that powers the vehicle

Commercial GPS units are used by consumers in a host of applications, including aviation, geosciences, marine applications, mapping, survey, outdoor recreation, vehicle tracking, automobile navigation, and wireless communications. Since there are a number of companies involved in GPS, only four of these will be reviewed. Companies that are selling their GPS services for other than space support include Garmin, which is headquartered in Olathe, Kansas, and has subsidiary offices in the United Kingdom and Taiwan. Garmin sells navigation receivers that are portable and have brought navigation to the masses for hiking, motor boat operation, and other recreational vehicle arenas.

Another large company competing in the manufacture of receivers is Magellan Navigation, Ltd., located in Santa Clara, California. Magellan brought into market the world's first handheld commercial receiver for ordinary uses. In 1995, Magellan introduced the first hand-held GPS receiver under $200, which led to even greater market expansion. In 2011, Magellan offers a wide variety GPS devices for various applications, such as the Magellan RoadMate and Maestro series for vehicle navigation.

Trimble Navigation Limited, located in Sunnyvale, California, offers services very similar to those of Garmin and Magellan. Trimble also has a subsidiary in the United Kingdom. Trimble offers GPS positioning devices for a variety of endeavors, including for construction and engineering work, agricultural applications, and technical field work (e.g., utility company workers). For fiscal year 2010, Trimble had worldwide revenues of $1.3 billion.

Motorola Corporation has been very cooperative in their affiliation with universities putting payloads* on satellites and on balloons. Using Motorola GPS units such as the MotoNav™, university students have tracked balloon payloads over 390 kilometers (240 miles) and have used the navigation information to determine the jet stream speed and balloon altitudes over the United States. As the GPS system continues, so too, will ideas from small companies about how to use this information commercially, thus developing industries that people can only dream about at this time in our history.

Other global or regional navigation satellite systems, besides the U.S. GPS and the European Union's Galileo, are being developed by Japan, India, China. The Russians already have a satellite navigation system, called GLONASS (short for Global Navigation Satellite System). Much like its American counterpart (the NAVSTAR GPS system), the GLONASS system was initially conceived and funded to meet the navigational needs of the military. Due to economic problems, the GLONASS system fell into disrepair in the late 1990s and throughout much of the 2000s. But the Russian government made new investments in order to resurrect GLONASS, and has projected that the system would have its full complement of 24 primary navigation satellites, along with several spares, by the end of 2011.

Japan's satellite navigation system is named the Quasi-Zenith Satellite System (QZSS), which is designed to use a minimum of three satellites in orbits around Earth so that at least one of the three is at or near its zenith (highest altitude above Earth) at any one time. The QZSS system is designed to be supplemented by America's GPS system, which means that the QZSS system will only have to use a few satellites for regional coverage. By taking data from both the QZSS and GPS navigational systems, users can obtain much higher positioning data in Japan and the Pacific region than compared to using just one of the systems alone. Japan hopes to have its navigation system operational by 2013 using four satellites.

China is working to develop its own satellite navigation system called Beidou, named after the Big Dipper constellation. As of late 2011, the Beidou-2 navigation system consists of eight satellites. More satellites will be added, so that the full Beidou-2 system (also called the COMPASS system) will consist of at least 30 satellites, with hopes of offering worldwide navigational services by 2020.

India's system is called the Indian Regional Navigational Satellite System (IRNSS), and is designed to operate with a total of seven satellites. As of late 2011, Indian government officials were expecting the IRNSS system to be fully operational by 2014.

 See also **Military Customers (Volume 1) • Navigation from Space (Volume 1) • Reconnaissance (Volume 1) • Remote Sensing Systems (Volume 1)**

Resources

Books and Articles

Hinch, Stephen W. *Outdoor Navigation with GPS.* 3rd ed. Birmingham, AL: Wilderness Press, 2010.

McNamara, Joel. *GPS For Dummies.* Hoboken, NJ: Wiley, 2008.

Misra, Pratap and Per Enge. *Global Positioning System: Signals, Measurements, and Performance.* Lincoln, Mass.: Ganga-Jamuna Press, 2010.

Sellers, Jerry Jon, et al. *Understanding Space: An Introduction to Astronautics.* Boston: McGraw-Hill Custom Pub., 2004.

Websites

Coursey, David. "Air Force Responds to GPS Outage Concerns." *ABC News.* <http://abcnews.go.com/Technology/AheadoftheCurve/story?id=7647002&page=1> (accessed October 26, 2011).

Global Positioning System. <http://www.gps.gov/> (accessed October 26, 2011).

Magellan. Magellan Navigation, Inc. <http://www.magellangps.com/> (accessed October 26, 2011).

Strom, Steven R. *Charting a Course Toward Global Navigation.* The Aerospace Company. <http://www.aero.org/publications/crosslink/summer2002/01.html> (accessed October 26, 2011).

Trimble. Trimble Navigation, Ltd. <http://www.trimble.com/index.aspx> (accessed October 26, 2011).

Goddard, Robert Hutchings
American Inventor and Educator
1882–1945

Robert Hutchings Goddard was born in Worcester, Massachusetts, on October 5, 1882. After reading science fiction as a boy, Goddard became excited about exploring space. At his high school graduation, in 1904, he gave the valedictorian speech and spoke one sentence in particular that is now often associated with his life work: "It has often proved true that the dream of yesterday is the hope of today, and the reality of tomorrow." He pioneered modern liquid-fueled rocketry in the United States and founded a field of science and engineering. He is often referred to as the Father of Space Exploration. Goddard received a bachelor's of science degree from Worcester Polytechnical University in 1908, and a master's degree in physics and a philosophy (Ph.D.) degree from Clark University, also in Worcester, Massachusetts, in 1910 and 1911, respectively. In 1912, he accepted a research fellowship position at Princeton University (New Jersey).

As a physics graduate student, Goddard conducted static tests with small solid-fuel rockets and, in 1912, he developed the mathematical theory of rocket propulsion. In 1916, the Smithsonian Institution provided funds for his work on rockets and, in 1919, published his research as "A Method of Reaching Extreme Altitudes." Goddard argued that rockets could be used to explore the upper atmosphere and suggested that with a velocity* of 11.18 kilometers per second (6.95 miles per second), without air resistance, an object could escape Earth's gravity and head into space, such as toward the Moon or other celestial bodies. This became known as Earth's escape velocity.

Goddard wrote a book in 1920 called *A Method of Reaching Extreme Altitudes* (published by the Smithsonian Institution), which described how rockets were a viable way to send objects out into space, maybe even to the Moon. Unfortunately, Goddard's ideas were ridiculed by some people in the popular press, probably the most stinging of them was the January 13, 1920 newspaper editorial in the *New York Times,* which came one day after an article was written in the newspaper about Goddard and his book. This criticism prompted him to become secretive about his work. However, he continued his research with liquid oxygen and liquid-fueled rockets. On March 16, 1926, while at his aunt's farm in

* **payload** any cargo launched aboard a rocket that is destined for space, including communications satellites or modules, supplies, equipment, and astronauts; does not include the vehicle used to move the cargo or the propellant that powers the vehicle

Auburn, Massachusetts, Goddard launched his first liquid-fueled rocket (named "Nell") that lifted itself only 12.5 meters (41 feet) off the ground for a mere 2.5 seconds. However, this event heralded the age of modern rocketry as Goddard proved that liquid propellants can be used as fuel for rockets. Today, the launch site is a National Historic Landmark. Then, on July 17, 1929, Goddard flew the first instrumented payload*, consisting of an aneroid barometer, a thermometer, and a camera. This was the first instrument-carrying rocket. After rising about 27 meters (90 feet), the rocket turned and struck the ground 52 meters (171 feet) away, causing a large fire.

Charles A. Lindbergh, the first pilot to make a non-stop solo flight from New York to Paris and an international celebrity, visited Goddard and was sufficiently impressed to persuade philanthropist Daniel Guggenheim to

* **transonic barrier** the aerodynamic behavior of an aircraft changes dramatically as it moves near the speed of sound, and for early pioneers of transonic flight such changes were deemed dangerous, leading some to hypothesize there was a "sound barrier" where drag became infinite

provide Goddard with funding, with which Goddard set up an experiment station near Roswell, New Mexico. From 1930 to 1941, Goddard launched rockets of ever-greater complexity and capability. The culmination of this effort was the launch of a rocket to an altitude of 2,743 meters (9,000 feet) in 1941. Late in 1941, Goddard entered naval service and spent the duration of World War II (1939–1945) developing a jet-assisted takeoff rocket to shorten the distance required for heavy aircraft launches. This work led to the development of the throttleable Curtiss-Wright XLR25-CW-1 rocket engine that later powered the Bell X-1 and helped overcome the transonic barrier* in 1947.

Goddard died in Baltimore, Maryland, from throat cancer on August 10, 1945. His many patents (over 200) include: a signal application device with a vacuum tube, a multi-stage rocket, and a rocket fueled with gasoline and liquid nitrous oxide. After his death, in 1959, the NASA Goddard Space Flight Center, in Greenbelt, Maryland, was named in his honor. In addition, in recognition of his work in rocketry, the Robert H. Goddard Library was established on the Clark University campus, and Goddard Hall houses the Department of Chemical Engineering at Worcester Polytechnic Institute. The impact crater Goddard on the Moon was also named after him. Goddard was vindicated on July 17, 1969, by the *New York Times* when it stated its error about Dr. Goddard's book *A Method of Reaching Extreme Altitudes* within its editorial forty-nine years earlier. His book is now considered one of the most important books on rocketry ever written.

 See also **Careers in Rocketry (Volume 1) • Rocket Engines (Volume 1) • Rockets (Volume 3)**

Resources

Books and Articles

Clary, David A. *Rocket Man: Robert H. Goddard and the Birth of the Space Age.* New York: Hyperion, 2003.

Goddard, Esther C., ed., and G. Edward Pendray, associate ed. *The Papers of Robert H. Goddard.* New York: McGraw-Hill, 1970.

Lehman, Milton. *This High Man: The Life of Robert H. Goddard.* New York: Farrar, Straus, 1963.

Winter, Frank H. *Prelude to the Space Age: The Rocket Societies, 1924–1940.* Washington, DC: Smithsonian Institution Press, 1983.

Websites

Banks, Phyllis Eileen. *Robert H. Goddard, Space Pioneer.* Southern-NewMexico.com. <http://www.southernnewmexico.com/Articles/People/RobertH.Goddardspacepione.html> (accessed October 13, 2011).

Dr. Robert Hutchings Goddard. Clark University, Archives and Special Collections. <http://www.clarku.edu/research/archives/goddard/> (accessed October 13, 2011).

Robert H. Goddard. New Mexico Museum of Space History, International Space Hall of Fame. <http://www.nmspacemuseum.org/halloffame/detail.php?id=11> (accessed October 13, 2011).

Rocket Scientist Robert Goddard. Time Magazine. <http://www.time.com/time/magazine/article/0,9171,990613,00.html> (accessed October 13, 2011).

* **payload** any cargo launched aboard a rocket that is destined for space, including communications satellites or modules, supplies, equipment, and astronauts; does not include the vehicle used to move the cargo or the propellant that powers the vehicle

* **orbit** the circular or elliptical path of an object around a much larger object, governed by the gravitational field of the larger object

Ground Infrastructure

The ground-based infrastructure for a satellite is responsible for a number of support functions, such as commanding the spacecraft, monitoring its health, tracking the spacecraft to determine its present and future positions, collecting the satellite's mission data, and distributing these data to users. A key component of the infrastructure is the ground station, which is an Earth-based point of contact with a satellite and a distributor of user data.

Spacecraft and payload* support consists of maintaining a communications link with the satellite to provide satellite and payload control. The ground station collects satellite telemetry (transmitted signals) to evaluate its health, processes state of health information, determines satellite orbit* and altitude, and issues satellite commands when required.

Mission data receipt and relay is a vital function of the ground station. This includes receiving mission data and payload telemetry. The ground station computers process these data into a usable format and distribute them to the users by way of electronic communication lines such as satellite or ground-line data link or even the Internet.

A generic ground station consists of an antenna to receive satellite signals; radio frequency receiving equipment to process incoming raw electronic signals; and mission data recovery equipment, computers, and data interface equipment to send data to users. Additionally, telemetry, tracking, and control equipment monitors the spacecraft's health status, and radio frequency transmitting equipment sends commands to the satellite via the station antenna.

Station Personnel

A large satellite control station has several types of centers staffed by a diverse range of qualified personnel. The Control Center (CC) accomplishes overall control of the ground station, and all other station centers are responsible to the CC. A senior individual who has several years' experience working with the satellite systems and the station leads this center.

This complex in Madrid, Spain, is one of three complexes that comprise the NASA Deep Space Network, which spans the world and tracks spacecraft throughout the solar system and, eventually, beyond. *Courtesy NASA/JPL/Caltech.*

The Satellite Operations Control Center (SOCC) is responsible for satellite health. A senior engineer leads this team, which tracks the satellite and its telemetry. The Payload Operations Control Center (POCC) is responsible for the payload's status and health. A senior engineer also leads this team, which monitors the spacecraft's payload and the quality of data it is collecting. The SOCC and POCC must also determine causes if the spacecraft malfunctions and corrections that might be required. The Mission Control Center, such as the National Aeronautic and Space Administration (NASA) Mission Control Center (MCC) at the Johnson Space Center in Houston, Texas, is responsible for reviewing mission data to ensure their quality for the users. Because it is the last link before the users obtain their data, an expert such as a scientist leads this center. The Station Control Center is responsible for the ground station upkeep, such as distributing power, providing cooling for the computers, and taking care of general maintenance on all other station equipment. This team is usually led by a civil engineer with a number of years of maintenance experience.

For a smaller satellite operation many of these jobs are combined. One computer might run the CC and the MCC functions while another accomplishes the SOCC and POCC jobs. If the operation is small, three people and two computers can run the entire ground station.

Commercial Satellite Ground Stations

Depending upon the size, commercial satellite ground stations are available from several corporations with different price structures. This section examines three major corporations, one in the United States, Raytheon; one in Sweden, SSC; and one in Russia, RDC ScanEx.

Raytheon Corporation of Aurora, Colorado, offers large ground stations including antennas, satellite command and control, mission planning, management, front-end processing, and terminal equipment. One example of their extensive capability for satellite ground station operations is their software, which uses over three million lines of code in operations activities. With their 2,000 Aurora-based employees, Raytheon has built and supplied more than 40 international ground stations. The large ground stations are prohibitively expensive for any organizations other than governments or very large universities. Three recent Raytheon contracts were issued for a GPS-aided geosynchronous augmentation system (GAGAN) being developed by the government of India (2009), and NASA's Common Ground System for use with the agency's Joint Polar Satellite System (JPSS) and the ground infrastructure system for the FAISAT (Final Analysis, Inc. Satellite) global wireless data system (2011).

The Swedish Space Corporation (Rymdbolaget) was founded in 1972 to develop products and programs in telecommunications, earth observation, meteorology, and navigation. The company changed its name to SSC in April 2011. It made a major foray into the field of satellite ground stations in 2009 when it acquired the U.S.-based company Universal Space Network (USN). USN was founded in 1996 by former NASA astronaut Charles "Pete" Conrad, Jr. (1930–1999). The company began on a small scale with the construction of ground stations in Alaska and Hawaii. With funding from the Swedish Space Corporation and the equity firm of Warburg Pincus, it then expanded to other locations worldwide. USN's largest operation now is a system called PrioraNet, which consists of ground stations around the world that operate 24 hours a day 365 days a year to download and transmit data from satellites to users worldwide. The system currently has ground stations at Dongara, in Western Australia; Esrange, in northern Sweden; Inuvik, Canada; North Pole, Alaska; Poker Flat, Alaska; Santiago, Chile; South Point, Hawaii; Riverside, California; Miami, Florida; and Punta Arenas, Chile. The last three of these stations are expected to begin operations in the third quarter of 2012.

RDC ScanEx was founded in 1989 by Vladimir Gershenzon, a graduate of the Moscow Institute of Physics and Technology. The company's mission was to develop and sell inexpensive personal technology for the retrieval and processing of data from space satellites. The company, headquartered in Moscow, Russia, focuses on personal computers to acquire, track, and download data from several different satellites such as the National Oceanic & Atmospheric Administration polar satellites and the Russian Earth remote sensing satellites. As of 2009, RDC ScanEx provides data for numerous Earth remote sensing programs. Some of them are: IRS and

RADARSAT-1 (in Canada), IKONOS and GEOEYE-1 (in the United States), Resourcesat-1 and Cartosat-1 (in India), ALOS (in Japan), EROS (in Israel), and ENVISAT-1 (in the European Union).

As of late 2011, the two major components of the ScanEx product line are the UniScan and Alice-SC systems. The UniScan system makes use of 2.4 (UniScan-24) and 3.1 meter (UniScan-36) reflector disks that transmit data to personal computers with a spatial resolution ranging from 0.7 m to 1 km. Each disk covers an area of 12 million square kilometers. The Alice-SC system uses a 1.2 meter reflector that weighs less than 50 kilograms and also connects directly to a personal computer. Its resolution is in the range of about 1,100 meters.

 See also **Communications for Human Spaceflight (Volume 3) • Communications Satellite Industry (Volume 1) • Satellite Industry • (Volume 1) Tracking of Spacecraft (Volume 3) • Tracking Stations (Volume 3)**

Resources

Books and Articles

Fleeter, Rick. *The Logic of Microspace.* El Segundo, CA: Microcosm Press, 2000.

Larson, Wiley J., and James R. Wertz, eds. *Space Mission Analysis and Design,* 3rd ed. Torrance, CA: Microcosm Press, 1999.

Ley, Wilfried, Klaus Wittmann, and Willi Hallmann, eds. *Handbook of Space Technology.* Chichester, UK: Wiley, 2009.

Martin, Donald H., Paul R. Anderson, and Lucy Bartamian. *Communication Satellites,* 5th ed. El Segundo, CA: Aerospace Press, 2007.

Olla, Phillip. *Commerce in Space: Infrastructures, Technologies, and Applications.* Hershey, PA: Information Science Reference, 2008.

Sellers, Jerry Jon. *Understanding Space.* Boston: McGraw-Hill, Inc., 2004.

Websites

Deep Space Network. Jet Propulsion Laboratory, National Aeronautics and Space Administration. <http://deepspace.jpl.nasa.gov/dsn/> (accessed October 13, 2011).

Mechanisms of Earth Observation Satellites. <http://www.gds.aster.ersdac. or.jp/gds_www2002/seminer_e/e.o.s_e/e.o.s_e.html> (accessed October 13, 2011).

SanEx. <http://www.scanex.ru/en/index.html> (accessed October 13, 2011).

SSC Group. <http://www.sscspace.com/> (accessed October 13, 2011).

H

Human Spaceflight Program

The first human to reach space, Soviet cosmonaut Yuri Gagarin, made a one-orbit, ninety-minute flight around Earth on April 12, 1961. In total, he was in space for about one hour, forty-eight minutes. As of October 11, 2011, thirty-five countries have sent people into space. These countries are: Afghanistan, Austria, Brazil, Bulgaria, Belgium, China, Cuba, Czechoslovakia, France, Germany, Hungary, India, Israel, Italy, Japan, Malaysia, Mexico, Mongolia, the Netherlands, Poland, Romania, Russia (and the former Soviet Union), Saudi Arabia, Slovenia, South Korea, South Africa, South Korea, Spain, Sweden, Switzerland, Syria, the Ukraine, the United Kingdom, the United States, and Vietnam. Of these countries, a total of 521 men and women have flown in space, according to rules set by the Fédération Aéronautique Internationale (FAI), the international governing body for astronautics and aeronautics.

Most of these individuals were a mixture of career astronauts, trained either to pilot space vehicles or to carry out a changing variety of tasks in orbit, and payload* specialists, who also went through extensive training in order to accompany their experiments into space.

In addition, there were a few people who got the opportunity to go into space because of their jobs on Earth (e.g., U.S. politicians). Some individuals paid their own way into space as spaceflight participants, sometimes commonly called space tourists. Other individuals flew into space because their country or company had paid for access to space, thereby getting the right to name someone to participate in the spaceflight in exchange for that funding. For example, a prince from Saudi Arabia—Prince Sultan Salman Abdelaziz Al-Saud (1956–)—flew into space on June 17, 1985, aboard the space shuttle *Discovery* (STS-51G), and helped launch a Saudi communications satellite. A Japanese journalist (Toyohiro Akiyama (1942–), representing his television network, went aboard the Soviet space station *Mir* in December 1990.

Citizens in Space

The United States began a "private citizen in space" program in the 1980s. The goal of the program was to identify ordinary individuals who could communicate the experience of spaceflight to the public. The first person selected was a teacher, Christa McAuliffe, who was the finalist of the agency's Teacher in Space Program. Unfortunately, she and six astronauts were killed when the external tank attached to the space shuttle *Challenger* exploded

* **payload** any cargo launched aboard a rocket that is destined for space, including communications satellites or modules, supplies, equipment, and astronauts; does not include the vehicle used to move the cargo or the propellant that powers the vehicle

Human spaceflight is becoming increasingly diverse, with an emphasis on international cooperation. Clockwise from right are Curtis L. Brown Jr., commander; Steven W. Lindsey, pilot; Stephen K. Robinson, mission specialist; Pedro Duque, mission specialist representing the European Space Agency; Chiaki Mukai, payload specialist representing Japan's National Space Development Agency (currently the Japan Aerospace Exploration Agency); Scott F. Parazynski, mission specialist; and U.S. Senator John H. Glenn Jr. (Democrat-Ohio), payload specialist. *NASA.*

▶

* **low Earth orbit** an orbit between 300 and 800 kilometers (185 and 500 miles) above Earth's surface

seventy-three seconds after liftoff on January 28, 1986. After the disaster, the United States abandoned the idea of taking ordinary people into space, and limited trips aboard the space shuttle to highly trained specialists. The restriction was relaxed in 1998, when U.S. Senator John Glenn (1921–), the first American to orbit Earth, was assigned as a payload specialist for the STS-95 mission. Part of his mission assignment, at the age of 77 years, was to study how the microgravity environment of outer space affects the elderly. However, when the space shuttle *Columbia* was destroyed (and all of its crewmembers killed), in 2003, over Texas on its way back from space, the restriction was once again re-instituted. The only country that allows private citizens to go into space aboard its spacecraft is Russia, which has sent several wealthy space tourists to space onboard its Soyuz spacecraft.

The space shuttle *Challenger* and *Columbia* disasters are vivid reminders that taking humans into space is a difficult and risky undertaking. It is also very expensive; the launch of a space shuttle, for example, was priced at several hundred million U.S. dollars during its lifetime from 1981 to 2011. Only three countries, the United States, Russia (from 1922 to 1991 known as the Soviet Union), and China have developed the expensive capabilities required for human spaceflight. As the last country to send a manned spacecraft into orbit, China accomplished this feat on October 15, 2003, when it sent its *Shenzhou-5* spacecraft into orbit with astronaut Yang Liwei onboard for a fourteen-orbit mission about Earth.

Low Earth Orbit

Human spaceflight since 1961 has been limited to low Earth orbit* (LEO) with the exception of the period from December 1968 to December 1972, when twenty-seven U.S. astronauts (three per mission) traveled to the vicinity of the Moon during Project Apollo. Of these astronauts, twelve actually landed on the Moon's surface (within six Apollo missions) and carried out humanity's first human exploration of another celestial body.

Project Apollo was the result of a 1961 Cold War political decision by U.S. President John F. Kennedy to compete with the Soviet Union in space. Since the end of Apollo, advocates have argued for new exploratory manned missions to the Moon and especially to Mars, which is considered the most interesting accessible destination in the solar system. However, the lack of a compelling rationale for such difficult and expensive missions has meant that no nation, or group of nations, has been willing to provide the resources required to complete such an enterprise. However, as of 2011, several countries are in the developmental stages of manned missions to the Moon or design stages of manned missions to Mars. These countries include the United States, Russia, China, India, and Japan, along with the member-nations of the European Union's European Space Agency (ESA).

The Soviet Program

During the 1960s the Soviet Union attempted to develop the capability to send people to the Moon, but abandoned its efforts after three test failures with no crew aboard. Beginning in the early 1970s the Soviet Union launched a series of Salyut space stations, which were capable of supporting several people in space for many days. Then, in March 1986, the Soviet Union launched the larger *Mir* space station. *Mir* was continuously occupied for most of the subsequent fifteen years until it reached the end of its lifetime; it was de-orbited into the Pacific Ocean in March 2001. Soviet cosmonaut Valery Polyakov spent 437 days, 18 hours (launched on January 8, 1994 and returned home on March 22, 1995) aboard *Mir,* the longest spaceflight by any person, a record that continues to stand as of October 11, 2011. After the United States and Russia began cooperating

Canadian mission specialist Chris A. Hadfield stands on the remote manipulator system (RMS) of the space shuttle *Endeavour,* attached to the portable foot restraint. *NASA.*

in their human spaceflight activities in 1992, U.S. astronaut Shannon Lucid visited *Mir,* and spent 188 days in orbit, the longest spaceflight by an American up to that point.

Human Spaceflight at the Beginning of the Twenty-first Century

Human spaceflight at the beginning of the twenty-first century is no longer a government monopoly. The possibility of privately operated, profit-oriented human spaceflight activities, long advocated by a variety of groups and individuals, is now a reality. A significant step in the direction of private spaceflight occurred in April 2001 when American millionaire Dennis Tito coordinated with space tourism company Space Adventures, Ltd. (Vienna, Virginia) to pay Russia to send him for a six-day visit to the International Space Station onboard one of its Soyuz space capsules. South African Mark Shuttleworth, who made a similar Russian trip beginning on April 25, 2002, followed Tito into space as the second space tourist. The third spaceflight participant was Gregory Olsen (from the United States) in 2005. Anousheh Ansari (Iran/U.S.) in 2006, Charles Simonyi (Hungary/U.S.) in 2007, and Richard Garriott (U.S.) in 2008, followed Olsen into space. Simonyi became the first spaceflight participant to fly twice into space when he returned in March 26, 2009. Canadian businessperson Guy Laliberte was launched into space on September 30, 2009, for a 12-day visit to the International Space Station (ISS).

In 1996 the United States' National Aeronautics and Space Administration (NASA) began to turn over much of the responsibility for operating the space shuttle to a private company, United Space Alliance (USA), which is jointly owned by Boeing and Lockheed Martin, the two largest U.S. aerospace firms. However, NASA limited United Space Alliance's freedom to market space shuttle launch services to nongovernment customers, and NASA retained control over which people could fly aboard the shuttle. However, after Tito's flight, this policy was re-evaluated, and NASA decided to accept applications from paying customers for such flights. A privately funded corporation called MirCorp worked with Russia to try to keep the *Mir* station in operation, perhaps by selling trips to the space station for tens of millions of dollars to wealthy individuals or by other forms of private-sector use of the facility. Tito was MirCorp's first customer, but he could not be launched in time to travel to *Mir* before it was de-orbited. After the February 2003 space shuttle *Columbia* disaster the NASA decision to accept paying customers was rescinded.

However, in the near future, civilians will get a chance to go into space onboard spacecraft operated by private enterprises such as Virgin Galactic. Using spacecraft developed by Scaled Composites, the company is expecting to begin sub-orbital flights in or around 2013. Already, the first two commercial astronauts have gone into space. On June 21, 2004, the first sub-orbital flight of the privately funded spacecraft occurred when Michael Melvill piloted Scaled Composites' *SpaceShipOne* into space. Brian Binnie became the second commercial astronaut when he flew another Scaled Composite flight into space on October 4, 2004.

Other companies developing private space flights for humans and/or cargo include Space Exploration Technologies Corporation, Masten Space Systems, SpaceDev/Sierra Nevada, and Blue Origin. For instance, on May 26, 2010, the first successful flight of a vertical take-off, vertical landing (VTVL) vehicle was performed by aerospace startup company Masten Space Systems, which is based in Mojave, California. The company is developing a line of VTVL spacecraft for unmanned suborbital research flights and, eventually, for unmanned orbital flights.

Further, two more spacecraft—the Cygnus by Orbital Sciences Corporation and the Dragon by Space Exploration Technologies Corporation (SpaceX)—are being developed to fly back and forth to the International Space Station. They will initially be unmanned, only carrying cargo for resupply missions to the space station, but eventually the two companies hope to modify them for manned missions.

The International Space Station and Beyond

The International Space Station (ISS), developed and funded by an international partnership—the United States (National Aeronautics and Space Administration), Canada (Canadian Space Agency), Russia (Russian Federal Space Agency), Japan (Japan Aerospace Exploration Agency), and the member-nations of the European Union (European Space Agency: which includes as its major contributors to the ISS: France, Germany, Italy, Denmark, Belgium, Switzerland, Sweden, Norway, Spain, and the Netherlands)—will offer opportunities for privately financed experiments in its various laboratories. It was possible in the past for those carrying out such experiments to pay NASA to send their employees to the space station, but the U.S. space shuttle program was terminated on August 31, 2011, and the space shuttle fleet retired. Since then, the only way to and from the International Space Station is through Russia's Soyuz space capsules.

If research or other activities aboard the International Space Station prove to have economic benefits greater than the cost of operating the facility, it is conceivable that it could be turned over to some form of commercial operator in the future. However, such a proposal is highly unlikely. If the International Space Station were fully or partially commercialized, and the various ways of transporting experiments, supplies, and people to and from the station were operated in whole or part by the private sector, the future could see the overall commercialization of most activities in low Earth orbit. If this were to happen, governments could act as customers for the transportation and on-orbit services provided by the private sector on a profit-making basis. As of 2011, the International Space Station is expected to be supported financially through 2020, and possibly to 2028, by its government sponsors.

The most exciting vision for the future is widespread public space travel, sometimes called space tourism. If this vision were to become reality, many individuals, not just millionaires or those with corporate sponsorships, could afford to travel into space, perhaps to visit orbiting hotels or other destinations. In recent years, only wealthy people could travel into space as tourists with estimated costs of between US$20 and $40 million. However, even that opportunity was halted in 2010 when the Russian Federal Space Agency stopped such trips through U.S.-based Space Adventures to take wealthy space tourists to the International Space Station (ISS) aboard Russian Soyuz spacecraft. The Russians temporarily halted the program because it did not have the capacity to send tourists into space along with its other obligations. Specifically, the size of the ISS Expedition crews was to increase from three to six beginning in 2011 (which increased the number of Russian flights annually to the ISS), along with the fact that the U.S. space shuttle program was ending in that same year (which placed all the capability of sending people to the ISS with the Russians). Russia hopes to reinstitute its spaceflight participant program in 2012 or 2013.

In the meantime, Space Adventures is offering suborbital flights through Armadillo Aerospace, which is developing a suborbital commercial spacecraft that will take off vertically and land vertically, what is called a vertical takeoff, vertical landing (VTVL) craft. The company is expecting to offer suborbital flights to space in the near future.

There is still much to develop before less expensive flight are possible. However, as mentioned earlier, several private space tourist ventures are now developing ways to send large numbers of private citizens into space. For instance, Bigelow Aerospace is developing inflatable habitat modules that one day could expand into space hotels. The U.S. company is currently testing the *Genesis I* and *Genesis II* prototypes, already orbiting in space. The company is also expecting to launch its *BA 330* module in 2014 or 2015. The space habitation module, expandable to 12,000 cubic feet (330 cubic meters), will directly support zero-gravity scientific and manufacturing research, along with indirect activities involving space tourism.

In addition, Bigelow is designing a *BA 2100* module that would be six times larger than the *BA 330*. The company is also developing a series of inflatable modules, under the name of Bigelow Next-Generation Commercial Space Station. The complex will include various spacecraft modules, along with a central docking node, solar arrays, propulsion, and crew capsules. The company hopes to launch the components in 2014, with leasing options starting in 2015. So far, Bigelow has contracts with organizations in seven countries interested in leasing its space station modules.

 See also **Astronauts, Types of (Volume 3) • Career Astronauts (Volume 1) • Cosmonauts (Volume 3) • History of Humans in Space (Volume 3) • International Space Station (Volumes 1 and 3) • Tourism (Volume 1)**

Resources

Books and Articles

DK Publishing. *Space Shuttle: The First 20 Years: The Astronauts' Experiences in Their Own Words.* New York: DK Publishing, 2002.

Harland, David M. *Exploring the Moon, The Apollo Expeditions.* New York: Springer, 2008.

Haugen, David, and Zack Lewis, editors. *Space Exploration.* Detroit: Greenhaven Press, 2012.

Landfester, Ulrike, et al., editors. *Humans in Outer Space—Interdisciplinary Perspectives.* Berlin: Springer, 2011.

Maurer, Eva, et al., editors. *Cosmic Enthusiasm: The Cultural Impact of Soviet Space Exploration since the 1950s.* New York: Palgrave Macmillan, 2011.

McCurdy, Howard E. *Space and the American Imagination.* Baltimore: Johns Hopkins University Press, 2011.

Websites

The Apollo Program. National Aeronautics and Space Administration. <http://spaceflight.nasa.gov/history/apollo/index.html> (accessed October 14, 2011).

Bonsor, Kevin. *How Space Tourism Works.* How Stuff Works. <http://science.howstuffworks.com/space-tourism.htm> (accessed October 14, 2011).

"China's First Astronaut Crowned 'Space Hero'."*Xinhua News Agency.* <http://www.china.org.cn/english/features/etdz/79480.htm> (accessed October 14, 2011).

Civilians in Space FAQs. CBC News. <http://www.cbc.ca/news/background/space/spacetourism.html> (accessed October 11, 2011).

European consortium plans suborbital vehicle (Project Enterprise). Space Fellowship. <http://spacefellowship.com/News/?p=4096> (accessed October 11, 2011).

First Russian space tourist will not lift off until 2009. RIA Novosti. <http://en.rian.ru/russia/20071005/82599456.html> (accessed October 11, 2011).

History Home. National Aeronautics and Space Administration. <http://history.nasa.gov> (accessed October 11, 2011).

International Space Station. European Space Agency. <http://www.esa.int/esaHS/iss.html> (accessed October 11, 2011).

Virgin Galactic. <http://www.virgingalactic.com/> (accessed September 28, 2011).

Virgin Galactic to Offer Space Travel Adventures. Fox Business. <http://www.foxbusiness.com/personal-finance/2011/10/03/virgin-galactic-to-offer-space-travel-adventures/> (accessed October 4, 2011).

Virgin Galactic Unveils Suborbital Spaceliner Design. Space.com. <http://www.space.com/news/080123-virgingalactic-ss2-design.html> (accessed October 14, 2011).

I

Insurance

Every few weeks or so another rocket carrying a telecommunications satellite thunders heavenward. These satellites might be destined to become part of the international telephone network, provide direct-to-home television, or provide a new type of cellular phone service or Internet backbone link. Regardless of each satellite's eventual purpose, a wide range of people around the world will be focused on its progress as its fiery plume streaks upward. Flight controllers monitor positions while the manufacturers of the satellites check critical systems. The satellite owners wait anxiously to see if their critical investments will successfully reach their orbital destinations. But there is another group of people, often overlooked, who also intensely monitor the fate of the rockets and satellites—the space insurance community.

According to the Satellite Industry Association (SIA), in June 2011 there were 365 commercial communications satellites orbiting Earth. The production, launch, and ground control of those and other commercial satellite types is a major industry, with annual revenues of nearly 170 billion dollars in 2010. However, the commercial space industry would not exist today without the space insurance industry. For example, in the case of a satellite slated to launch, unless the owners of and investors in the satellite are able to obtain insurance, the satellite will never be launched. A typical telecommunications satellite costs upwards of around $200 million. Another $80 million to $100 million is needed to launch this satellite into its proper orbit. Given the historic 10 to 15% failure rate of rockets, very few private investors or financial institutions will place this amount of money at risk without insurance to cover potential failures. In fact, the National Aeronautics and Space Administration (NASA) found that in the 1990 and 2000s, the GOES (Geostationary Operational Environmental Satellites) series of satellites had a 15% failure rate for launch and deployment. Thus, insurance is an essential part of the financing for any commercial space venture.

The Growth and Development of the Space Insurance Industry

The growth of the commercial space industry and the growth of the space insurance industry go hand in hand. National governments did not require insurance at the beginning of the space age, so it was not until the early 1970s, when companies decided to build the first commercial

It can cost hundreds of millions of dollars to build and launch a satellite. Because there is an approximate 10 to 15% rocket failure rate, insurance is a critical element of space business. © *Corbis.* ▶

* **expendable launch vehicles**
  launch vehicles, such as a
  rocket, not intended to be reused

satellites for the long-distance phone network, that space insurance was born. In these early years the few space insurance policies were usually underwritten as special business by the aviation insurance industry.

The destruction of the space shuttle *Challenger* in 1986 marked a dramatic new phase in space insurance. NASA decided that the shuttle would no longer be used to launch commercial satellites. This forced commercial satellites onto expendable launch vehicles*, which had a higher risk of failure than the relatively safe shuttle. Owners and investors actively sought insurance to protect their satellite assets, and this growing demand established space insurance as a class of insurance of its own.

The success of commercial satellites has led to sustained growth in the space insurance industry. In 2010, the worldwide space insurance industry took in nearly 900 million dollars in premiums, while making payouts for losses of 351 million dollars. The ability to obtain insurance against launch and satellite failures has enabled investors to achieve commercial financing for space projects. In turn, this has stimulated the growth of the commercial space industry such that commercial spending on space projects was able to reach parity with government spending by 1998. The role of space insurance in securing commercial financing is so well-established that government agencies, such as NASA and the European Space Agency (ESA), now also insure selected projects. This trend will continue and space insurance will play a central role in unlocking financing for new commercial ventures on the International Space Station and in other avenues in low-Earth orbit and beyond.

Insurance—A Global Industry

Space insurance, like other forms of insurance, is a global industry. The largest concentration of companies is in London (United Kingdom), where space insurance and insurance in general originated. Major companies exist all over the world, however, including in the United States, Germany, France, Italy, Australia, and Japan. Virtually every country that has commercial satellites participates in the space insurance industry, either through direct underwriters or through reinsurers. Reinsurers insure the insurance companies, agreeing to accept some of the risk for a fee. This spreading of risk is crucial, as it is difficult for one company or country to absorb a loss in the hundreds of millions of dollars from a single event. The global spreading of risk takes advantage of worldwide financial resources and is a fundamental aspect of the insurance industry.

Insurance premiums can vary from 4 to 25% of the project's cost depending on the type of rocket and satellite, previous history, new technology being used, and policy term. Accurate and up-to-date information is essential in setting rates, and a major issue is government restrictions on the flow of technical information. Market forces and recent losses also affect rates. Rates were increasing in the early twenty-first century as companies adjusted to the very high insurance losses incurred from 1998 to 2000. The year 2000 was an especially troubling time for the space insurance industry following the disasters incurred with the companies Iridium and Globalstar. Nearly forty launches per year had been expected just prior to 2000, but with the demise of these two rivals, the launch count was under thirty per year immediately afterwards. Since that time, launch numbers not only rebounded, but have shown steady growth to meet the demand for more commercial satellite capacity. Statistics from the SIA indicate that worldwide revenues for the satellite industry posted an annual growth of slightly over eleven percent from the years 2005 through 2010.

Brokers and Underwriters

The space insurance industry is essentially made up of two types of companies: the brokers and the underwriters. The broker's task is to put together an appropriate insurance program for the satellite owner, while the underwriter puts up security in the form of insurance or reinsurance. The broker identifies the client's insurance needs over the various phases of a satellite's life: manufacture, transport to the launch site, assembly onto the rocket, launch, in-orbit commissioning, and in-orbit operations. The broker then approaches the underwriting companies and asks how much coverage they will provide and what premium rate they will charge. Most underwriting companies will not take more than $50 million for any one risk. Hence, the broker must often contact several underwriters to place the client's total risk at consistent rates. Once the package is agreed to, legal contracts complete the arrangements.

Jobs in Space Insurance

The space insurance industry can offer fascinating work for those interested in space. There are two main roles: the broker, who has a business development role—finding clients and negotiating insurance programs; and the underwriter, who leads the complex task of establishing the insurance rates. Experience in the space insurance industry is essential for both of these roles, and often a business, legal, financial, or technical background is required. Companies also have technical experts who understand satellites and rockets, a legal department for writing contracts, a finance department handling the accounts and money transfers, and a claims department for assessing losses and processing claims. Actuaries, who generally have a mathematics background, model the expected losses and help the underwriter and the technical experts set the rates.

As permanent habitation of space through the International Space Station leads to the discovery of new commercial opportunities, space insurance will evolve to cover them. In addition, as more space debris (junk) continues to build up in orbit about Earth, there is more risk of collisions with satellites. For example, on February 9, 2009 a defunct Russian satellite (*Kosmos-2251*) collided with an operational *Iridium 33* satellite operated by Iridium Satellite LLC about 790 kilometers (490 miles) over the sky of northern Siberia in Russia. Both satellites were destroyed, placing even more space junk in orbit around Earth and making it even more risky for spacecraft and satellites to travel around the planet.

As of mid-2010, the U.S. Strategic Command was tracking about 22,000 objects with diameters of ten centimeters (about four inches) or larger in orbit around the Earth. However, according to the European Space Agency's Meteoroid and Space Debris Terrestrial Environment Reference, more than 600,000 objects larger then one centimeter across are in orbit, most of which are not being tracked. Thus, the use of insurance remains an essential ingredient for commercial space business and will continue to play a vital role in humanity's growing commercial exploration of space.

The prospect of space tourists going up into space on a regular basis is producing a new avenue of additional business for the space insurance industry. By late 2011, 450 people had booked a flight with Virgin Galactic to ride on their advertised sub-orbital flights as early as 2013. The cost of each ticket is $200,000, with a down payment of $20,000. Space insurance underwriter Lloyd's of London has been participating in negotiations with Virgin Galactic. The insurance will likely involve a combination of an aviation risk and a space risk while operating their flights both in Earth's atmosphere and above it. Talks with the pioneering space tourist company are also ongoing concerning third-party liability protection against claims from people living near the New Mexico launch site, such as claims from pollution caused by launches or from falling debris in the unlikely event of a crash over their homes. As space is no longer the sole domain of professional astronauts and scientists, and as more people get the chance to go into space, the space insurance industry will no doubt see new revenues coming its way.

 See also **Communications Satellite Industry (Volume 1) • Launch Vehicles, Expendable (Volume 1) • Launch Vehicles, Reusable (Volume 1) • Reusable Launch Vehicles (Volume 4) • Satellite Industry (Volume 1) • Search and Rescue (Volume 1)**

Resources

Books and Articles

Maral, Géard and Michel Bousquet. *Satellite Communications Systems: Systems, Technologies, and Technology.* 5th ed. Chichester, U.K.: John Wiley & Sons, 2009.

Seedhouse, Erik. *Tourists in Space: A Practical Guide.* Berlin: Springer, 2008.

Websites

GOES-News. Goddard Space Flight Center. National Aeronautics and Space Administration. <http://goes.gsfc.nasa.gov/text/goesnew.html> (accessed October 21, 2011).

Hessel, Evan. "Collision 500 Miles Above The Earth." *Forbes.* <http://www.forbes.com/2009/02/11/communications-satellite-iridium-technology-wireless_0211_iridium.html> (accessed October 21, 2011).

Whalen, David J. "Communications Satellites: Making the Global Village Possible." *National Aeronautics and Space Administration.* <http://history.nasa.gov/satcomhistory.html> (accessed October 21, 2011).

International Space Station

The International Space Station (ISS) is a scientific and technological wonder. It is a dream being realized by a multinational partnership. The ISS provides a permanent human presence in space and a symbol of advancement for humankind. There is great promise and discovery awaiting those who will commercially and scientifically use the space station.

Just as the global explorers of the fifteenth century circled the globe in their square-sailed schooners in search of riches—gold, spices, fountains of youth, and other precious resources—so too is the space station a wind-jammer plowing the waves of space, exploring the riches it holds. The space station brings together the adventure of fifteenth-century explorers with twenty-first-century technology and industry. The space station is, at its essence, an infrastructure that will facilitate and transmit new knowledge, much like that provided through the virtual world of the Internet.

Boeing technicians slide a systems rack into the U.S. Laboratory module intended for the International Space Station. The laboratory will contain 13 racks with experiments and 11 racks with support systems, with each rack weighing nearly 1,200 pounds (on Earth). *NASA*.

* **microgravity** the condition experienced in freefall as a spacecraft orbits Earth or another body; commonly called weightlessness; only very small forces are perceived in freefall, on the order of one-millionth the force of gravity on Earth's surface

Concerning the space station, the question is: From where will the value of this virtual world come? Or, put in terms of the fifteenth-century explorers, "What is the spice of the twenty-first century in the new frontier of space?"

At this stage, no one can guess what the most valuable and profound findings from space station research will be. Space research done to date, however, does point the way to potential areas of promise that will be further explored on the ISS. The space environment has been used to observe Earth and its ecosphere, explore the universe and the mystery of its origin, and study the effects of space on humans and other biological systems, on fluid flow, and on materials and pharmaceutical production. In all, many different types of experiments occur on the ISS, including life sciences, physical sciences, human research, biology and biotechnology, materials sciences, technology development, and Earth and space science.

The space station creates a state-of-the-art laboratory to explore human subjects and the world. The discoveries that scientists have made to date in space, while significant, are only the foundation for what is to come. Research in microgravity* is in its infancy. Throughout the thousands of years that physical phenomena have been observed, including the relatively recent four hundred years of documented observations, it has been only in the period since the 1960s that experiments in microgravity of more than a few seconds have been observed; only since the late 1990s has there been a coordinated set of microgravity experiments in space. For instance, the Active Rack Isolation System-ISS Characterization Experiment (ARIS-ICE), conducted from March 2001–June 2002, was an ISS experiment designed to characterize the effectiveness of the ARIS facility. ARIS was made to detect and isolate on-orbit vibrations found on the International Space Station. Sensitive experiments might be adversely affected by such vibrations, so the ARIS absorbs the shock before it can harm such devices and materials.

History is rife with failed predictions. Nevertheless, perhaps the best way to try to predict the future is to look at the evolution of the past, using history and the current situation as a jumping-off point, while recognizing the challenge in predicting the future. The following represents an attempt

The view from space shuttle *Atlantis* following its undocking from the International Space Station, after successfully outfitting the station for its first resident crew. *NASA.*

* **telecommunications infrastructure** the physical structures that support a network of telephone, Internet, mobile phones, and other communications systems

to peer into the future, to see what promise lies ahead for the space station by looking at the past and the present.

Previous Advances from Space Activities

In the twentieth century, space exploration had a profound impact on the way humans viewed itself as a species and the world affording life to each individual. Viewing planet Earth from space for the first time gave humans a unique perspective of Earth as a single, integrated whole. Observations of Earth's atmosphere, land, and oceans have allowed scientists to better understand the planet as a system and, in doing so, the human role in that integrated whole. Many aspects of human lives that are now taken for granted were enabled, at least in part, by investments in space. Whether making a transpacific telephone call, designing with a computer-aided design tool, using a mobile phone, wearing a pacemaker, or going for an MRI (magnetic resonance imaging) scan, humans are using technology that space exploration either developed or improved.

In the early twenty-first century, commercial interests offer a myriad of products and services that either use the environment of space or the results of research performed in microgravity. Just a few examples include:

- *Satellite communications:* Private companies have operated communications satellites for decades. Today, private interests build, launch, and operate a rapidly expanding telecommunications infrastructure* in space. The initial investment in space of the United States helped fuel the information revolution that spurs much of the nation's economy today. For instance, the ISS is often used as a "communications satellite" as it circles above Earth. ISS science officer Don Pettit, while on the space station as part of the Expedition 6 crew, conducted the "Saturday Morning Science" program. Pettit recorded different videos to explain concepts in science for schoolchildren. In addition, the Amateur Radio on the International Space Station (ARISS) program has allowed the astronauts and cosmonauts aboard the space station to communicate with Amateur Radio Clubs and ham radio operators around the world, along with talking with school children, teachers, parents, and whole communities.

- *Earth observation/remote sensing:* A growing market for Earth imagery is opening up new commercial opportunities in space. Private interests now sell and buy pictures taken from Earth orbit. Land-use planners, farmers, and environmental preservationists can use the commercially offered imagery to assess urban growth, evaluate soil health, and track deforestation. For instance, the ISS Agricultural Camera (ISSAC) takes visible light and infrared light images of the Earth in order to study dynamic Earth processes such as ecological changes caused by seasonal progressions or human impacts. For instance, the northern Great Plains has been imaged with respect to its grasslands, forests, and croplands. Ranchers can determine which lands are being overgrazed and undergrazed for better utilization of their animals.

■ *Recombinant human insulin:* The Hauptman-Woodward Medical Research Institute, in collaboration with Eli Lilly and Company, has used structural information obtained from crystals grown in space to better understand the nuances of binding between insulin and various drugs. Researchers there are working on designing new drugs that will bind to insulin, improving their use as treatments for diabetic patients. On the International Space Station, biology and biotechnology occupy much of the experiments performed in the orbiting laboratory, from animal biology, cellular biology, and microbiology, to plant biology, protein crystal growth, and the study of anemia in astronauts returning to Earth from long flights in space. The study, called *Effects of Microgravity on the Haemopoietic System: A Study on Neocytolysis (Neocytolysis)*, found that astronauts returning from space flights often had anemia (the decrease of red blood cells caused by their destruction while in space). Results of the study led to more effective treatments of anemia.

What all these discoveries have in common is that they use space as a resource for the improvement of human conditions. Efforts aboard the ISS will continue this human spirit of self-improvement and introspection. In addition, the twenty-first century holds even greater promise with the advent of a permanent human presence in space, allowing that same spirit to be a vital link in the exploration process. The space station will maximize its particular assets: prolonged exposure to microgravity and the presence of human experimenters in the research process.

Potential Space Station-Based Research

The ISS will provide a laboratory setting that can have profound implications on human health issues on Earth. Many of the physiological changes that astronauts experience during long-term flight resemble changes in the human body normally associated with aging on Earth. Bone mass is lost and muscles atrophy*, and neither appear to heal normally in space. By studying the changes in astronauts' bodies, scientists might begin to understand more about the aging process. Scientists sponsored by the National Aeronautics and Space Administration (NASA) are collaborating with the National Institutes of Health (NIH) in an effort to explore the use of spaceflight as a model for the process of aging. This knowledge may be translated into medical wonders, such as speeding the healing of bones and thereby reducing losses in productivity. By beginning to understand the process by which bones degenerate, scientists might be able to reverse the process and expedite the generation of bone mass.

The microgravity environment offers the opportunity to remove a fundamental physical property—gravity—in the study of fluid flow, material growth, and other phenomena. The impacts on combustion, chemistry, biotechnology, and material development are promising and exciting. The combustion process, a complex reaction involving chemical, physical, and thermal properties, is at the core of modern civilization, providing over 85% of the world's energy needs. By studying this process on the ISS,

* **atrophy** condition that involves withering, shrinking, or wasting away

commercial enterprises could realize significant savings by introducing new-found efficiencies.

Researchers have found that microgravity provides them with new tools to address two fundamental aspects of biotechnology: the growth of high-quality crystals for the study of proteins and the growth of three-dimensional tissue samples in laboratory cultures. On Earth, gravity distorts the shape of crystalline structures, while tissue cultures fail to take on their full three-dimensional structure.

The microgravity environment aboard the ISS will therefore provide a unique location for biotechnology research, especially in the fields of protein crystal growth and cell/tissue culturing. Protein crystals produced in space for drug research are superior to crystals that can be grown on Earth. Previous research performed on space-grown crystals has already increased knowledge about such diseases as AIDS (acquired immune deficiency syndrome), emphysema, influenza, and diabetes. With help from space-based research, pharmaceutical companies are testing new drugs for future markets.

In addition to these scientific findings, the ISS serves as a real-world test of the value of continuous human presence in space. There are already companies focused on space tourism and the desire to capitalize on the human presence in space. A myriad of future scenarios are possible, and the imagination of entrepreneurs will play a key role.

Inevitably, private interests will move to develop orbital infrastructure and resources in response to a growing demand for space research and development, along with tourism and commercialism. The permanent expansion of private commerce into low Earth orbit* will be aided as the partners of the ISS commercialize infrastructure and support operations such as power supply and data handling. This trend is already under way with several commercial payloads having flown on past space shuttle missions and on the ISS.

The ISS is an unparalleled, international collaborative venture. In view of the global nature of the ISS, the international partners (the United States, Russia, Japan, Canada, and member-countries of the European Space Agency) recognize the value of consulting on and coordinating approaches to commercial development. Each international partner retains the autonomy to operate its own commercial program aboard the ISS within the framework of existing international agreements, and mechanisms of cooperation are possible where desired.

 See also **Aging Studies (Volume 1)** • **Crystal Growth (Volume 3)** • **History of Humans in Space (Volume 3)** • **International Space Station (Volume 3)** • **Made with Space Technology (Volume 1)** • **Microgravity (Volume 2)** • **Mir (Volume 3)** • **Skylab (Volume 3)**

Resources

Books and Articles

Catchpole, John E. *The International Space Stations: Building for the Future.* Berlin: Springer, 2008.

Committee for the Decadal Survey on Biological and Physical Sciences in Space, Space Studies Board, Aeronautics and Space Engineering Board, Division on Engineering and Physical Sciences, National Research Council of the National Academies. *Life and physical sciences research for a new era of space exploration: an interim report.* Washington, D.C.: National Academies Press, 2010.

Jamison, David E., editor. *The International Space Station.* New York: Nova Science, 2010.

Reference Guide to the International Space Station. Washington, D.C.: National Aeronautics and Space Administration, 2010.

Websites

Active Rack Isolation System — ISS Characterization Experiment (ARIS-ICE). National Aeronautics and Space Administration. <http://www.nasa.gov/mission_pages/station/research/experiments/ARIS-ICE.html> (accessed October 25, 2011).

Amateur Radio on the International Space Station (ARISS). National Aeronautics and Space Administration. <http://www.nasa.gov/mission_pages/station/research/experiments/ARISS.html> (accessed October 25, 2011).

Effects of Microgravity on the Haemopoietic System: A Study on Neocytolysis (Neocytolysis). National Aeronautics and Space Administration. <http://www.nasa.gov/mission_pages/station/research/experiments/Neocytolysis.html> (accessed October 25, 2011).

International Space Station. National Aeronautics and Space Administration. <http://www.nasa.gov/mission_pages/station/main/index.html> (accessed October 25, 2011).

International Space Station Agricultural Camera (ISSAC). National Aeronautics and Space Administration. <http://www.nasa.gov/mission_pages/station/research/experiments/ISSAC.html> (accessed October 25, 2011).

International Space Station: Canada's Contribution. Canadian Space Agency. <http://www.asc-csa.gc.ca/eng/iss/canada.asp> (accessed October 25, 2011).

International Space Station: Human Spaceflight and Exploration. European Space Agency. <http://www.esa.int/esaHS/iss.html> (accessed October 25, 2011).

Japan Aerospace Exploration Agency. <http://www.jaxa.jp/index_e.html> (accessed October 25, 2011).

NASA's Plans to Explore the Moon, Mars and Beyond. National Aeronautics and Space Administration. <http://www.nasa.gov/exploration/home/index.html> (accessed October 25, 2011).

Office of the Chief Technologist. National Aeronautics and Space Administration. <http://www.nasa.gov/offices/oct/home/index.html> (accessed October 25, 2011).

Russian Federal Space Agency. <http://www.roscosmos.ru/index.asp?Lang=ENG> (accessed October 25, 2011).

Saturday Morning Science. National Aeronautics and Space Administration. <http://issresearchproject.grc.nasa.gov/saturdayScience.php> (accessed October 25, 2011).

International Space University

As space programs become increasingly international and commercial in nature, the education of the space-sector workforce needs to continually adapt to maintain this pace. The International Space University (ISU) is dedicated to meeting this challenge by training the world's next generation of professionals who will lead the way into space.

The ISU provides postgraduate training for students of all nations who are interested in space. Each year, the ISU conducts a two-month summer program in various locations around the world. Examples of past ISU Summer Session locations include Australia, Canada, Chile, Germany, Japan, Sweden, Spain, Thailand, and the United States. In addition, the ISU provides a master's degree in space studies (its original degree program started in 1995) and a master's degree in space management (which was implemented in 2004) at its central campus in Illkirch-Graffenstaden, a large suburb in the city of Strasbourg, in the northeastern part of France.

Because specialist knowledge alone is no longer sufficient to meet the challenges of complex space programs implemented by governments and companies from around the world, the ISU presents an interdisciplinary, international, and intercultural approach. The university provides participants with a thorough appreciation of how space programs and space business work. This is accomplished through extensive coursework in science, engineering, law and policy, business and management, and other space-related fields. In addition, all ISU students participate in a student design project that allows them to integrate their classroom learning in a complex, hands-on, practical exercise. Because this design project is conducted with classmates from all over the world, it allows ISU students to master the challenges of working with international teammates in an intercultural environment.

Since the first summer session in 1988, more than 3,000 students from more than 100 countries have reaped the benefits of an ISU education. Most of these students have gone on to work in successful careers in space and related fields. Universally, ISU alumni credit their professional success to the broad intellectual perspectives gained at the university as well as to their extensive international network of contacts in the space community. The university's interactive, international environment

At an International Space University summer session, students learn about spacesuit design. *International Space University.*

provides its students with continuous opportunities to forge professional relationships with colleagues and with its faculty, who also come from many different countries.

The Origins of the International Space University

The ISU was the brainchild of three young men (Todd B. Hawley, Peter Diamandis, and Robert D. Richards) in their early twenties who, as college students, became interested in space exploration. Passionate about space, they proposed a university dedicated to a broad range of space-related subjects for graduate students from all parts of the world. With their enthusiasm, they succeeded in winning over important players in the space field, including science fiction author Arthur C. Clarke, who became chancellor of ISU. The founders' vision for the university was that it would be an institution dedicated to a peaceful, prosperous, and boundless future through the study, exploration, and development of space for the benefit of all humanity.

Hawley served as the first chief executive officer of ISU. He was also the chair of the organization Students for the Exploration and Development of Space from 1983 to 1985. Hawley also conducted research on solar power satellites before his death of HIV/AIDS-related causes in 1995. Diamandis is currently the founder and chair of the X PRIZE Foundation, a non-profit organization whose mission is to provide competitions to benefit humankind, such as prizes for the building of private spaceships. Richards is the founder of Odyssey Moon Limited (based in The Isle of Man), a commercial lunar enterprise company. He is also involved with Optech Inc., a company dealing with airborne laser systems, three-dimensional imaging systems, and other such technologies.

The ISU began to materialize in the summer of 1987 as participants in the first ISU summer session gathered at the Massachusetts Institute of Technology in Cambridge. Four years after this initial session, Strasbourg, France, was selected as the site for the ISU central campus. The first master's program, which entailed classes in space studies, was initiated in September 1995 following the move of the campus to Europe. The second master's program, in space management, was started in 2004.

In 2011, ISU launched its first program in the Southern Hemisphere, a joint module in its executive MBA program with the University of South Australia in Mawson Lakes. The module was designed to explore ways in which space research can benefit those who are living in the Southern Hemisphere to the degree that it already has for those in the Northern Hemisphere.

 See also **Careers in Space Science (Volume 2) • Education (Volume 1)**

Resources

Websites

International Space University. <http://www.nasa.gov/centers/ames/ISU/> (accessed November 2, 2011).

International Space University. <http://www.isunet.edu/> (accessed November 2, 2011).

Paths to Progress. <http://www.unisa.edu.au/itee/spaceprogram/images/SHSSP-2011-WhitePaper.pdf> (accessed November 2, 2011).

L

Launch Industry

The process by which satellites, space probes, and other craft are launched from the surface of Earth into space requires equipment, machinery, hardware, and a means, usually a rocket vehicle of some sort, by which these materials are lifted into space. The businesses that exist to service these launching needs form an international space transportation industry. Historically, that industry has been dominated by firms in the United States and Russia (and, previously, the Soviet Union). More recently, launch companies have been created in a number of other nations, including Brazil, China, France, India, and Japan, as well as by the European Space Agency. With the end of the space shuttle program in the United States in 2011, the fundamental character of the launch industry worldwide has undergone a fundamental change with a shift of responsibility for launch programs from governmental agencies to private companies.

The Origins of the Industry

The commercial launch industry has its roots in the Cold War missiles that formed the launching rockets of the early space age. Following the launching of the world's first artificial satellite, *Sputnik 1,* by the Soviet Union on October 4, 1957, the governments of the Soviet Union and the United States were the only entities that possessed space launch vehicles. The fleets of intercontinental ballistic* missiles developed by the two nations served as the basis for the only rockets big and powerful enough to lift satellites into space. The launching rockets were used mainly to place government spacecraft into Earth orbit or towards the Moon or other planets. During the period of the late 1950s to early 1960s commercial satellites were not in existence for which commercial launchers were needed. All spacecraft were owned by the civil or military part of the governments of the Soviet Union and the United States.

When foreign nations that were aligned with either country needed a satellite launch in the mid–1960s, the satellite was shipped to the launching site of the United States or the Soviet Union and launched. Eventually the technology of satellites matured to the point where other nations sought to have their own space programs. Europe, the United Kingdom, France, and Germany each developed varying types of spacecraft that were launched by the United States from sites at either Cape Canaveral in Florida or Vandenberg Air Force Base in southern California. Japan, Brazil, and China also developed satellite technology. Nations aligned

* **ballistic** the path of an object in unpowered flight; the path of a spacecraft after the engines have shut down

Launching this rocket requires equipment, machinery, and hardware, all of which are provided by businesses within the launch industry. *NASA.*

with the Soviet Union included Mongolia, China, and other Southeast Asian nations. France and China were among the first nations other than the United States and the Soviet Union to develop their own independent space-launching rockets. Japan developed a version of the U.S. Delta space booster under a licensing agreement with McDonnell Douglas Corporation, although the technology of the launching rocket was carefully protected by the United States.

Evolution of the U.S. Launch Industry

A dominant theme for most of the first half century of the space age was research. Rockets, satellites, space probes, and other types of spacecraft were launched in order to collect data about virtually every part of the solar system. A host of space missions were designed to study the Moon, the planets, asteroids, comets, and the interstellar space. In some respects, the acme of that research was the Apollo program in the United States that landed humans on the Moon in six missions between 1969 and 1972. The two launch vehicles used in the Apollo program were the Saturn IB and Saturn V rockets, built, respectively by the Chrysler Corporation and the Douglas Aircraft Company, and by Douglas, Boeing and North American Aviation. Both before and after the Apollo program, the two workhorses of the U.S. launch program were the Delta and Atlas rockets. The first Delta rockets were built by Douglas, and the first Atlas rockets, by the Convair Division of General Dynamics. Both rockets were used extensively for satellite launches through the 1960s and 1970s, but were expected to be in low demand once the space shuttle program got underway in 1981. National Aeronautics and Space Administration (NASA) anticipated using the shuttles for the delivery of essentially all spacecraft into orbit. The throwaway expendable rockets, all based on earlier generations of ballistic missiles, were to be completely replaced by the shuttles. However, the development of the shuttle took longer and was more expensive and less reliable than was promised. By the mid–1980s the U.S. Department of Defense, a major customer for space shuttle launches, elected to continue production of the Delta and Titan launching rockets as alternates to the shuttles for launches of military spacecraft. In 1985, President Ronald Reagan announced a shift in U.S. space policy that would allow both civil and government satellites to be launched on these expendable rockets as well as the shuttles.

Following the destruction of the space shuttle *Challenger* in January 1986, a further shift in policy diverted satellite launches away from shuttles entirely. The shuttles would launch only those cargos that could not be flown aboard any other vehicle. This policy shift served to reinvigorate the U.S. space launch industry, which had lost most of its share of space-launching services during the period when the shuttles were taking over the launching of all U.S. satellites.

Commercial versions of the Delta and Atlas rockets, partially funded by the Defense Department and NASA, soon entered service. McDonnell

Douglas began offering a commercial Delta II rocket as a launching vehicle. General Dynamics Corporation and Convair Astronautics began selling larger Atlas rockets. A third commercial rocket based on the huge military Titan III booster was also sold commercially but only for a brief period. Its builder, Martin Marietta Corporation, could find only two customers before it was phased out. Titans were relegated to missions for the U.S. military.

Following the end of the Cold War in the early 1990s, the U.S. aerospace industry contracted through a series of mergers and acquisitions. Martin Marietta acquired the line of Atlas rockets, adding them to its stable of military Titan boosters. The Boeing Company acquired McDonnell Douglas and its Delta boosters. Eventually, Martin Marietta was itself acquired by Lockheed Corporation, forming Lockheed Martin Corporation. By the mid–1990s the U.S. launch industry was comprised of Boeing, Lockheed Martin, and a smaller firm called Orbital Sciences Corporation, which sold two smaller rockets, the winged Pegasus and the Taurus.

As of 2011, both Delta and Atlas rockets continue to play a crucial role in the space vehicle launch industry. All members of the two series are now built by United Launch Alliance, a joint venture of the Boeing and Lockheed Martin corporations. Between 2002 and the end of 2011, the Atlas V has been used for 27 missions, primarily the placement of communications satellites into orbit. It has also been used for the deployment of the Mars Reconnaisance Orbiter, the New Horizons spacecraft, and the Juno Jupiter orbiter. Two versions of the Delta rocket are now in use. The Delta II rocket was used for 146 launches between 1989 and the end of 2011, while the Delta IV rocket was used for 17 launches between 2002 and the end of 2011. As with the Atlas, the primary missions for the Delta rocket have been the launching of satellites, although they have also been used to deliver the GRAIL lunar orbiter and the WISE and GLAST space telescopes into orbit.

The first decade of the twenty first century saw more significant changes in the U.S. launch industry. The first of these changes was a consequence of the termination of NASA's space shuttle program in August 2011. With the end of that program, the only vehicle available for the delivery of personnel and goods to the International Space Station (ISS) was Russia's Soyuz rocket. To broaden the options for contact between Earth and the ISS, NASA decided to promote the development of launch vehicles and associated activities by private companies. By the end of 2011, NASA had contracted with two private corporations, Space Exploration Technologies (SpaceX) and Orbital Sciences Corporation to develop rockets for the delivery of cargo to the ISS by 2012. A second new trend in the development of launch vehicles was the increasing pace of manned spacecraft development. The first ship to carry human tourists into space was SpaceShipOne, developed by Mojave Aerospace Ventures, in 2004. SpaceShipOne used a form of launch that does not make use of rockets.

The craft was carried into the air by a jet airplane, called White Knight, which released SpaceShipOne at an altitude of about 50,000 feet. At that point, SpaceShipOne's own engines fired, taking the craft higher into the atmosphere, from which it eventually returned to Earth in a trajectory like that of any commercial airliner. SpaceShipOne was retired immediately after its successful voyage, with plans for two second-generation vehicles, ShapeShipTwo and WhiteKnightTwo, to be developed by The Spaceship Company, a joint venture of the Virgin Group and Scaled Composites. As of September 2011, The Spaceship Company reported that it had received more than 450 reservations, at $20,000 per person, for flights on SpaceShipTwo. Another consequence of the forthcoming privatization of space vehicle launching is an increasing interest in the development of new spaceports in the United States. As of late 2011, there were eight federal and eight non-federal spaceports, with additional spaceports proposed for Alabama, Hawaii, Indiana, Texas, Wisconsin, and Wyoming.

Russian Launchers

The changes wrought by the end of the Cold War affected the Soviet Union's space-launching programs as well. Commercial sales of the Proton rocket were conducted initially by a government-industry partnership called Glavcosmos in the early 1980s. However, after the Cold War ended, Protons were made commercially available as part of the rocket catalog being assembled by Lockheed Martin. A separate firm called International Launch Services (ILS) was established in June 1995 to sell both Atlas and Proton rockets on the world's commercial market. ILS, jointly owned by the U.S. firm Lockheed Martin and the Russian rocket design companies Khrunichev, and Energyia, reported a backlog of $3 billion worth of rocket contracts in 2000 and conducted six Proton and eight Atlas launches during that year. As of late 2011, the Proton rocket had been used for more than 360 launches since its maiden voyage in 1965. Of this number, 130 had been conducted by ILS. Proton rockets are launched from Baikonur Cosmodrome, near the small town of Tyuratam in the deserts of Kazakhstan. The launch site is the largest of its type in the world.

A second launch system that evolved out of Soviet Union technology is a modified version of the Russian Zenit rocket now being manufactured by the Yuzhnoye Design Bureau of Ukraine. The rocket is launched from a floating launching pad towed to the mid–Pacific Ocean. As of late 2011, the most popular version of the rocket, the Zenit-3SL, had been launched thirty times, with two total failures and one partial failure.

Yet a third commercial rocket, the R-7, was also developed originally in Russia. Variants of the R-7 have been used to launch more spacecraft than any other rocket in the world. Among the most historic of these launches were the first Earth satellite, *Sputnik 1*, launched in 1957, and the world's first human in space, Yuri Gagarin, in 1961. Among the R-7 variants that have since been discontinued are the Luna, Vostok, Molniya, Polyot, and Semyorka rockets. The last remaining R-7 variant in production is the Soyuz rocket, currently being manufactured by the Russian firm of S.P. Korolev Rocket and Space Corporation Energia, whose headquarters are in Moscow. A French company, Starsem, which calls itself "the Soyuz Company," is responsible for all launches of the Soyuz rocket, which take place at Baikonur Cosmodrome. Starsem was created in 1996 as the commercial arm of Soyuz launches, as well as other launch vehicles and accessory equipment. The Soyuz rocket has been used to carry astronauts to the International Space Station—and also to carry space tourists to the space station, such as Dennis Tito in April 2001, and Charles Simonyi in 2007 and 2009.

The Development of Arianespace

By the mid–1970s, Europe sought to develop a commercial means of launching European-made military, civil, and commercial satellites. The United States had begun the development of the partially reusable space shuttle during this period, and France was left out of the shuttle's development. In partial response, France and several European countries came together to form an entity called the European Launcher Development Organization (ELDO). Among their first objectives was the development of a commercial space-launching rocket that would be entirely made within the nations that were part of the space organization. Eventually ELDO was merged with the European Space Research Organization in order to establish the European Space Agency (ESA). The company that emerged from that effort was named Arianespace, and its family of expendable rockets was called Ariane.

Ariane was unlike the Soviet R-7 and Proton rockets and U.S. Delta and Atlas rockets because it was not based on an existing ballistic missile design but was instead created from the start as a commercial launching vehicle. The company was established in March 1980 and conducted its first launch on May 24, 1984. The rocket has been produced in five major variants which, as of late 2011 had experienced more than 250 successful launches since its maiden flight in December 1979. Ariane now consists of 21 companies in ten European countries. Its corporate headquarters are in Courcouronnes, France, where it employs about 300 people. Its launches are conducted from the Centre Spatial Guyanais launch site in Kourou, French Guiana, on Earth's equator.

Japan in Space Launch

Japan entered the space age in August 1963 with the attempted launch of the first in its series of Lambda (L) rockets, based on the earlier Kappa (K) sounding rockets developed by the country's space agency. The L rockets were not very successful and were replaced by the far more successful Mu (M) series, that flew from 1966 to 1985. M rockets placed into orbit two radio astronomy satellites, two x ray astronomy satellites, an asteroid probe, and a solar observatory. Simultaneously with the M series, the Japanese received a license to begin producing a version of the

Arianespace of Europe is one of three firms that currently dominate the global launch services industry. The Ariane 5 rocket has replaced the smaller Ariane 4, seen here. *European Space Agency/Photo Researchers, Inc.*

▶

* **payload** any cargo launched aboard a rocket that is destined for space, including communications satellites or modules, supplies, equipment, and astronauts; does not include the vehicle used to move the cargo or the propellant that powers the vehicle

U.S. Delta rocket. The first rocket in that Nu (N) series, the N-1 was successfully launched in September 1975. (N and H series rockets are also designated with Roman numerals, as N-I and H-I.) It was used for seven missions (one of which was a failure) before being replaced by the N-2 in 1981 and the H-1 in 1986. The H-1 was then replaced by a much larger rocket, the H-2. Although the H-2 was designed primarily for the delivery of satellites and other payloads*, a commercial version was also planned. Development of the H-2, however, was slowed by launch failures, and the program was terminated in 1999. It was replaced by the more advanced—and cheaper—H-2A. Made by Mitsubishi Heavy Industries (MHI), for the Japan Aerospace Exploration Agency (JAXA), the H-2A made its first test flight in 2001. Between then and the end of 2011, the rocket has been used for 19 launches, all of which have been successful. Japanese rockets are launched from the Tanegashima Space Center on Tanegashima Island, about 70 miles south of Kyushu Island. In 2007, the management of the H-2A was transferred from JAXA to MHI. Now a private concern, the first launch by MHI of its H-2A rocket was the Japanese SELENE (Selenological and Engineering Explorer), spacecraft launched in September 2007 for the Moon. In 2009, MHI received its first order for the launch of a non-Japanese satellite, Korea Multipurpose Satellite 3 (KOMPSAT-3), scheduled for delivery in early 2012.

China Competes

China is also developing commercial launch vehicles, with all of them based on the nation's early missile design programs. The Chang Zheng (Long March) series of rockets, available in different sizes, each capable of lifting different types of satellites, became available in April 1970 with the launch of Chang Zheng 1, carrying the Dong Feng Hong satellite into space. Since that time, 15 versions of the original rocket have been developed, of which eight are still in service [series 2 (2C, 2D, 2F), series 3 (3A, 3B, 3C), and series 4 (4B, 4C)]. Four additional versions of the rocket [2E(A), 3B(A), and 5] are also being developed. The Chang Zheng series has been marked by some spectacular successes and a few dramatic failures. The successes include 75 consecutive successful launches between August 1996 and August 2009, the successful launch of the first space-craft to carry a Chinese astronaut into space (*Shenzhou 5* in October 2003), and the first three-man mission and EVA (first extravehicular activity) flight (*Shenzhou 7* in September 2008). Design and development of Chinese rocket vehicles is the responsibility of the China Academy of Launch Vehicle Technology. Launches take place at one of three sites: Jiuquan Satellite Launch Center, in the Gobi Desert; Taiyuan Satellite Launch Center, in Kelan County, Xinzhou Prefecture, Shanxi Province; Xichang Satellite Launch Center, about 40 miles northwest of Xichang City, in Liangshan Yi Autonomous Prefecture, Sichuan Province.

India Enters the Fray

The Indian government has also been developing a family of space launch vehicles that will also be offered for commercial sales. The Polar Satellite Launch Vehicle (PSLV) and the Geosynchronous Satellite Launch Vehicle (GSLV) expendable rockets were successfully tested for launching satel-lites into polar orbits* as well as launching larger commercial communi-cations craft. The first successful launch of a PSLV occurred on May 26, 1999 when it launched the OceanSat 1 payload. As of late 2011, there have been 19 attempted launches of the PSLV, one of which (the first) failed, and one of which was a partial success. The GSLV has been less successful, with only two completely successful launches between April 2001 and December 2010. Design and development of Indian launch

* **polar orbits** orbits that carry a satellite over the poles of a planet

vehicles is the responsibility of the Indian Space Research Organisation, which has six research sites and five construction and launch facilities. India uses two launch sites for most of its missions, the Vikram Sarabhai Space Centre, at Thiruvananthapuram, and the Satish Dhawan Space Centre (Sriharikota), in Andhra Pradesh.

Other Competitors in the Industry

The South Korean government, through its Korea Aerospace Research Institute (KARI), has developed the Korea Space Launch Vehicle (KSLV). The maiden voyage of South Korea's first launch vehicle, the Naro-1, occurred on August 25, 2009 from the Naro Space Center in Goheung County, South Jeolla.

Brazil is developing and hopes to eventually market a smaller commercial rocket called the VLS (Satellite Launch Vehicle). Two prototype launch failures, however, in the late 1990s, placed the project's future in doubt. Then, on August 22, 2003, the VLS-1 V03 prototype rocket exploded on the launch pad at the Alcântara Launching Center (northern Brazil) only a few days before its planned launch date. The fourth prototype, VLS-1 V04, is still under testing, and an actual launch is not scheduled until 2012 or later.

Israel also has developed a commercial rocket, adapted in part from its ballistic missile program. Thus far, launches of the rocket, called the Shavit, have been limited to Israeli satellites. Efforts to base the Shavit at launch sites other than in Israel have not been successful. Its first launch, with an experimental payload, occurred on September 19, 1988. The first operational payload was launched successfully on April 5, 1995. Between September 1988 and June 2010, nine launches were attempted, of which five were successful. Launches all take place at Palmachim Air Base, located near the cities of Rishon LeZion and Yavne on the Mediterranean Sea.

Ukraine launched its own payload (Sich-1) with its own launch system (Tsyklon) on August 31, 1995. From 1991 to 2007, the country made 97 launches. Its launch vehicles include Tsyklon, Zenit, Dnepr, and Mayak. In addition, Sea Launch (which is partially owned by Ukrainian companies) uses the Ukrainian Zenit 3SL launch system for its launches from a mobile sea platform.

Pakistan and Indonesia have expressed interest in developing space-launching rockets, but neither has yet developed a final configuration for commercial sales. The Indonesian government made an announcement in 2000 that development of a commercial space booster was a major priority, but no reports have been seen as to the project's fate. In 2001, Pakistan announced that it was in the developmental stage of building the country's first Satellite Launch Vehicle (SLV). It hopes to launch its own launch vehicle by 2014.

North Korea claimed a test flight of a satellite launcher on August 31, 1998. Many Western observers believe, however, that the rocket is a ballistic missile and is not yet in a launch vehicle design configuration.

On April 5, 2009, North Korea claimed it launched its second satellite. This claim has yet to be confirmed by other space-faring countries.

Russia has several new designs of expendable rockets on the drawing boards, including a rocket called Angara that may replace the Proton for commercial sales. In 2001, the United States began fielding a new family of rockets called the evolved expendable launch vehicle (EELV). Both Boeing and Lockheed Martin were selling EELV versions, called the Delta IV and Atlas V, respectively. Both companies were awarded developmental contracts for their Boeing Delta IV rocket and Lockheed Martin Atlas V rocket.

With changes continuing to affect the space industries of the world, and new technologies in development, no one can predict the future direction of the space launch industry. The people and equipment seeking rides to space are as varied—and unpredictable—as the evolution of the rocket itself.

 See also **Augustine, Norman (Volume 1) • Emerging Space Businesses (Volume 1) • Global Industry (Volume 1) • Launch Facilities (Volume 4) • Launch Sites (Volume 3) • Launch Vehicles, Expendable (Volume 1) • Satellites, Types of (Volume 1) • Spaceports (Volume 1)**

Resources

Books and Articles

Cliff, Roger, Chad J. R. Ohlandt, and David Yang. *Ready for Takeoff: China's Advancing Aerospace Industry.* Santa Monica, CA: Rand National Security Research Division, 2011.

Commercial Space Transportation: Industry Trends and Key Issues Affecting Federal Oversight and International Competitiveness. [Washington, DC]: United States Government Accountability Office, 2011.

Hunley, J. D. *The Development of Propulsion Technology for U.S. Space-launch Vehicles, 1926-1991.* College Station, TX: Texas A & M University Press, 2007.

———. *U.S. Space-launch Vehicle Technology: Viking to Space Shuttle.* Gainesville, FL: University Press of Florida, 2008.

Kuczera, Herbert, and Peter W. Sacher. *Reusable Space Transportation Systems.* Chichester, UK: Springer, 2011.

National Research Council. *Revitalizing NASA's Suborbital Program: Advancing Science, Driving Innovation, and Developing Workforce.* Washington, D.C. National Academies Press 2010.

Websites

Encyclopedia of Science. <http://www.daviddarling.info/encyclopedia/ETEmain.html> (accessed October 9, 2011).

International Launch Services. <http://www.ilslaunch.com/> (accessed October 9, 2011).

Launch Vehicles Guide. <http://www.aerospaceguide.net/launchvehicles/index.html>(accessed October 9, 2011).

Satlaunch.Net - Satellite Launch Info. <http://www.satlaunch.net/> (accessed October 9, 2011).

Starsem. <http://www.starsem.com/> (accessed October 9, 2011).

Virgin Galactic. <http://www.virgingalactic.com/> (accessed October 9, 2011).

Launch Services

Commercial launch services are used to place satellites in their respective orbits. Launch services represent, by far, the most lucrative aspect of the launch-for-hire business. The European Space Agency (ESA), United States, Russian Federation (Russia), People's Republic of China (China), India, and other countries and some international organizations supply or plan to supply launch vehicles for the purpose of placing satellites in orbit. Although the majority of launch vehicles are used for broadcast satellites, satellites are also launched for wireless telephony. For example, the mobile communications systems Globalstar and Iridium Satellite LLC are satellite phone networks for global use. Today, the launch service industry is dominated by a handful of large companies, including Arianespace and Starsem in Europe, United Launch Alliance in the United States, and International Launch Services in Russia. In other parts of the world, nations are turning to governmental, quasi-governmental, and private companies to provide satellite launch services. Examples of this trend have been the creation of Earth2Orbit and the Antrix Corporation in India and the rise of small, innovative private companies such as SpaceX in the United States. This essay describes the major service launch providers (SLP) as of late 2011.

Europe

Arianespace is a French company established in 1980 for the purpose of placing commercial satellites into Earth orbit, the first company of its kind in the world. The company offers three rockets for use in its services, the Ariane 5, Soyuz 2, and Vega rockets. The three are designed, respectively, for heavy, medium, and light loads. As of late 2011, the Ariane 5 and Soyuz 2 launchers had been used successfully 214 times, with the first Vega launch planned for early 2012. The company is the leader in launching satellites into geostationary transfer orbit. The first stage of an Ariane 5 launch vehicle has a Vulcain liquid propellant engine that uses a fuel tank

A Delta rocket lifts off at Cape Kennedy carrying the world's first commercial communications satellite, *Intelsat 1* or Early Bird. The satellite, launched in 1965, was intended to operate as a switchboard relaying radio, television, teletype, and telephone messages between North America and Europe.
© *AP Images.*

that contains about 130 tons of liquid oxygen (LOX) and approximately 25 tons of liquid hydrogen (LH_2). Two solid rocket boosters (SRBs) are attached to the sides of the giant fuel tank, with each weighing about 277 tons when full. The second stage has one engine that uses one of two fuels. The Ariane 5G has the Aestus engine that uses monomethylhydrazine (MMH) and nitrogen tetroxide, while the Ariane 5 ECA has the HM7-B engine that uses liquid oxygen and liquid hydrogen. The heavy-lift Ariane 5 rocket is used mostly for commercial use but can be used for other purposes: the launching of scientific and military spacecraft, for example.

Space Sciences, 2ⁿᵈ Edition

Four versions have so far been built for the active Ariane 5 rocket: Ariane 5G, Ariane 5G+, Ariane 5 ECA (Evolution Cryotechnique type A), and Ariane 5 ES-ATV (Evolution Storable–Automated Transfer Vehicle). Ariane G and Ariane ES can send a payload into low-Earth orbit (LEO), while Ariane G, Ariane G+, Ariane 5 ECA, and Ariane ES-ATV are able to send payloads into geostationary transfer orbit (GTO). For instance, an Ariane ES-ATV launch vehicle placed the European Space Agency's ATV *Jules Verne* into LEO in March 2008 so that it could resupply the crew of the International Space Station.

A second major launch service provider in Europe is the French company Starsem. It is responsible for all commercial launches of the Soyuz rocket, which take place at Baikonur Cosmodrome. Starsem was created in 1996 as the commercial arm of Soyuz launches, as well as other launch vehicles and accessory equipment. The Soyuz rocket has been used to carry astronauts to the International Space Station—and also to carry space tourists to the space station, such as Dennis Tito (1940–) in April 2001, and Charles Simonyi (1948–) in 2007 and 2009. As of late 2011, the company's most recent missions have included two resupply flights to the International Space Station (ISS), one crew replacement flight to the ISS, and the deployment of two Globalstar communications satellites, the tenth flight for that company.

United States

The major LSP in the United States is United Launch Alliance, a company formed in December 2006 as the result of a merger of the launch services of Boeing and Lockheed Martin. The company uses three rocket launchers with long histories in the U.S. space industry, the Delta II, Delta IV, and Atlas V rockets. The Delta II rocket was first launched in 1989 and has a record of 148 successful launches in 150 attempts. The Delta IV rocket was first launched in 2003 and has had 16 successful launches out of 17 attempts. The Atlas 5 rocket was first launched in 2002 and has had 26 successful launches out of 27 attempts. The Delta rockets can have a number of configurations to provide the lifting power needed for payloads of various sizes. For example, three or six or nine solids may be attached to the first stage of the Delta II.

Two companies of growing importance in the United States are the Orbital Sciences Corporation and Space Exploration Technologies Corporation, better known as SpaceX. Orbital Sciences Corporation was founded in 1982 for the construction of launch vehicles and other space hardware. Since that time, it has built more than 570 launch vehicles and, as of late 2011, held about 60 percent of the small launch vehicle market worldwide. SpaceX was founded in 2002 to develop reusable space vehicles for commercial and other purposes. The company has developed two such vehicles, the Falcon 1 and Falcon 9 rockets, neither of which had, as of late 2011, yet flown commercial missions. However, the company appears to be successful so far, with the number of employees doubling every year since its founding. In 2011, the company was awarded the Space Foundation's Space Achievement Award.

Russia

The first launch service provider in Russia was established in February 1992 with the creation of a consortium of three existing companies, Lockheed Martin, from the United States, and the Khrunichev and Energia companies from Russia. That company was given the name of International Launch Services (ILS). It continues to be the primary LSP in Russia today. ILS relies primarily on the Proton rocket, one of the oldest and most successful of all Soviet/Russian rockets. As of late 2011, the Proton had been used successfully for more than 360 launches, of which 130 had taken place under the auspices of the ILS. With the formation of United Launch Alliance in 2006, Lockheed Martin continued its affiliation with ILS, limiting its participation to non-US launches.

Eurockot is a joint venture of the European Aeronautic Defence and Space Company (EADS) and the Russian company, Khrunichev State Research. The company uses a modified Russian SS-19 ICBM for its launches. The rocket is used to place payloads in low Earth orbit*. Since its first launch in 2000, Eurockot has performed 12 of 13 successful launches, the most recent of which have been the European Space Agency's Gravity Field and Steady-State Ocean Circulation Explorer (GOCE) satellite in March 2009, its SMOS/PROBA-2 (Soil Moisture and Ocean Salinity Satellite/PROBA2) satellite in November 2009, and the Japanese SERVIS-2 satellite in June 2010.

China

The People's Republic of China has developed a series of rockets, all called Chang Zheng (Long March). Unlike the situation in most other countries, launch services are provided by a governmental agency, the China Academy of Launch Vehicle Technology (CALVT), which itself is a subsidiary of China Aerospace Science and Technology Corporation (CASC). Although nominally a state agency, CALVT is to some extent autonomous in its daily activities. The company consists of some 27,000 employees located in 13 research facilities. The marketing arm of CASC is the China Great Wall Industry Corporation (CGWIC).

China's efforts to expand its launch services have been hampered by political differences with the United States. During the period from 1998 to early 2000, very few U.S.-built satellites were exported to China for launching because of changes in U.S. export controls for space hardware. Since then, the *Shenzhou* spacecraft, first launched on October 15, 2003, and the *Chang'e 1* lunar orbiter, launched on October 24, 2007, were both launched through China's own Long March rockets. With the future completion of the heavy-lift Long March 5 launch vehicle, the Chinese will have more ability to launch payload weight into space. The maximum capacity for a payload is around 25,000 kilograms (55,100 pounds

* **low Earth orbit** an orbit between 300 and 800 kilometers (185 and 500 miles) above Earth's surface

The European Infra-Red Space Observatory (ISO) was launched by the Ariane 44P from Kourou space center in French Guiana in 1995. The launch was the rocket's eightieth. © *AP Images/ESA*.

of equivalent weight), to low-Earth orbit, and 14,000 kilograms (30,800 pounds of equivalent weight), to geostationary transfer orbit. The first launch of the Long March 5 is expected in 2014. As of late 2011, CGWIC had experienced 136 launches, of which 129 were successful. All of the 35 commercial flights in that record had been successful, delivering satellites for a number of nations, including Australia, Brazil, France, Germany, Hong Kong, Japan, Nigeria, Pakistan, Sweden, and the United States.

Other International Organizations

A number of other launch service providers are also available worldwide. The Japanese government, for example, relies on Mitsubishi Heavy Industries for both the production of its launch vehicles and for the launches themselves. Its most popular rocket, the H-2A, has had 19 successful launches since its maiden voyage in 2001. All of its payloads have been Japanese spacecraft and satellites with the exception of one Australian satellite in 2002; a South Korean satellite is scheduled for launch in early 2012. The country of Ukraine together with a division of Boeing (Boeing Commercial Space), the Russian Energia Company, and the Norwegian company Kvaerner Maritime (a ship and oil rig builder) have formed Sea Launch. Sea Launch consists of an assembly and command ship (*Sea Launch Commander*), a semisubmersible launch platform called Ocean Odyssey (a former oil rig converted to a seagoing launching platform), a two-stage rocket called Zenit, and a third stage for the Zenit. The platform and the ship are based at Long Beach, California, and travel to longitude 154° west on the equator for launching. This procedure takes advantage of Earth's rotation speed, thus adding 465 meters per second (1,524 feet per second) at the equator for the Zenit 3SL, either reducing the amount of propellant needed or permitting an increase in payload weight. As of April 2009, Sea Launch had launched 31 rockets, with two failures and one partial failure.

Reusable Launch Vehicles

The only operational reusable rocket to date is the U.S. space shuttle, which ended in August 2011. The system that launched the shuttle was only partially reusable because the large external fuel tank holding liquid propellant was discarded after each launch, and the solid rocket boosters that were recovered from the ocean needed extensive refurbishment at considerable expense before their reuse. Reusability of rockets is a much desired but not a realized concept. There have been research studies, but nothing has been fully tested, leaving the commercial worth of reusable launch vehicles still in question for lifting payloads into space.

Currently, several companies are developing reusable launch vehicles to take paying customers on suborbital flights. These space tourism companies are hoping to make money on people's desire to go into space. The Scaled Composites' WhiteKnightTwo launch vehicle is the closest thing to a reusable rocket that humans have to date. Virgin Galactic is buying a fleet of WhiteKnightTwo launch vehicles and SpaceShipTwo spacecraft so it can send tourists on suborbital flights into space beginning in 2012. Both the launch vehicle and spacecraft return to Earth for repeat flights. Other companies competing in this new space tourism market with reusable launch vehicles are Armadillo Aerospace, Blue Origin, and Spaceplane Kistler.

 See also **Emerging Space Businesses (Volume 1)** • **Launch Facilities (Volume 4)** • **Launch Industry (Volume 1)** • **Launch Sites (Volume 3)** • **Launch Vehicles, Expendable (Volume 1)** • **Launch Vehicles, Reusable (Volume 1)** • **Reusable Launch Vehicles (Volume 4)** • **Rockets (Volume 3)** • **Spaceports (Volume 1)**

Resources

Books

Smith, Lesley Jane, and Ingo Baumann. *Contracting for Space: Contract Practice in the European Space Sector.* Farnham, Surrey, UK: Ashgate, 2011.

Websites

2011 Commercial Space Transportation Forecasts. Federal Aviation Administration (FAA). <http://www.faa.gov/about/office_org/headquarters_offices/ast/media/2011%20Forecast%20Report.pdf> (accessed October 13, 2011).

Arianespace. <http://www.arianespace.com/index/index.asp> (accessed October 13, 2011).

Eurockot. <http://www.eurockot.com/joomla/index.php> (accessed October 13, 2011).

International Launch Services. http://www.ilslaunch.com/ (accessed October 9, 2011).

United Launch Alliance. <http://www.ulalaunch.com/site/default.shtml> (accessed October 13, 2011).

Virgin Galactic. <http://www.virgingalactic.com/> (accessed October 13, 2011).

Launch Vehicles, Expendable

Expendable launch vehicles (ELVs) are rockets that carry satellites, people, and space probes but are not recovered or reused. These rockets are expendable, meaning that they are thrown away after their flights are completed. Expendable rockets can take many forms, but the most commonly used ones are powered by either liquid fuel or solid fuel and use multiple stages to propel their cargoes of spaceships, probes, or satellites into outer space.

Each of the rocket's stages consists of a self-contained rocket engine or motor and the fuel such as hydrogen, kerosene, or a solid fuel that looks a lot like the eraser on a pencil. Along with the engine and fuel are tanks to hold the materials, lines and pumps, and electrical systems to move the engine while in flight. Once the fuel in the stage has been used up, the stage is usually dropped away and the next stage ignited. These rockets continue to burn stage by stage until the correct altitude or speed for its designated space mission has been reached.

The Soviet Soyuz spacecraft as photographed from the American Apollo spacecraft in Earth orbit in July 1975. The three major components of the Soyuz craft are easily visible: the Orbital Module (sphere-shaped), the Descent Vehicle (bell-shaped), and the Instrument Assembly Module (cylinder-shaped). © *AP Images/NASA.*

* **orbit** the circular or elliptical path of an object around a much larger object, governed by the gravitational field of the larger object

Evolution from Military Missiles

Expendable rockets are used today by Brazil, China, the European Space Agency, France, India, Iran, Israel, Italy, Japan, North Korea, South Korea, Russia, Ukraine, the United Kingdom, and the United States to place satellites into Earth orbit* and toward the Moon and planets. Most of the rockets now in use as launching vehicles evolved from missiles developed during military conflicts such as World War II (1939–1945). Beginning with a German missile called the V-2, these weapons were created to carry large high-energy explosives to hit cities or troop encampments. Later, after World War II ended, larger and more powerful missiles were created to carry nuclear weapons to targets on the other side of Earth.

Two nations, the United States and the Soviet Union, were the first to develop these large bomb-carrying missiles. After World War II ended, the German scientists who designed the V-2 missile fled Germany. Some of them surrendered to the United States, while others escaped to the Soviets. Each group of German scientists, engineers, and technicians sought to continue the development of rocketry and missiles in their new countries. In the Soviet Union, missile development was made a top priority by the Communist government. Part of the reason for the emphasis on missiles was the Soviet Union's lack of jet-powered bombers that could carry atomic bombs from Russia to targets in the United States. If large missiles could be built successfully, they could carry the bombs to their targets.

* **ballistic** the path of an object in unpowered flight; the path of a spacecraft after the engines have shut down

* **payload** any cargo launched aboard a rocket that is destined for space, including communications satellites or modules, supplies, equipment, and astronauts; does not include the vehicle used to move the cargo or the propellant that powers the vehicle

In the United States, Wernher von Braun (1912–1977) headed the group of German rocket and missile experts. Von Braun was the head of the German missile program and was considered the most advanced expert on rocketry designs during and after the war. In the Soviet Union, his counterpart was Sergei Korolev (1906–1966), who was designated by the Soviet government as the "Chief Designer" of human-carrying and large expendable rockets.

Korolev became the head of a large central design bureau, and his "customers" were the specific missions assigned to his bureau. These included the design, manufacture, and test of the first Soviet intercontinental ballistic* missile, the R-7; the first-and second-generation human-carrying space capsules, called Vostok and Soyuz; and a series of larger and more advanced liquid-powered expendable rockets, called Proton and N-1. These latter rockets were to be used in the Soviet lunar-landing program. The original purpose for the Proton, however, was as a very large missile that could fly from Russian bases and attack targets in the United States. The missile version of Proton was never developed, and instead it became a launching rocket for heavy payloads* and space probes to the planets. Proton was also designated as the carrier rocket for the Soviet piloted lunar space capsule, called Zond.

In the United States of America, von Braun developed a series of liquid-fueled rockets called Juno and a large army missile called Redstone. These rockets were adapted for scientific space missions by von Braun's team working at the Redstone Arsenal in Huntsville, Alabama. By emplacing a small basket of solid rockets on the nose of the Juno II, von Braun was able to insert a small U.S. satellite into Earth orbit on January 31, 1958, marking the first U.S. artificial Earth satellite. This was after Korolev had done the same with the R-7, having launched the Sputnik satellite three months earlier, on October 4, 1957.

Improvements to U.S. and Soviet Rockets

Increasingly, both the United States and the Soviet Union made improvements to their missiles that made them capable satellite and space capsule launchers. The U.S. equivalent to the Soviet R-7 was an intercontinental ballistic missile named Atlas. Developed initially for the U.S. Air Force to carry atomic warheads to targets in the Soviet Union, the Air Force and the newly created National Aeronautics and Space Administration (NASA) modified the missile's design to replace its bomb-carrying nosecone. In the nosecone's place, with reinforced nose sections, the Atlas could carry an additional liquid-fueled rocket stage that could send payloads—satellites, capsules, or space probes—into orbit or to the Moon or Mars.

In 1959, the Atlas missiles were modified to carry Project Mercury one-person space capsules, just as Korolev had done with the R-7 and its Vostok and Soyuz capsules. Eventually the Atlas and R-7 each received more powerful engines and larger upper stages. While the Mercury and

A Delta IV Heavy rocket belonging to United Launch Alliance blasts off on November 10, 2010. It is carrying a National Reconassiance Office satellite into orbit. © *AP Images/PRNewsFoto/ Pratt & Whitney Rocketdyne/Pat Corke.*

Vostok projects have long since ended, both the Atlas and R-7 rockets are still in service, using advanced subsystems and powerful upper stages. Both are being sold today on the commercial space launch market, competing with each other for commercial sales.

The Atlas and R-7 were not the only throwaway rockets to evolve during the Cold War. The United States took a smaller and more limited intermediate range rocket called the Thor and adapted it for launching scientific space probes, beginning in 1960. Eventually the Thor grew to become, under the name Delta, one of the most reliable space launchers in history. In its Delta II and IV variants, it is still in government, military, and commercial use today.

Von Braun also developed the only U.S. throwaway rocket that was created from scratch and not evolved from missiles. From October 1961 until May 1973, NASA used three versions of a rocket called Saturn to support the man-on-the-Moon program called Project Apollo. The larger of these rockets, the Saturn V, sent Apollo astronauts to the surface of the Moon from 1969 to 1972 and lifted America's first space station, *Skylab,* in May 1973. *Skylab* itself was developed from the upper stage of the Saturn V rocket.

Briefly, the Soviet Union developed an expendable rocket called Energyia that was not developed from a missile. It flew in 1988 and 1989, in one flight carrying the unpiloted *Buran* space shuttle. The collapsing economic situation in Russia forced the abandonment of the Energyia program after those two flights.

The Evolution of Other Nations' Expendable Rockets

The launch vehicles used by China also evolved from ballistic missile designs. In the past, the country has flown Feng Bao and Kaituozhe rockets but, today, are dedicated to Long March rockets. However, the expendable rockets developed and flown by Japan (Lambda, Mu, N, H, where HII-A is active), Brazil (Satellite Launch Vehicle, VLS), and India (Polar Satellite Launch Vehicle, PSLV, and Geosynchronous Satellite Launch Vehicle, GSLV) are all new designs that had no direct missile ancestor, although all were strongly influenced by missile systems in use at the time. Israel's expendable rocket, a small booster named Shavit, is thought to have evolved from that nation's Jericho missile program. Iran flies the Safir launch vehicle, while North Korea launches the Baekdusan and South Korea flies the Korea Space Launch Vehicle (KSLV, now called the Naro). The Ukraine continues to launch its Zenit launch system, while the United Kingdom retired its Black Arrow rocket in the 1970s and currently uses foreign made rockets to launch its payloads into space, mostly through the European Space Agency.

The European Space Agency's (ESA's) fleet of commercial Ariane rockets, derived from the Europa rocket, were also created entirely apart from any missile project. Since 1979, Ariane rockets have been launching commercial satellites for customers worldwide, French military and government payloads, and payloads for the European Space Agency and the French Space Agency, CNES.

Today, expendable rockets are the mainstay of civil, military, and commercial satellite launches. The R-7 is still flying as the booster that lifts the Soyuz piloted space capsules. A commercial version is also for sale, flying from the same launching pad where the first satellite, *Sputnik 1,* and the first human, aboard *Vostok 1,* were launched in 1957 and 1961, respectively. The U.S. Atlas and Delta rockets are expected to be flying well into the twenty-first century, as are the Chinese missile-derived space boosters. The era of the expendable rocket may prove to be a long one in the evolving history of the space age.

 See also **Emerging Space Businesses (Volume 1) • Launch Facilities (Volume 4) • Launch Vehicles, Reusable (Volume 1) • Reusable Launch Vehicles (Volume 4) • Rockets (Volume 3) • Spaceports (Volume 1) • von Braun, Wernher (Volume 3)**

Resources

Books and Articles

Baker, David. *The Rocket: The History of Rocket and Missile Technology.* New York: Crown Publishers, 1978.

Darrin, Ann Garrison, Beth Laura O'Leary, eds. *Handbook of Space Engineering, Archaeology, and Heritage.* Boca Raton: CRC Press, 2009.

Lennick, Michael *Launch Vehicles: Heritage of the Space Race.* Burlington, Ontario: Apogee Books, 2006.

Ley, Willy. *Rockets, Missiles, and Men in Space.* New York: Viking Press, 1967.

Websites

Arianespace. <http://www.arianespace.com/index/index.asp> (accessed November 15, 2011).

China National Space Agency. <http://www.cnsa.gov.cn/n615709/cindex.html> (accessed November 15, 2011).

European Space Agency. <http://www.esa.int/esaCP/index.html> (accessed November 15, 2011).

Indian Space Research Organization. <http://www.isro.org/> (accessed November 15, 2011).

Israeli Space Agency. <http://www.most.gov.il/English/Units/Israel+Space+Agency/default.htm> (accessed November 15, 2011).

Japan Aerospace Exploration Agency. <http://www.jaxa.jp/index_e.html> (accessed November 15, 2011).

National Aeronautics and Space Administration. <http://www.nasa.gov/> (accessed November 15, 2011).

Russian Federal Space Agency. <http://www.federalspace.ru/?lang=en > (accessed November 15, 2011).

* **payload** any cargo launched aboard a rocket that is destined for space, including communications satellites or modules, supplies, equipment, and astronauts; does not include the vehicle used to move the cargo or the propellant that powers the vehicle

Launch Vehicles, Reusable

A reusable launch vehicle (RLV) is a craft designed to place payloads* or crews into Earth orbit, and then return to Earth for subsequent launches. RLVs are designed to reduce launch costs by reusing the most expensive components of the vehicle rather than discarding them and building new ones for each mission (as is the case with expendable launch vehicles, known as ELVs). The definition of RLVs does not include reusable craft launched from expendable launch vehicles. As of 2011, the only operational RLV has been the U.S. space shuttle and its RLV status is classified as only partially reusable because its external fuel tank was thrown away while its solid rocket boosters were refurbished after each flight. As of August 31, 2011, the space shuttle fleet has been retired by the U.S. space agency National Aeronautics and Space Administration (NASA). However, a number of concepts are being studied or developed. Some are partially reusable, while others are totally reusable. Most employ rockets, while others used jet engines or aircraft.

RLVs may be categorized by whether the vehicle takes off horizontally or vertically and whether it lands horizontally or vertically; for instance, a VTHL vehicle is one that has a vertical takeoff, horizontal landing. An RLV may also be described as single-stage-to-orbit (SSTO) or two-stage-to-orbit (TSTO). Vehicles such as the space shuttle, which took off vertically using a two-stage system and landed horizontally, have the easiest design because horizontal takeoff involves more demanding flight loads,

The aim of the X-34 program was the development of low-cost space access through reusable launch vehicles. *NASA.*

vertical landing requires the craft to carry enough propellants to land, and SSTO requires a higher ratio of propellant to vehicle weight. Nevertheless, the economics of preparing a single stage, rather than two stages, have kept space engineers interested in SSTO designs. Future RLVs also are expected to employ more advanced, reliable systems, making them safer than expendable launch vehicles, and thus allow launches from inland sites (i.e., no stages to splash down into the ocean), perhaps even airports, where weather is less of a concern than at coastal spaceports.

Early Concepts

Because it was easier to adapt existing military missiles, which are designed for a single flight, most launchers have been expendable. Nevertheless, space visionaries have often focused on RLVs. One of the most significant early concepts was a three-stage vehicle designed by German-born American engineer Wernher von Braun (1912–1977) in 1952, and popularized in his book *Across the Space Frontier*. The first two stages would parachute into the ocean for recovery while the winged third stage, carrying crew and cargo, landed like an airplane. In addition, the 1951 movie *When Worlds Collide* depicted a rocket-powered sled that gives a vehicle its initial boost.

Several RLV concepts were advanced in the 1960s. Notable among these was a reusable design by U.S. engineer Philip Bono (1921–1993), then working with the Douglas Aircraft Company. His innovative design comprised a core vehicle holding a payload bay, liquid oxygen tank, and a ring of small rocket engines around the base. Liquid hydrogen was carried in external tanks that could be hinged outward to enhance atmospheric control during entry. This unique engine arrangement followed the *aerospike* concept developed by Rocketdyne. In this approach, the pressure from the shock wave produced by the vehicle's high-speed ascent becomes the outer wall of the engine nozzle from which the exhaust streams. The resulting exhaust appears to be a spike of hot gas, thus leading to the nickname "aerospike."

In the late 1960s, the U.S. aerospace industry offered a number of reusable designs as the National Aeronautics and Space Administration (NASA) sought ways to reduce the cost of space launches. U.S. aerospace engineer Maxwell Hunter (1922–2001), then with Lockheed Missiles & Space Company, proposed in 1984 a wedge-shaped reusable vehicle, called the X-rocket, with main engines in its tail, and a large external tank that was shaped like an inverted V and was wrapped around the nose of the vehicle. The tank would be discarded after its propellants had been consumed, leaving the main body to return to Earth. Hunter's concept, later renamed Space Ship eXperimental (SSX), eventually led to the Delta Clipper (DC-X), which was developed by the McDonnell Douglas Corporation for the U.S. Department of Defense (DoD). The experimental craft was tested between 1993 and 1995 by the DoD, and later taken over by NASA in 1996. The privately funded company Blue

Origin, with its experimental spacecraft New Shepard, is now using the DC-X design. Amazon.com founder Jeff Bezos is funding the project at a Culberson County, Texas space facility. The first successful test flight of its spacecraft *Goddard* occurred on November 13, 2006. The company hopes to fly an unmanned suborbital New Shepard vehicle in 2012. However, on August 24, 2011, the company lost its test vehicle during the fifth test flight when instabilities within the craft prompted ground personnel to destroy the vehicle.

Space Shuttle

Following the *Apollo 11* Moon landing in 1969, NASA proposed a space program that would provide the basic building blocks in support of a wide range of human space missions: a space shuttle, a space station, a space tug, and a nuclear interplanetary stage. In this plan, the space shuttle was a fully reusable vehicle. The booster would fly back to the launch site after launch, and the orbiter at the end of the mission; both would be quickly prepared for the next mission. NASA soon realized that such a massive craft would cost more than it could afford. A series of redesign efforts traded the high development cost and low per-flight cost of the original design for a lower development cost and a higher per-flight cost. Dozens of variations were studied before arriving at the final design. One interesting variation employed two piloted fly-back boosters and a piloted orbiter outwardly identical to the boosters. The concept was to reduce costs by designing one airframe for two purposes. This design meant, however, that three vehicles had to be prepared for each launch. The Soviet Union largely copied the final space shuttle design for its *Buran* shuttle, which flew only once.

Advanced RLVs

Even before the shuttle started flying, designers continued to look at advanced reusable concepts, such as North American Rockwell's immense winged Star Raker, which was envisioned as taking off and landing like a jetliner. The SSX (Space Ship eXperimental) proposed by Hunter in 1984 was based on an earlier design of a passenger vehicle. Hunter's efforts helped lead to a DoD project that opened the current reusable era. The DoD's purpose was to design a single-stage vehicle that could orbit military replacement satellites during a national emergency.

McDonnell Douglas Corporation was contracted to build and test fly the DC-X, a one-third-scale suborbital model of the Delta Clipper, a larger version that would launch satellites on short notice. While not capable of spaceflight, the DC-X incorporated many of the technologies needed for an SSTO vehicle, including highly automated systems enabling a quick turnaround (just twenty-six hours) between launches. It made eight successful test flights between August 18, 1993, and July 7, 1995, and then was taken over by NASA and flown four times as the DC-XA between May 18 and July 31, 1996. It was destroyed on its last

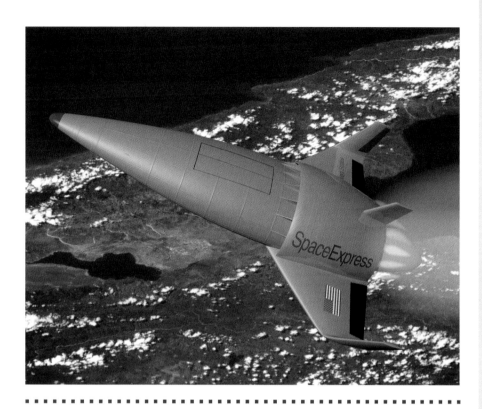

NASA and several private ventures have initiated the development of reusable launch vehicles, such as this artist's concept of an express rocket, for the promotion of space tourism. *NASA.*

flight when one landing strut failed to deploy and the vehicle tipped over at landing.

The DC-X led NASA to a broader launch vehicle technology program to reduce the cost of putting a payload in space from $22,000 per kilogram of mass ($10,000 per pound of weight on Earth) to $2,200 per kilogram ($1,000 per pound of weight) or less. The principal programs were the Lockheed Martin X-33 and the Orbital Sciences Corporation X-34. The X-33 was a one-third-scale test model of the Lockheed Martin concept for VentureStar an automated vehicle capable of launching up to 18,650 kilograms (50,000 equivalent pounds, in weight). In operation, VentureStar would launch, orbit, and land much as the shuttle did, but without discarding boosters or tanks. Other major differences include systems that can be readied for re-flight with less maintenance (or no maintenance) than the shuttle required. Significant structural and other problems raised the cost of the X-33 project. In 2001, NASA canceled the project. Also canceled was the X-34, a demonstration vehicle built largely from commercially available parts. The vehicle, if it had been made operational, would have been launched from a jumbo jet, flown to an altitude of 76,200 meters (250,000 feet), and then glided to Earth for landing. It, too, encountered severe technical problems and was canceled by NASA in 2001.

In place of the X-33 and X-34 programs, NASA initiated the Space Launch Initiative (SLI) program to study more conventional two- or three-stage-to-orbit second-generation RLV, possibly using the aerospike engine concept, which looked promising in the X-33 project.

The important underlying features would be new electronics and materials that would allow automated preparation and checkout of vehicles and more rapid launches, and highly automated manufacturing processes. Goals include reducing the risk of crew loss to once per 10,000 missions, and the cost of launches to less than $1,000 per pound of payload in orbit. The SLI program ended essentially at the conclusion of the X-43 program in 2004, after several successful tests of the unmanned hypersonic aircraft. In 2006, the U.S. Air Force announced the Boeing X-51 scramjet vehicle. The first "free-flight" test flight for the X-51, part of the Waverider program, occurred on May 26, 2010 when it achieved its longest flight time to date (over 200 seconds) at speeds exceeding Mach 5, or five times the speed of sound. The second test flight occurred on June 13, 2011. However, the flight was terminated earlier than expected when a fuel problem occurred.

Beyond the second-generation RLV, NASA is looking at advanced space transportation concepts that could realize the earlier dreams of combining jet rocket combustion cycles in a single power plant, use electromagnetic railways as an Earthbound booster stage, or even laser-and microwave-powered craft. However, as of October 2011, a Single Stage to Orbit manned vehicle has yet to prove its ability to fly regularly back and forth between outer space and Earth.

In addition to NASA's efforts, several private ventures have initiated activities to develop RLVs for business, including space tourism. Many have stalled or failed for lack of financial backing. However, one in particular seems to be on the brink of successfully bringing space tourism to a large number of people. Using spacecraft developed by Scaled Composites, Virgin Galactic, headed by British entrepreneur Richard Branson, is expecting to begin sub-orbital flights in or around 2013. The primarily reusable spacecraft, *SpaceShipTwo* vehicle, is a series of manned spacecraft that will be launched with the help of one of a series of manned *WhiteKnightTwo* launching vehicles. Scaled Composites, founded by Burt Rutan, builds both vehicles.

The Spaceship Company, a joint venture of Virgin Galactic and Scaled Composites, owns the technology involved in the project. Virgin Galactic will have the *WhiteKnightTwo* take *SpaceShipTwo* from launch to about 18 kilometers (11 miles) above Earth. At that time, *SpaceShipTwo* will continue under its own power while the carrier vehicle returns to Earth. The spacecraft then completes the trip to space, about 100 kilometers (62 miles) above Earth. Test flights have taken place in 2011 with more scheduled for 2012. The first commercial flight is scheduled for sometime in 2013.

Other companies developing private flights into space with the use of reusable launch vehicles include Space Exploration Technologies Corporation (SpaceX, developing partially reusable Falcon 1, Falcon 9, and Falcon Heavy launch vehicles), Blue Origin (New Shepard), and Masten Space Systems (Xaero). For instance, a SpaceX Falcon 9 launch vehicle launched from the Cape Canaveral Air Force Station on June 4,

2010, for its first successful orbital insertion. Then, the second launch of a Falcon 9 rocket with a Dragon space capsule as payload—the first time this has been done by a private company—was accomplished on December 8, 2010. A two-orbit flight occurred of the Dragon spacecraft, which then splashed down in the Pacific Ocean. In September 2011, Elon Musk, the founder of SpaceX, announced that the company was designing a Falcon 9 launch vehicle in which its boosters would fly back to Earth, making the launch system a completely reusable launch vehicle.

Blue Origin is working on its VTVL New Shepard spacecraft, which (as mentioned earlier) is based on the technology for the McDonnell Douglas DC-X. It hopes to fly its spacecraft unmanned sometime in 2012, and manned soon thereafter. Further, Masten Space Systems, headquartered in Mojave, California, is developing a VTVL spacecraft for unmanned suborbital flights. It currently is developing its Xaero reusable VTVL launch vehicle as part of NASA's Flight Opportunities Program.

XCOR Aerospace is developing its Lynx suborbital spaceplane that will be the approximate size of a private airplane. Carrying a pilot and one passenger, it would take off and land from a conventional runway, and be capable of flying into space several times per day. XCOR envisions its Lynx, and its next-generation Mark II, to provide transportation for various space tourism companies. XCOR hopes to have its Lynx operational by 2012 or later.

In the United Kingdom, Reaction Engines Limited is developing the single-stage-to-orbit Skylon, an unmanned spaceplane with a hybrid air- and liquid oxygen-breathing rocket motor. The craft uses an air-breathing rocket engine to launch from a conventional runway and fly through the atmosphere, where it would then switch the engine to use liquid oxygen to take it into orbit about Earth. Once in orbit, the spaceplane could deploy its payload or perform another mission objective, and eventually return to a runway back on Earth to be used again.

Reusable commercial launch vehicles are being developed in the United States to such an extent that the Federal Aviation Administration publicly issued new regulations in 2006 regarding the maintenance and operations of commercial reusable launch vehicles (in the United States and its territories) for suborbital and orbital flights.

 See also **Emerging Space Businesses (Volume 1)** • **Launch Services (Volume 1)** • **Launch Vehicles, Expendable (Volume 1)** • **Reusable Launch Vehicles (Volume 4)** • **Rockets (Volume 3)** • **Spaceports (Volume 1)** • **von Braun, Wernher (Volume 3)** • **X Prize (Volume 1)**

Resources

Books and Articles

Bono, Philip, and Kenneth Gatland. *Frontiers of Space.* New York: Macmillan, 1969.

Dick, Steven J. *America in Space: NASA's First Fifty Years.* New York: Abrams, 2007.

Jenkins, Dennis R. *Space Shuttle: The History of Developing the National Space Transportation System.* Cape Canaveral, FL: Motorbooks, 1997.

McCurdy, Howard E. *Space and the American Imagination.* Baltimore: Johns Hopkins University Press, 2011.

Neal, Valerie, Cathleen S. Lewis, and Frank H. Winter. *Spaceflight: A Smithsonian Guide.* New York: Prentice Hall Macmillan, 1995.

Parker, Martin, and David Bell, editors. *Space Travel and Culture: From Apollo to Space Tourism.* Malden, MA: Wiley-Blackwell/Sociological Review, 2009.

Sparks, James C. *Winged Rocketry.* New York: Dodd Mead, 1968.

von Braun, Wernher, Frederick I. Ordway III, and Dave Dooling. *Space Travel: A History.* New York: Harper Collins, 1985.

Wilson, Andrew. *Space Shuttle Story.* London: Deans International Publishing, 1986.

Websites

Bergin, Chris. *X-33/VentureStar—What really happened.* NASAspaceflight.com. <http://www.nasaspaceflight.com/2006/01/x-33venturestar-what-really-happened/> (accessed October 26, 2011).

Blue Origin. <http://www.blueorigin.com/> (accessed October 3, 2011).

Guide to Commercial Reusable Launch Vehicle Operations and Maintenance. Federal Aviation Administration. <http://www.faa.gov/about/office_org/headquarters_offices/ast/licenses_permits/media/RLV_OM_Guidelines_revD_032905_final.pdf> (accessed October 26, 2011).

Masten Space Systems. <http://masten-space.com/> (accessed October 7, 2011).

Rose, Charlie. *A conversation with Amazon.com CEO Jeff Bezos.* CharlieRose.com. <http://www.charlierose.com/view/interview/8784> (accessed October 26, 2011).

Scaled Composites. <http://www.scaled.com/> (accessed October 7, 2011).

The Space Launch Initiative: Technology to Pioneer the Space Frontier. Marshall Space Flight Center, National Aeronautics and Space Administration. <http://www.nasa.gov/centers/marshall/news/background/facts/slifactstext02.html> (accessed October 26, 2011).

SpaceX Working on Reusable Falcon 9 with Fly-back Boosters. SpaceNews. com. <http://www.spacenews.com/launch/110930-spacex-falcon-flyback-boosters.html> (accessed October 26, 2011).

Star-Raker. Astronautix.com. <http://www.astronautix.com/lvs/staraker. htm> (accessed October 26, 2011).

Thisdell, Dan. *Spaceplane engine tests under way.* FlightGlobal.com. <http://www.flightglobal.com/news/articles/spaceplane-engine-tests-under-way-361501/> (accessed October 26, 2011).

Virgin Galactic. <http://www.virgingalactic.com/> (accessed October 26, 2011).

Virgin Galactic Unveils Suborbital Spaceliner Design. Space.com. <http://www.space.com/news/080123-virgingalactic-ss2-design.html> (accessed October 26, 2011).

XCOR Aerospace. <http://www.xcor.com/> (accessed October 3, 2011).

X-33 VentureStar. Federation of American Scientists. <http://www.fas.org/programs/ssp/man/uswpns/air/xplanes/x33.html> (accessed October 26, 2011).

Law of Space

The law of space is the field of the law that relates to outer space or outer space-related activities conducted by governments, international organizations, and private individuals. International space law governs the activities of states and international organizations, whereas domestic, or national, space law governs the activities of individual countries and their citizens. Both areas of space law govern the activities of private persons and businesses.

International Space Law

Soon after the launch of the Soviet satellite *Sputnik 1* in 1957, the United Nations became active in the creation, development, and implementation of a system for studying the legal problems that may result from the exploration and use of outer space. Since the establishment of the United Nations Committee on the Peaceful Uses of Outer Space in 1958, five major multilateral treaties and conventions have been adopted or ratified by many countries to establish the framework to address concerns about outer space matters:

1. The Treaty on Principles Governing the Activities of States in the Exploration and Use of Outer Space, including the Moon and Other Celestial Bodies (1967), commonly known as the Outer Space Treaty, established basic principles regarding outer space and its exploration and use, including the conduct of activities pursuant to international law;

* **orbit** the circular or elliptical path of an object around a much larger object, governed by the gravitational field of the larger object

2. The Agreement on the Rescue of Astronauts, the Return of Astronauts and the Return of Objects Launched into Outer Space (1968), commonly known as the Rescue Agreement, safeguarded the prompt return of astronauts to the host country;

3. The Convention on International Liability for Damage Caused by Space Objects (1972), commonly known as the Liability Convention, established an international legal regime to assess liability and compensation for damage, injury, or death resulting from space activities and objects;

4. The Convention on Registration of Objects Launched into Outer Space (1976), commonly known as the Registration Convention, provides for a centralized registry of space objects maintained by the Secretary General of the United Nations; and

5. The Agreement Governing the Activities of States on the Moon and Other Celestial Bodies (1979), commonly known as the Moon Agreement, declares the Moon to be the "common heritage of all mankind." It has not been ratified by states that have or are likely to develop the ability to orbit* around or land on the Moon.

These treaties provide the framework for international space law. Their principles and provisions relate to the exploration and use of outer space. They are binding upon the countries that have ratified them. Together, these principles and provisions guide countries that have not ratified the international treaties. Space law includes international arms control treaties that prohibit or restrict the deployment or use of certain rocket and missile weapon systems. Space law also includes agreements, treaties, conventions, rules and regulations of nations and international organizations, executive and administrative orders, and judicial decisions.

As of November 1, 2011, 100 nations have ratified the Outer Space Treaty (the largest number of ratifying counties of the five treaties) while another 26 countries have signed it but not yet ratified it. As of that same date, 92 countries have ratified the Rescue Agreement, while 24 countries have signed it and two international intergovernmental organizations (the European Space Agency [ESA] and the European Organization for the Exploration of Meteorological Satellites [EOEMS]) declared their acceptance of the rights and obligations provided for in this Convention. On that same date, 90 countries have ratified the Liability Convention, while another 23 have signed it and the ESA, the EOEMS, and the European Telecommunications Satellite Organization (ETSO) have declared their acceptance.

As of November 1, 2011, 55 countries have ratified the Registration Convention, four have signed it, and the ESA and EOEMS have declared their acceptance. As of that same date, thirteen nations have ratified the Moon Agreement, while another four countries have signed it, but not ratified it.

The International Telecommunications Union (ITU), an agency of the United Nations, is the world's regulatory body for the coordination and

regulation of the radio frequency spectrum and the space-based and land-based facilities that provide global telecommunications services. Nearly every country in the world is a member of the ITU. The launching of commercial satellites, particularly those that provide telecommunications services, is the most common space activity. The ITU's management of the orbital positions of these satellites and prevention of harmful interference caused by radio frequency transmissions are accomplished by the adherence of the member countries to the provisions of the International Telecommunication Convention (1973).

In the United States the Federal Communications Commission is the regulatory authority for private, commercial, and state and local government use of the radio frequency spectrum. The commission controls the licensing of space satellites and ground Earth stations and regulates their use of radio frequencies* to ensure that telecommunications services are free from interference by other transmissions.

Domestic Space Law

Launches of commercial satellites for telecommunications, remote sensing*, and global positioning systems* are governed by the regulatory regimes of the countries that conduct those launches. As the launching

The United Nations has developed international treaties that act as space law for most countries. © *Joseph Sohm, ChromoSohm Inc./Corbis.*

* **frequencies** the number of oscillations or vibrations per second of an electromagnetic wave or any wave

* **remote sensing** the act of observing from orbit what may be seen or sensed below on Earth or another planetary body

* **global positioning systems** a system of satellites and receivers that provide direct determination of the geographical location of the receiver

state, a country that has launched a rocket or missile is liable under the international treaties for any damage caused by the launch activity to third parties. Each launching state's domestic regulations also address the responsibilities and liabilities of the launching entity.

The conduct of space launches from U.S. territory or by U.S. citizens or businesses from anywhere outside the national territory is governed by the Commercial Space Launch Act of 1984. This act supplies the legal framework for the relationship between the government and the commercial space launch industry. The act directs the secretary of transportation to regulate space transportation and streamline the licensing process for commercial space activities pursuant to international treaty obligations and national policy. The Office of the Associate Administrator implements this statute for Commercial Space Transportation of the Federal Aviation Administration. Commercial launch providers are required by the act to obtain insurance for death, injury, or property damage suffered by third parties. If an accident occurred during the launch activity, the U.S. government would pay any international claims in excess of the insurance level.

Other laws of the United States apply to the operation of private remote sensing satellites (the Land Remote Sensing Commercialization Act of 1984); the encouragement of commercial activity on the space station and government purchases of commercially produced scientific data (the Commercial Space Act of 1998); the protection of intellectual property and the patenting of inventions made, used, sold, or practiced in space (Public Law 101–580 on Inventions in Outer Space, 1990); and attempts to minimize the amount of orbital debris (the National Aeronautics and Space Administration Authorization Act, 1994 and 1995).

As the space industry has expanded from a niche market benefiting a limited high-tech scientific and military base to a global concern affecting geographically dispersed industries and consumers, the scope of space law has grown. From the initial steps of establishing the principles for the exploration and use of outer space by nations and governments, space law in the twenty-first century encompasses rational and reasonable approaches to representing the demands of persons in virtually every part of the world for enhanced communications, education, entertainment, environmental, and transportation services. As space commerce grows, space law will continue to address the unique problems posed by commercial activities in space.

▶ See also **Careers in Space Law (Volume 1) • Law (Volume 4) • Legislative Environment (Volume 1) • Regulation (Volume 1) • Spaceports (Volume 1)**

Resources

Books and Articles

Bhat, B. Sandeepa. *Space Law in the Era of Commercialisation.* New Delhi: Eastern Book Company, 2010.

Jakhu, Ram. *International Space Law.* Dordrecht; London: Springer, 2011.

Jakhu, Ram, Tommaso Sgobba, and Paul Stephen Dempsey. *Do We Need an International Regulatory Framework for Space?* Vienna; London: Springer, 2011.

Mardon, Austin A., et al. *Space Rescue Systems in the Context of International Laws.* Edmonton: Golden Meteorite Press, 2011.

Pelton, Joseph N., and Ram S. Jakhu. *Space Safety Regulations and Standards.* Oxford, UK; Burlington, MA: Butterworth-Heinemann, 2010.

Websites

Agreement Governing the Activities of States on the Moon and Other Celestial Bodies. Office of Outer Space Affairs, United Nations. <http://www.unoosa.org/oosa/SpaceLaw/moon.html> (accessed October 23, 2011).

Agreement on the Rescue of Astronauts, the Return of Astronauts and the Return of Objects Launched into Outer Space. Office of Outer Space Affairs, United Nations. <http://www.unoosa.org/oosa/SpaceLaw/rescue.html> (accessed October 23, 2011).

Convention on International Liability for Damage Caused by Space Objects. Office of Outer Space Affairs, United Nations. <http://www.unoosa.org/oosa/SpaceLaw/liability.html> (accessed October 23, 2011).

Convention on Registration of Objects Launched into Outer Space. Office of Outer Space Affairs, United Nations. <http://www.unoosa.org/oosa/SORegister/regist.html> (accessed October 23, 2011).

Treaty on Principles Governing the Activities of States in the Exploration and Use of Outer Space, including the Moon and Other Celestial Bodies. Office of Outer Space Affairs, United Nations. <http://www.oosa.unvienna.org/oosa/SpaceLaw/outerspt.html> (accessed October 23, 2011).

Legislative Environment

When Congress created the National Aeronautics and Space Administration (NASA) from the National Aeronautics and Space Act on July 29, 1958, it did not consider the need to commercialize space. At the time, the United States was in a conflict and competition with the Soviet Union called the Cold War. The purpose of the new space agency was to bring

geostationary orbit a specific altitude of an equatorial orbit where the time required to circle the planet matches the time it takes the planet to rotate on its axis. An object in geostationary orbit will always remain over the same geographic location on the equator of the planet it orbits

together America's many space programs to compete with the Soviet Union, which had launched the first satellite into orbit.

NASA's missions, as defined by Congress, were to expand human knowledge of Earth and space, to preserve America's role as a leader in "aeronautical and space science and technology," and to cooperate with other nations in the pursuit of these goals. Not until 1984 did Congress add the requirement for NASA to "seek and encourage to the maximum extent possible the fullest commercial use of space."

The Legal Framework to Support Commercial Space Activities

The legal framework to support commercial space enterprises gradually evolved in response to the development of space technologies and the maturing of space industries. The U.S. Defense Department set the first major space business into motion by launching the Satellite Communication by Orbit Relay Equipment (*SCORE*) in 1958. The orbiting vehicle could receive and record audio signals from ground stations, then rebroadcast them to other locations around Earth, creating a rudimentary global communications delivery system.

British author and futurist Arthur C. Clarke (1917–2008) was the first to suggest using space to facilitate communications. In 1945, in a paper titled "Extraterrestrial Relays," Clarke proposed a three-satellite constellation, placed in geosynchronous* orbit 35,800 kilometers (22,300 miles) above Earth, in which the satellites could send and receive audio signals. Although SCORE operated only two weeks before its batteries died, it validated the concept of space communications and opened the way for private industry to build a network in space.

Private industry subsequently developed two communication satellites, *Telstar 1* and *Relay 1,* which NASA launched in 1962. The satellites were powered by solar cells and relayed telephone, television, and data signals between ground stations on Earth. The success of the new technology led Congress to pass the Communications Satellite Act—the first commercial space legislation. The act established the Communications Satellite Corporation (COMSAT, sometimes also written as Comsat), a partnership between government and private industry that in turn brought eleven countries into a consortium called Inter-Governmental Organization (IGO) that began to fund and operate a nonprofit global communication satellite network dubbed INTELSAT, short for International Telecommunications Satellite Organization. In 2001, INTELSAT was turned into a private organization (with over one hundred participating countries), with its name changed from INTELSAT to Intelsat. Then, in 2005, the company was sold to a private-equity consortium comprised of Madison Dearborn Partners (United States), Permira (United Kingdom), Apax Partners (United Kingdom), and Apollo Management (United States).

The creation of additional satellite communication networks followed in the 1970s, including Eurosat, Arabsat (Arab Satellite Communication Organization), and Inmarsat (International Maritime Satellite Organization).

Government was involved in all these systems. Not until 1974 did private industry place into orbit* a purely commercial satellite—*Westar 1*—which was financed by Western Union. The satellite offered video, audio, and data services. Soon after, RCA Americom (short for Radio Corporation of America's American Communications) launched *Satcom 1,* (Satellite Communications 1) in 1975, designed with solid-state transmitters.

As of September 1, 2011, according to the Union of Concerned Scientists, there were 965 active satellites orbiting Earth, of which 443 were American. In addition to communications, satellites are used for such purposes as remote sensing, navigation, and military uses. The Federal Communications Commission (FCC) licenses communication satellites within the United States.

Facilitating Commercial Payloads

Space launch vehicles, like satellites, were originally developed in partnership with the federal government. Until 1984 U.S. private industry relied on NASA to boost commercial payloads* to space. With the passage of the Commercial Space Act (of 1998), NASA was removed from the launch business—except for the space shuttle—and the industry was allowed to operate as a commercial enterprise. Thereafter, satellite producers could

The U.S. Congress is ultimately responsible for setting the policies and budget for the American space program. © *AP Images/Ron Edmonds.*

* **orbit** the circular or elliptical path of an object around a much larger object, governed by the gravitational field of the larger object

* **payload** any cargo launched aboard a rocket that is destined for space, including communications satellites or modules, supplies, equipment, and astronauts; it does not include the vehicle used to move the cargo nor the propellant that powers the vehicle

* reusable launch vehicles
launch vehicles, such as the
space shuttle, designed to be
recovered and reused many
times

contract directly with launch providers to deliver payloads to space. To ensure public safety, the government requires all launches to be licensed. Originally, this was the responsibility of the Department of Transportation (DOT), but it was later transferred to the Office of Commercial Space Transportation at the Federal Aviation Administration (FAA).

Commercial space enterprises have grown to include remote sensing. From space, satellites can collect spectral data that has commercial applications in such fields as natural resource management, urban planning, and precision agriculture. The licensing of remote sensing satellites is the responsibility of the National Oceanic & Atmospheric Administration (NOAA).

The Commercial Space Act of 1998

In 1998, Congress passed a series of amendments to the Commercial Space Act to further promote the commercial development of space. One provision authorized the licensing of space vehicles that re-enter the atmosphere—a response to the development of reusable launch vehicles*. The legislation encouraged the commercial purchase of remote sensing data for science applications, instead of reliance on government employees to build and gather such data. It also required NASA to establish a market-based price structure for commercial enterprises aboard the International Space Station (ISS).

NASA and the Department of Defense

Congress annually enacts legislation to fund NASA and space programs in the Department of Defense (DoD). These bills often contain policy directives. For instance, legislation in 1999 to fund NASA included a provision to allow the space agency to retain funds generated from commercial activities on the ISS.

Jurisdiction over NASA and DoD space programs is spread among several congressional committees. The Armed Services Committees in the House and the Senate consider policy issues involving defense space programs. Oversight of NASA is the responsibility of the Science Committee in the House and the Commerce Committee in the Senate. The Appropriations Committees in both the House and Senate provide funding for space programs.

Finally, the president issues legal directives that affect the development of space commerce. Presidential order has increasingly opened to the public the Global Positioning System (GPS), a constellation of a global navigation satellite system (GNSS) built originally for military purposes. For instance, on May 1, 2000, President Bill Clinton directed the military to stop scrambling GPS signals, thereby improving the accuracy and marketability of commercial GPS devices.

The U.S. president and Congress annually determine how much the federal government will spend on space research and development. The presidential administration submits to Congress comprehensive budget plans for NASA and the DoD space programs that contain proposed funding for individual research and development projects. Congress reviews and often changes the amounts requested. For instance, in seven of its eight years, the Clinton administration cut NASA's budget. Congress, during the Clinton years, generally supported additions to the budget, but not enough to reverse the general downward trend.

NASA's budget has fluctuated over the half century of its existence. Its first year allocation in 1958 was only $89 million dollars, less than 0.1 percent of the total federal budget. That situation changed significantly during the 1960s as the result of the nation's effort to place a human on the Moon before the end of the decade. It reached $5.933 billion dollars in 1966, 4.41 percent of the federal budget, the highest point in the agency's history. As the Apollo program came to an end, the agency's budget fell once again to less than one percent of the federal budget, settling at about 0.8 percent of the federal budget through most of the 1980s. NASA experienced a modest increase in prosperity in the 1990s, when its budget finally reached just more than one percent of the federal budget in the period 1991 to 1993. U.S. presidents and the Congress then started to gradually reduce the agency's budget (less than 0.8 percent in 2000, less than 0.7 percent in 2003, and less than 0.6 percent in 2006). According to President Barack Obama's projected budget, the agency is anticipating an annual increase of about 6 percent in the period 2011 to 2015, with a total allotment over the five years of just over $100 billion. The largest share of that budget is allocated to research in earth and planetary science, astrophysics and heliophysics ($5.814 billion in 2015), with lesser amounts going to space exploration programs ($5.179 billion), space operations (such as the International Space Station; $4.130 billion), cross-agency support ($3.462 billion), and aeronautics and space research and technology ($1.818 billion).

In 1998, revenue generated from space commerce eclipsed, for the first time, the investment in space from government. By October 2007, space commerce was generating over $100 billion per year in revenue, with estimates that this number would reach in excess of $137 billion by 2009. This trend suggests that government is increasingly moving to the sidelines and private industry is now leading the way in the commercial development of space. However, new regulations and legislation will be needed in the future to provide a clear legal framework for the expansion of space commerce.

 See also **Clarke, Arthur C. (Volume 1) • Law (Volume 4) • Law of Space (Volume 1) • Regulation (Volume 1) • Spaceports (Volume 1)**

Resources

Books and Articles

Duggins, Pat. *Final Countdown: NASA and the End of the Space Shuttle Program.* Gainesville, FL: University Press of Florida, 2007.

National Research Council. Committee on the Rationale and Goals of the U.S. Civil Space Program. *America's Future in Space: Aligning the Civil Space Program with National Needs.* Washington, DC: National Academies Press, 2009.

Sundahl, Mark J. *Space Tourism and Export Controls: A Prayer for Relief.* Dallas: Southern Methodist University School of Law, 2010.

Websites

Commercial Space Act of 1998. Union of Concerned Scientists. <http://geo.arc.nasa.gov/sge/landsat/sec107.html> (accessed October 24, 2011).

Enactment of Title 51—National and Commercial Space Programs. <http://uscode.house.gov/cod/t51/Pub%20L%20111-314%20(auth%20version%20from%20GPO).pdf> (accessed October 24, 2011).

Future of Human Space Flight, Government Officials. <http://www.c-spanvideo.org/program/293473-1> (accessed October 24, 2011).

Text of H.R. 970: Federal Aviation Research and Development Reauthorization Act of 2011. <http://www.govtrack.us/congress/billtext.xpd?bill=h112-970> (accessed October 24, 2011).

UCS Satellite Database. Union of Concerned Scientists. <http://www.ucsusa.org/nuclear_weapons_and_global_security/space_weapons/technical_issues/ucs-satellite-database.html> (accessed October 24, 2011).

Whalen, David J. *Communications Satellites: Making the Global Village Possible.* National Aeronautics and Space Administration. <http://www.hq.nasa.gov/office/pao/History/satcomhistory.html> (accessed October 24, 2011).

Licensing

All commercial launches (or re-entries or landings) conducted by U.S. companies are regulated by the Commercial Space Launch Act (CSLA) of 1984 and its 1988, 1998, and 2004 amendments. Under the CSLA, each launch (or re-entry or landing) must have a license. This is true even when launching offshore, as is the case with Space Exploration Technologies Corporation (SpaceX), based in Hawthorne, California, launching its Falcon 1 rocket at the Reagan Test Center on the island atoll of Kwajalein in the Pacific Ocean; or Sea Launch, a venture composed of U.S. company Boeing Commercial Space, Russian company Energia, Norwegian company Aker Solutions, and Ukrainian company

SDO Yuzhnoye/PO Yuzhmash that launches from a ship near the equator in the Pacific Ocean.

These regulations are an outcome of the United Nations' 1967 Treaty on Principles Governing the Activities of States in the Exploration and Use of Outer Space, including the Moon and Other Celestial Bodies (commonly called the Outer Space Treaty), which places responsibility for any liabilities that might result from a space launch and/or reentry (for instance, if a person or building is hit by a spent rocket stage) on the launching state.

To assure "the public health and safety, safety of property, and the national security and foreign policy interests of the United States," Congress enacted the CSLA and established the Office of the Associate Administrator Commercial Space Transportation (AST, or sometimes FAA/AST), which is part of the Federal Aviation Administration (FAA). Its mission statement is "… to ensure protection of the public, property, and the national security and foreign policy interests of the United States during commercial launch or reentry activities, and to encourage, facilitate, and promote U.S. commercial space transportation." Consequently, its purpose is to provide the relevant rules, laws, regulations, and documents necessary to obtain a launch license.

FAA/AST is organized as the Office of the Associate Administrator (the headquarters office of AST); the Space Systems Development Division (SSDD); the Licensing and Safety Division (LASD); and the Systems Engineering and Training Division (SETD). SSDD develops regulations and policy and provides engineering support and forecasting; LASD is the organization that actually evaluates applications and issues licenses; and SETC defines safety standards for current and new space launch and re-entry systems and sites while defining methods to assure and verify those standards are met. Helping to advise the FAA/AST is an industry group, the Commercial Space Transportation Advisory Committee (COMSTAC), which the FAA/AST established and sponsors. COMSTAC meets quarterly. The FAA/AST also licenses spaceport operators (and their spaceports).

The FAA/AST has issued active licenses for the following spaceports (launch site operators, their locations, and license expirations):

- California Spaceport (Spaceport Systems International, at Vandenberg Air Force Base, September 18, 2016)
- Oklahoma Spaceport (Oklahoma Space Industry Development Authority, Burns Flat, June 11, 2016)
- Spaceport Florida (Space Florida, Cape Canaveral Air Force Station, June 30, 2015)
- Cecil Space Port (Jacksonville Aviation Authority (Florida), Cecil Airport, January 10, 2015)
- Mojave Air and Space Port (East Kern Airport District, Mojave, California, June 16, 2014)

* **payload** any cargo launched aboard a rocket that is destined for space, including communications satellites or modules, supplies, equipment, and astronauts; it does not include the vehicle used to move the cargo nor the propellant that powers the vehicle

■ Spaceport America (New Mexico Spaceflight Authority, Las Cruces, New Mexico, December 15, 2013)

■ Kodiak Launch Complex (Alaska Aerospace Corporation, Kodiak Island, September 24, 2013)

■ Mid-Atlantic Regional Spaceport (Virginia Commercial Space Flight Authority, Wallops Flight Facility, December 18, 2012)

FAA/AST officials encourage those seeking a launch license to meet with the organization in "pre-licensing consultations" before submitting an actual license application. The FAA/AST conducts a policy review, a payload* review, a safety evaluation, an environmental review, and a financial responsibility determination based on the data in the license application. It contacts the applicant if it needs more data or if it requires the applicant to change something to qualify for the license.

Once the official application arrives, the FAA/AST has 180 days to issue a license. Since 1984, officials have issued the license in almost every case. There have been only a few exceptions, and in these cases FAA/AST initially rejected the application because of technical lapses. Once applicants made corrections, the FAA/AST granted the license and the rocket flew. After the FAA/AST issues a license, it monitors the licensee through launch to assure compliance with regulations and requirements.

In 2004, the Commercial Space Launch Amendments Act (H.R. 5382) was enacted. The Amendment allows commercial tourism to take place in the United States; specifically, it allows paying customers to fly on suborbital launch vehicles. Several space tourism companies are actively pursuing the business of space tourism; specifically, for a set amount of money, sending private citizens into suborbital flights into space.

The first company likely to accomplish this goal is Virgin Galactic with the help of Scaled Composites. In early flights, Michael Melvill was issued his FAA commercial astronaut wings for his piloted Flight 15P of *SpaceShipOne* (above Mojave, California) on June 21, 2004. Brian Binnie became the second commercial astronaut for his Flight 17 (using the same spacecraft and at the same location) on October 4, 2004. Using the Scaled Composites-built *SpaceShipTwo*, the spacecraft carrying passengers into space, and *WhiteKnightTwo*, the carrier aircraft, Virgin Galactic hopes to begin commercial flights into space in 2013. These operational flights will begin after numerous test flights in 2011 and 2012 of the aircraft provided by Burt Rutan's Scaled Composites. Richard Branson, the founder of Virgin Group, backs this joint venture (called The Spaceship Company) that will launch six people per flight from the Mojave Spaceport in New Mexico, for the approximate 110-kilometer (68-mile) high trip into space. Initial flights will cost about $200,000, but later flights are estimated to cost only $20,000.

 See also **Law (Volume 4) • Law of Space (Volume 1) • Legislative Environment (Volume 1) • Regulation (Volume 1)**

Resources

Books and Articles

McCurdy, Howard E. *Space and the American Imagination.* Baltimore: Johns Hopkins University Press, 2011.

Websites

Active Commercial Space Licenses. Federal Aviation Administration. <http://www.faa.gov/data_research/commercial_space_data/current_licenses/> (accessed October 26, 2011).

Alaska Aerospace Corporation. <http://www.akaerospace.com/klc_overview.html> (accessed October 26, 2011).

Boyle, Alan. *Private-spaceflight bill signed into law.* MSNBC. <http://www.msnbc.msn.com/id/6682611/> (accessed October 26, 2011).

Commercial Space Launch Amendments Act of 2004. Federal Aviation Administration. <http://www.faa.gov/about/office_org/headquarters_offices/ast/media/PL108-492.pdf > (accessed October 26, 2011).

Mid-Atlantic Regional Spaceport. <http://www.marsspaceport.com/> (accessed October 26, 2011).

Mojave Air & Space Port. <http://www.mojaveairport.com/> (accessed October 26, 2011).

Office of Commercial Space Transportation. Federal Aviation Administration. <http://www.faa.gov/about/office_org/headquarters_offices/ast/> (accessed October 26, 2011).

Oklahoma Space Industry Development Authority. <http://www.okspaceport.state.ok.us/> (accessed October 26, 2011).

Spaceport America. <http://www.spaceportamerica.com/> (accessed October 26, 2011).

Spaceport Florida. <http://www.spaceportflorida.com> (accessed October 26, 2011).

Spaceport Systems International. <http://www.calspace.com/SSI/Welcome.html> (accessed October 26, 2011).

Treaty on Principles Governing the Activities of States in the Exploration and Use of Outer Space, including the Moon and Other Celestial Bodies. Office of Outer Space Affairs, United Nations. <http://www.unoosa.org/oosa/SpaceLaw/outerspt.html> (accessed October 26, 2011).

Literature

Long before *Sputnik 1* became humankind's first orbiting spacecraft in 1957 and before the first astronauts landed on the Moon in 1969, science fiction and science fact writers provided the theories, formulas, and ideas that gave birth to space travel. Some of these thinkers and storytellers wrote fancifully. Others expressed their ideas in precise mathematical equations with intricate scientific diagrams. All of them succeeded in helping to make space travel a reality by the mid-twentieth century.

Early Works of Science Fiction

Early works of science fiction relied more on whimsical solutions to spaceflight. During the seventeenth century, Francis Godwin's (1562–1633) *The Man in the Moon* employed a flock of swans to transport a voyager to the lunar surface. Frenchman Cyrano de Bergerac (1619–1655) wrote space travel novels that described bottles of morning dew lifting people into the sky.

Far more serious scientific thought went into the works of two nineteenth-century space fiction writers. In 1869, American Edward Everett Hale (1822–1909) wrote a novel called *The Brick Moon*. This book was the first that detailed the features and functions of the modern Earth-orbiting artificial satellite. French science fiction writer Jules Verne (1828–1905) penned two space travel works, *From the Earth to the Moon*, in 1865, and, five years later, *Round the Moon*. In both books, Verne chronicled the adventures of explorers from post-Civil War America who take a trip to the Moon. Although the form of propulsion was unrealistic (the explorers were shot into space by a gigantic cannon), many other aspects of the stories anticipated the actual U.S. lunar missions undertaken by the National Aeronautics and Space Administration (NASA) in the late 1960s and early 1970s. Verne correctly predicted everything from the phenomenon and effects of weightlessness in space to the shape of the capsule used by the Apollo astronauts. He even proved uncannily accurate in anticipating the Florida launch site, Pacific Ocean splashdown, and recovery by U.S. naval forces of the astronauts in the Apollo missions.

Twentieth-Century Rocket Pioneers

Verne's novels had a strong impact on the three most important rocket pioneers of the twentieth century. One of them was American physicist Robert H. Goddard (1882–1945). As a boy, Goddard was so inspired by the Frenchman's tales of lunar trips that he dedicated his life to achieving spaceflight. As a young physics professor in Massachusetts, Goddard designed and constructed solid-propellant-like rockets. In 1917, the Smithsonian Institution agreed to provide funding for his high-altitude rocket tests. Two years later, Goddard wrote a paper for the Smithsonian titled "A Method of Reaching Extreme Altitudes." This pamphlet discussed the mass required to propel objects beyond Earth's atmosphere—even to the Moon. He also theorized that liquid propellant made for a far more powerful and efficient fuel for rockets than solid propellant. Goddard launched the world's first liquid-fueled rocket in 1926. Ten years later, he published the results of this historic event in his second Smithsonian paper, "Liquid-Propellant Rocket Development."

The second father of modern rocketry was Russia's Konstantin Tsiolkovsky (1857–1935). His works focused on liquid-propellant rockets, kerosene as a fuel, and space station design. Tsiolkovsky's *Investigation of Universal Space by Means of Reactive Devices,* published in 1891, proposed the use of multistage rockets for space travel. Through his science fiction and mathematical study, he laid the foundation for the Soviet Union's successes in spaceflight, which began in the 1950s.

The third great pioneer of rocket theory was the Romanian-born Hermann Oberth (1894–1989). Like Goddard and Tsiolkovsky, Oberth espoused the virtues of liquid-fueled rockets for space voyages. In 1923, he wrote a book called *The Rocket into Planetary Space.* Besides advertising the use of liquid oxygen and alcohol for rocket fuel, he also stressed the importance of using strong yet lightweight alloys for constructing launch vehicles and spacecraft. His *Ways to Spaceflight,* written in 1929, discussed the possibility of building large orbiting space mirrors that could transmit energy to Earth and illuminate cities at night.

There were several other key spaceflight writers and theoreticians of the early twentieth century. In 1929, Hermann Noordung (1892–1929) of Croatia wrote *The Problem of Space Travel,* which discussed the engineering requirements for a space station. Eugene Sänger (1905–1964) of Austria developed basic concepts in rocketry and aerodynamics in his work *Rocket Flight Technology,* published in 1933. Sänger almost single-handedly invented the idea of an "aerospaceplane"—a direct ancestor of today's space shuttle. Germany's Fritz von Opel (1899–1971) also contributed much to the field of rocketry. In 1929, he made the first documented flight of a rocket-powered airplane.

Von Braun, Clarke, and Beyond

In the post-World War II era, two important science fact and science fiction authors stood out. The first was the German-born Wernher von Braun (1912–1977). As a gifted young rocket engineer, von Braun was instrumental in building the V-2 rockets that Germany fired at Britain and Belgium late in the war. After World War II ended in 1945, he moved to the United States where he directed the design and construction of NASA's Saturn rockets, which propelled astronauts into space and to the Moon. In between these two periods in his life, von Braun penned numerous books, essays, and articles about spaceflight. In 1952, he published *Prelude to Space Travel,* which greatly expanded upon Noordung's research in space station development. Four years earlier, he had written *The Mars Project* (published in 1962). In this book, von Braun detailed the first fully comprehensive plan for a human mission to Mars. During the early 1950s, he contributed to a popular series of space-related articles in *Collier's* magazine.

The other key literary figure during this period was Arthur C. Clarke (1917–2008). In 1945, the British-born writer published a paper in *Wireless World,* titled "Can Rocket Stations Give Worldwide Radio Coverage?" This was the first work to discuss the concept of communications satellites that stay in the same relative position above Earth. Such satellites made instant worldwide television, telephone, facsimile (fax), e-mail (electronic mail), and computer services possible. In 1968, the film based upon his science fiction book *2001: A Space Odyssey* captured the very mood and spirit of the space age.

Today, science fiction and fact authors continue the efforts begun by Verne, Goddard, Oberth, von Braun, and others. Through their imagination, knowledge, and words, the frontiers of space exploration are pushed forward.

 See also **Artwork (Volume 1) • Careers in Writing, Photography, and Film (Volume 1) • Clarke, Arthur C. (Volume 1) • Mars Missions (Volume 4) • Oberth, Hermann (Volume 1) • Rockets (Volume 1) • Sänger, Eugene (Volume 3) • Science Fiction (Volume 4) • Tsiolkovsky, Konstantin (Volume 3) • von Braun, Wernher (Volume 3)**

Resources

Books and Articles

Clarke, Arthur C. *Greetings, Carbon-Based Bipeds!* New York: St. Martin's Press, 1999.

McCurdy, Howard E. *Space and the American Imagination.* 2nd ed. Baltimore: Johns Hopkins University Press, 2011.

National Geographic Society. *Man's Conquest of Space.* Washington, DC: Author, 1968.

Noordung, Hermann. *The Problem of Space Travel.* Edited by Ernst Stuhlinger, J. D. Hunley, and Jennifer Garland. Washington, DC: National Aeronautics and Space Administration, NASA History Office, 1995.

Ordway, Frederick I., and Randy Liebermann, eds. *Blueprint for Space.* Washington, DC: Smithsonian Institution Press, 1992.

Stuhlinger, Ernst, and Frederick I. Ordway. *Wernher von Braun: Crusader for Space.* Malabar, FL: Krieger Publishing, 1994.

Websites

A Jules Verne Centennial, 1905–2005. Smithsonian Institution Libraries. <http://www.sil.si.edu/OnDisplay/JulesVerne100/> (accessed October 12, 2011).

Kluger, Jeffrey. "Rocket Scientist Robert Goddard." *Time.* <http://www.time.com/time/time100/scientist/profile/goddard.html> (accessed October 12, 2011).

Young, Anthony. *Remembering Wernher von Braun.* The Space Review. <http://www.thespacereview.com/article/656/1> (accessed October 12, 2011).

Lucas, George
American Screenwriter, Producer, and Director
1944–

Born on May 14, 1944, in Modesto, California, film director George Walton Lucas, Jr. studied film at the University of Southern California. His first feature film was *THX 1138.* The executive producer was Francis Ford Coppola, who would later gain fame directing *The Godfather* trilogy and *Apocalypse Now.* In 1973, Lucas co-wrote and directed *American Graffiti,* which won a Golden Globe and garnered five Academy Award nominations.

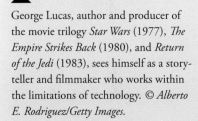
George Lucas, author and producer of the movie trilogy *Star Wars* (1977), *The Empire Strikes Back* (1980), and *Return of the Jedi* (1983), sees himself as a storyteller and filmmaker who works within the limitations of technology. © *Alberto E. Rodriguez/Getty Images.*

Within the space fraternity Lucas is recognized for the Star Wars movies. *Star Wars,* the first in the initial trilogy of tales about life and conflict in the universe, was released in 1977. The film broke box-office records and won seven Academy Awards. Lucas went on to write *The Empire Strikes Back* (1980) and *Return of the Jedi* (1983), and was executive producer for both. Lucas worked for twenty years developing a prequel to the trilogy, *Episode 1: The Phantom Menace,* released in 1999, for which he was writer, director, and executive producer. A second prequel, *Attack of the Clones,* was released in May, 2002, and the final, long-awaited third prequel, *Revenge of the Sith,* was released in May, 2005. This movie finally showed moviegoers how Anakin Skywalker transformed into the threatening Darth Vader.

Lucas sees himself as a storyteller and professes not to be particularly keen on technology. He admits that he has had to invent the necessary technology to tell his tales and believes the mark of a talented filmmaker is how well one works within the limitations imposed by the available technology.

In 1971, Lucas founded the film production company Lucasfilm Limited, which is headquartered in San Francisco, California. Industrial Light & Magic is one of its several divisions. As of 2011, he is the company's chairperson of the board. He also founded the George Lucas Educational Foundation in 1991, a nonprofit organization promoting innovation in education. In 2005, he was awarded the Life Achievement Award by the American Film Institute. Lucas continues to be one of America's most successful directors and producers.

▶ *See also* **Careers in Writing, Photography, and Film (Volume 1) • Entertainment (Volume 1) •** *Star Wars* **(Volume 4)**

* **orbit** the circular or elliptical path of an object around a much larger object, governed by the gravitational field of the larger object

Resources

Books and Articles

Salewicz, Chris. *George Lucas: The Making of His Movies.* New York: Thunder's Mouth Press, 1999.

Smith, Jim. *George Lucas.* London: Virgin Books, 2003.

Websites

AFI Life Achievement Award: George Lucas. American Film Institute. <http://www.afi.com/laa/laa05.aspx> (accessed October 13, 2011).

George Lucas. Lucasfilm Limited. <http://www.lucasfilm.com/inside/bio/georgelucas.html> (accessed October 13, 2011).

Lucasfilm Limited. <http://www.lucasfilm.com/> (accessed October 13, 2011).

Lunar Development

The spectacular advances of science and engineering in the twentieth century established the basis for creating permanent human settlements in space in the twenty-first century. Since the Moon is Earth's closest celestial neighbor and is in orbit* around Earth, it will logically be the next principal focus of human exploration and settlement. The Moon is an excellent platform for astronomical and other scientific investigations, for technological development, and for human habitation. It also has access to the virtually unlimited energy and material resources of space, which can be applied to the development needs of both the Moon and Earth. These opportunities, combined with the universal desire of humankind to explore and settle new lands, assure that the global transformation of the Moon into an inhabited sister planet of Earth will become a reality in this century.

Past Achievements and Near-Term Plans

Russia, the United States, Japan, China, and India have all launched spacecraft into orbit around the Moon. The European Space Agency (ESA), representing a host of European countries, has also launched spacecraft into lunar orbit. Two countries—Russia and the United States—have returned samples of lunar soil to Earth; only the United States has landed people on the Moon.

As of late 2011, the United States, China, India, and Japan have all publicly declared an intent to send astronauts to the Moon on or before the year 2020. Other countries, such as Russia or the European countries represented by the ESA, may also decide to send their own citizens to the lunar surface within the relatively near future. In contrast to the United State's Apollo program of the 1960s and early 1970s, in which astronauts only visited the Moon for brief periods, future manned missions to the Moon will most likely include plans for permanent occupation of lunar outposts.

The Present High Cost of Space Exploration

A major impediment to the exploration of space is the high cost of delivering cargoes from the surface of Earth into space. For example, the cost of launching a payload* into low Earth orbit* by the space shuttle was approximately $20,000 per kilogram ($10,000 per pound, equivalent weight on Earth), and that figure will be higher for missions to the Moon. Thus, it appears that lunar projects will continue to be extremely expensive into the near-term future, and that will hold true even if launch costs to low Earth orbit are reduced down to $2,000 per kilogram ($1,000 per equivalent pound on Earth).

In contrast to the current high cost of launching payloads into space, the exploration and development of the Moon will eventually be marked by a dramatic reduction in the cost of space exploration through the process known as in situ resource utilization, otherwise known as "living off the land." Industrial processes on Earth use energy, raw materials, labor, and machines to manufacture sophisticated products such as computers, medical imaging devices, rockets, and communication satellites. Eventually it will become feasible to use lunar materials to manufacture equally sophisticated products on the Moon.

The Moon has a reliable supply of energy in the form of sunlight, and the lunar regolith (Moon dirt) contains abundant supplies of iron, silicon, aluminum, and oxygen as well as traces of carbon, nitrogen, and other light elements. Arguably the most serious challenge to overcome may be providing water to human settlements. Water is essential to human life, and transporting adequate supplies of water from Earth to the Moon is probably out of the question. Human settlements are a realistic possibility, therefore, only if some mechanism can be found to obtain water from the Moon itself. For much of modern history, most scientists assumed that might not be possible because they did not believe the Moon had any—or at least sufficient—water supplies to meet the needs of human settlements. Some doubts about that assumption began to arise as a result of the U.S. *Lunar Prospector* mission to the Moon in 1998 and 1999. That mission found that hydrogen was present in low concentrations in the polar regions of the Moon, conceivably in the form of water ice. The amount of water detected was, however, far less than what would be needed by human settlements. No further evidence for the presence of water was produced until a decade later when two lunar missions initiated more aggressive searches for water ice on the Moon. The first of those missions was conducted by the Indian *Chandrayaan-1* spacecraft, which went into orbit around the Moon on November 8, 2008. A week later, a probe ejected by the spacecraft crashed onto the lunar surface, sending a large

* **payload** any cargo launched aboard a rocket that is destined for space, including communications satellites or modules, supplies, equipment, and astronauts; it does not include the vehicle used to move the cargo nor the propellant that powers the vehicle

* **low Earth orbit** an orbit between 300 and 800 kilometers (185 and 500 miles) above Earth's surface

plume of dust into the air. Instruments on the *Chandrayaan-1* mother ship analyzed the chemical composition of that material and found clear evidence for the presence of hydroxyl radicals (OH), present in all water molecules. The final report of that mission suggested that lunar soil may contain up to 1,000 ppm (parts per million; 0.1 percent) water. While those results were encouraging, they hardly assured researchers that the Moon could supply the water needed by colonizers from Earth.

Brighter prospects became known only two months after Indian scientists had reported their findings. On June 18, 2009, the U.S. National Aeronautics and Space Administration (NASA) had launched a lunar orbiter called *Lunar Reconnaissance Orbiter* (LRO), which carried with it a lunar probe called the *Lunar Crater Observation and Sensing Satellite* (LCROSS). NASA planned to use LCROSS in much the same way that *Chandrayaan-1* had used its probe. That is, LCROSS fired the upper stage of its structure into the lunar surface and then analyzed the plume of soil ejected during impact. The experiment produced essentially the same results as those observed by the Indian experiment, except more so. That is, rather than estimating a 0.1 percent moisture level for the lunar surface, the LCROSS data suggested that as much as five percent of the regolith might consist of water. Instead of being present in unusably small concentrations, the Moon's surface may actually hold millions or even billions of gallons of water. Some experts have estimated that the total amount of water on the Moon may be comparable to that present on Earth. If so, the problem of having water for lunar human settlements may not be as difficult as it has long appeared to be.

One more important finding about lunar water was announced in May 2010 by a group of researchers at the Carnegie Institution in Washington, D.C. These researchers were studying a group of rocks returned by astronauts during the Apollo 17 flight of 1972. They found that the rocks contained inclusions of water in volcanic rocks similar to those found on Earth. The water had apparenty been trapped in the rocks during the early stages of the Moon's life, some three billion years ago. If the samples examined by the Carnegie researchers are typical of most Moon rocks, the supply of water on the Moon may actually be very significant, sufficient to meet all the anticipated needs of a human colony. Of course, extracting the water from Moon regolith and rocks may not be a simple task. But, like other challenges facing Moon colonists, it does not necessarily have to be an insurmountable task.

For the initial large-scale development of lunar resources, tele-operated robots that have been delivered to the Moon will serve as the "labor" component for lunar industrial processes. Tele-operation is the process by which robots are controlled by scientists or technicians at remote locations using radio links and television monitors. Tele-operation procedures are widely used on Earth for diverse applications such as mining, undersea projects, and certain forms of surgery. It is fortuitous that the Moon

An artist's conception of a lunar mining facility illustrates possible exploration programs for the future. *Illustration by Pat Rawlings. NASA.*

always has the same face directed to Earth and that the round-trip time for communications between Earth and the Moon is less than three seconds. These conditions will allow Earth-bound operators of lunar robots to have a virtual presence on the Moon twenty-four hours per day, every day of the year.

Establishing a Lunar Base

The site for the first unmanned base will likely be on the Earth-facing side of the south polar region of the Moon. There are several sites in this region that always have Earth in view for continuous telecommunications and that receive as much as 340 days of sunlight per year for the generation of solar electric power. In addition, it is likely that a south polar base would have access to large concentrations of water (in the form of ice), which would be extremely useful for industrial operations as well as eventual human habitation. Not only is water an essential raw material for an inhabited lunar base, but the vitally important elements hydrogen and oxygen could be derived from the in situ ice as well.

Many countries have rockets that can be modified to place useful payloads on the Moon. In one scenario for the establishment of a lunar base, one or more of these existing rocket systems will be used to transport solar panels, communication systems, scientific equipment, and robots from

Earth to the south polar region of the Moon. When these components are in place, tele-operated rover vehicles will explore and analyze the lunar surface. Protocols for the preservation of unique features of the lunar environment will be observed, and scientific data will be obtained before local materials are utilized for research purposes. When surveys and analyses have been completed, tele-operated robots will then begin experiments with the production of various finished products, such as glass, bricks, wires, and transistors, from lunar dirt.

In the preceding scenario involving products that will be manufactured on the Moon, priority will be given to the fabrication of solar cells for the generation of electric power. The generation of electric power from the first solar cells created on the Moon—derived completely from lunar materials—will be a milestone in space exploration because it will demonstrate that human enterprises can be self-supporting in space. From that beginning, lunar-made solar cells will be added to the electric power system of the lunar base. As electric power levels grow, additional scientific and manufacturing equipment will be delivered from Earth, and the lunar base will expand in all of its capacities. Utilizing lunar ores to produce various metals, a rudimentary rail line could be constructed from the first base to other areas in the southern polar region, including the geographic South Pole. A "southern rail line" would greatly expand the ability to carry out exploratory missions and would facilitate the growth of lunar power and communication networks.

Humans Return to the Moon

During the build-up of the first unmanned lunar base, controlled ecological life support systems will undergo continued research and development on Earth and in space, on facilities such as a space station. Work will also commence on the development of reusable transportation systems that can ferry people between Earth and the Moon. When a reliable lunar electric power system is in place and underground chambers (for protection from radiation, temperature extremes, and micrometeorites) have been constructed, life support systems and agricultural modules will be delivered to the lunar base. Humans will then begin to inhabit the lunar base, with crews shuttling between Earth and the Moon every sixty to ninety days. With the presence of highly skilled workers, all aspects of lunar base activities will be expanded. The return of people to the Moon will most likely occur within a decade after establishment of the first unmanned lunar base.

As experience with lunar operations increases, the scientific and industrial capability of people on the Moon will trend toward parity with Earth, perhaps within two to three decades after the founding of the first manned base. Widely separated, permanent human settlements will be established, and the only cargoes that will need to be transported from Earth will be humans—the scientists, engineers, tourists, and colonists who will explore, develop, and inhabit the Moon.

Future Lunar Development

Geological expeditions will explore the mountain ranges, mares (plateaus), craters, and rills (narrow valleys) of the Moon, and investigate lava tubes that have been sealed for billions of years. Thousands of lunar-made telescopes will be placed at regular intervals on the Moon so that any object of interest in the universe may be observed continuously under ideal viewing conditions. People will live and work in large underground malls that have Earth-like living conditions. A rail system will provide high-speed access to all areas of the Moon and lunar tourism will be a growth industry. Millions of megawatts of low-cost, environmentally-sound electric power will be beamed from the Moon to Earth and other locations in space by the lunar power system.

By the mid-twenty-first century, spacecraft will be manufactured on the Moon and launched by electromagnetic "mass drivers" to all points of interest in the solar system, and robotic missions to nearby stars will be underway. Communication, power, and transportation systems will be built on the Moon and launched to Mars in support of the global human exploration and development of that planet. Asteroids and "burned out" comets in Earth's orbital vicinity, especially those that pose a threat of collision with Earth or the Moon, will be maneuvered out of harm's way and mined for their hydrocarbon, water, and mineral contents, which will then be delivered to Earth or the Moon.

The transformation of the Moon into an inhabited sister planet of Earth is an achievable goal that will be highly beneficial to the people of Earth, and will result in the following advancements:

- An expansion of scientific knowledge;
- Access to the virtually unlimited energy and material resources of space;
- Job and business opportunities;
- The advancement of all engineering disciplines;
- International cooperation;
- A greatly expanded program of solar system exploration; and
- The opening of endless frontiers.

The binary Earth-Moon planetary system will thus draw upon and benefit from the vast energy and material resources of space, and the spacefaring phase of humankind will be firmly established.

 See also **Lunar Bases (Volume 4) • Lunar Outposts (Volume 4) • Natural Resources (Volume 4) • Settlements (Volume 4) • Space Tourism, Evolution of (Volume 4) • Telepresence (Volume 4) • Tourism (Volume 1)**

Resources

Books and Articles

Benaroya, Haym. *Lunar Settlements.* Boca Raton, Fl: CRC Press, 2010.

Binienda, Wieslaw K., ed. *Proceedings of the 11th International Conference on Engineering, Science, Construction, and Operations in Challenging Environments 2008.* [Reston, Va.]: American Society of Civil Engineers, [2008].

Eckart, Peter. *The Lunar Base Handbook: An Introduction to Lunar Base Design, Development, and Operations,* 2nd ed. New York: McGraw-Hill, 2006.

Harris, Philip R. *Space Enterprise: Living and Working Offworld in the 21st Century.* Berlin: Praxis Publishing, 2009.

Schrunk, David G. *The Moon: Resources, Future Development and Settlement.* Berlin, Germany, and New York: Springer (Praxis Publishing), 2008.

Seedhouse, Erik. *Lunar Outpost: The Challenges of Establishing a Human Settlement on the Moon.* Berlin: Praxis Publishing, 2009.

Weeks, E. E. *Outsiders' Guide to Understanding Outer Space Development.* Philadelphia: Xlibris Corporation, 2004.

Websites

Al-Jammaz, Khaled, et al. *Elements for a Sustainable Lunar Colony in the South Polar Region.* <http://www.spacefuture.com/archive/elements_for_a_sustainable_lunar_colony_in_the_south_polar_region.shtml> (accessed October 9, 2011).

LUNA GAIA: A Closed-Loop Habitat for the Moon. International Space University. <http://www.isunet.edu/index.php?option=com_content&task=view&id=214&Itemid=251> (accessed October 9, 2011).

Lunar Exploration Themes and Objectives Development Process. National Aeronautics and Space Administration. <http://www.nasa.gov/exploration/home/why_moon_process.html> (accessed October 9, 2011).

Mendell, W. W. *The Second Conference on Lunar Bases and Space Activities of the 21st Century.* National Space Society. <http://www.nss.org/settlement/moon/library/lunar2.htm> (accessed October 9, 2011).

NASA-Funded Scientists Make Watershed Lunar Discovery. National Aeronautics and Space Administration. <http://www.nasa.gov/topics/moonmars/features/moon_water_prt.htm> (accessed October 9, 2011).

Open Luna. <http://openluna.org/> (accessed October 9, 2011).

M

Made in Space

History characterizes the various eras of civilization in terms of available materials technology, leading to the recognition of such eras as the Stone Age, Bronze Age, Steel Age, and Silicon Age. One of the areas of intense research in the present era has been the processing of materials in the space environment to develop new or improved products for use on and about Earth. In the 1960s, during the early phase of this effort, the advent of a new industry was predicted based on the promising initial results obtained, and it was anticipated that by the 1980s "made in space" would be a common label on a large number of products.

However, that new manufacturing industry based in space is still in the future, primarily because of the high cost of placing the carrier vehicles into orbit—around $22,000 to $77,000 per kilogram ($10,000 to $35,000 per equivalent pound of weight on Earth's surface) within the first decade of the twenty-first century. Advances in propulsion technology in the 2010s, however, could reduce the cost of transportation to space in the future. Currently the emphasis is on making better products here on Earth based on the knowledge and processes discovered through space research. To understand the great potential of space it is important to examine what it is that makes the space environment so unique in the processing of materials.

The Advantages of Space-Based Manufacturing

There are two primary effects on Earth that can be reduced to almost zero in the microgravity environment (nominally one-thousandth to one-millionth of Earth's gravitation) present in orbiting vehicles: sedimentation and thermal convection*. Sedimentation makes heavier liquids or particles settle at the bottom of a container, as when sugar added to coffee settles at the bottom of the cup. Thermal convection establishes currents where cooler fluids fall to the bottom and warmer ones rise to the top.

Because many chemical, fluid-physics, biological, and phase-change (e.g., changing from a liquid to a solid state) processes are affected by the effects of sedimentation and thermal convection, the form and size properties of materials formed under these influences are different in space compared to those formed under the influence of Earth's gravitation.

A study conducted for the National Aeronautics and Space Administration (NASA) in the early 1970s identified seventy-seven representative unique products or applications that can be obtained in space. Since that

* **convection** the movement of heated fluid caused by a variation in density; hot fluid rises while cool fluid sinks

* **DNA** deoxyribonucleic acid; the molecule used by all living things (and some viruses) on Earth to transmit genetic information

* **crystallography** the study of the internal structure of crystals

early study, the list of potential applications has at least doubled. Most of the items have been the subject of many investigations conducted on rockets, spacecraft such as the U.S. space shuttles, the Russian *Mir* space station, the International Space Station by scientists and engineering teams from many countries, particularly the United States; Germany, France, the United Kingdom, and Italy (within the European Space Agency); Russia; Canada; and Japan. The following two examples in the medical area serve to illustrate the current research.

Protein Crystal Growth

Growing crystals in microgravity has the advantage of virtually eliminating the thermal convection that produces poor crystal quality and increases the time required to grow a useful crystal. This advantage of microgravity is particularly important in obtaining crystals that have the size and high degree of structural perfection necessary to determine, through x-ray analysis, the three-dimensional structure of those complex organic molecules. For instance, long before the space era the structure of DNA* was determined using crystallography*, but there are many protein crystals that are difficult to grow under the influence of the gravitational force. Urokinase, a significant protein in cancer research, is an example of a protein that is difficult to grow on Earth and benefits from microgravity. Complete three-dimensional characterization, that is, determining the relative location of the approximately 55,000 atoms in this molecule, would permit pharmaceutical scientists to design drugs to counteract the harmful effects of urokinase in promoting the spread of cancer throughout the body, particularly in the case of cancerous breast tumors. In 2003, Instrument Technologies Associates (ITA) obtained permission to conduct an experiment on growing urokinase crystals aboard space shuttle *Columbia* flight STS-107. When that flight exploded upon its return to Earth, the experiment survived, but was not properly preserved by NASA officials. In 2009, ITA officials sued NASA for $8 million in compensation for losses suffered as a result of that neglect. As of late 2011, that case had still not been decided.

A more recent protein crystal growth experiment was conducted on the International Space Station (ISS) between 2008 and 2010 under the auspices of the European Space Agency (ESA). The experiment, called the Protein Crystallisation Diagnostic Facility, was assembled in two parts. The first part consisted of the Protein Crystallisation Diagnostics Facility (PCDF) Electronics Unit, transported to the space station in February 2008. The second component, the Processing Unit, was delivered to the station on the space shuttle *Discovery* in STS-119 in March 2009. The two components were then combined and put into operation in the ESA Columbus laboratory on the ISS, where crystal growth continued for three months. Protein crystals produced in that experiment were then returned to Earth on space shuttle *Endeavour* on flight STS-127 in June 2010.

Another group of protein crystal growth experiments are being conducted by the Japan Aerospace Exploration Agency (JAXA) in its ISS Kibo laboratory on the International Space Station. The JAXA project is being carried out in collaboration with Russia, which provides the rockets that carry the experiments to the ISS, and Malaysia, which provides planning and analysis of experiments for the project. As of late 2011, this consortium had delivered and retrieved more than a dozen protein experiments. Officials at JAXA see their experiments as providing a way of producing drugs for serious diseases with too few patients to attract private pharmaceutical corporate interest and for the production of enzymes needed for new waste disposal and energy generation technologies.

Microcapsules for Medical Applications

Experiments conducted in space have shown more spherical perfection and uniformity of size distribution in microcapsules, which are capsules with diameters of 1 to 300 micrometers (0.00004 to 0.012 inches). The size and shape of microcapsules are key factors in how effective they are at delivering medicinal drugs directly to the affected organs by means of injections or nasal inhalations. The feasibility of newly developed processes for producing multilayered microcapsules has been limited because of the effects of density differences in the presence of Earth-based gravity. In order to circumvent this, a series of experiments has been performed onboard the space shuttle to produce superior microcapsules in space. As of late 2011, a number of patents had been issued to individuals for methods of making and utilizing the use of microcapsules to deliver drugs. A 2011 patent, for example, described a microcapsule that consists of a drug for treating cancer and a magnetic material. When the microcapsule is administered to a patient and it reaches its target organ, a magnetic field is applied to the microcapsule. Heat produced by the application of the magnetic field on the magnetic material melts the capsule and releases the drug to the target organ without producing any harmful effects on surrounding tissue. In a variation of this invention, a drug and opaque material are embedded with a microcapsule, along with a magnetic material. When a magnetic field is administered to this microcapsule, it releases not only the drug but the opaque material, which allows a physician to observe the progress of the drug with standard imaging procedures.

Student Experiments in Space-Based Material Processing

Since the initial activity in materials processing in space, student participation has been an important part of the effort. Since the early space shuttle flights, the NASA-sponsored Getaway Special (GAS) program has provided experiment containers in the shuttle cargo bay capable of accommodating fifty to two hundred payloads* of 20 to 90 kilograms (50 to 200 equivalent pounds of weight on Earth's surface), with the primary focus on student experiments. Industry also plays an important role in student education. One U.S. space company pioneered a hands-on

* **payload** any cargo launched aboard a rocket that is destined for space, including communications satellites or modules, supplies, equipment, and astronauts; it does not include the vehicle used to move the cargo nor the propellant that powers the vehicle

EXAMPLES OF MATERIALS INVESTIGATIONS IN SPACE IN VARIOUS CATEGORIES

Category	Examples	Description
Materials Solidification	Vapor Deposition of Silicates	Vapor deposited on a substrate as a coating of metallic particles imbedded in a matrix.
	Crystal Growth	Organic or inorganic crystal growth in a liquid solution or through evaporation or osmosis.
	Directional Solidification of Metals	A metallic rod has a molten zone that is moved along the rod to produce a superior metal cell structure.
	Micro-encapsulation of Medicinal Drugs	Chemicals are combined in a chamber to form microcapsules containing various layers.
Chemical and Fluids Phenomena	Multiphase Polymers for Composite Structures	Very sensitive separation of cell group subpopulations using processes that do not work with the gravitation on Earth.
	Production of Catalysts	To use controlled gravitational acceleration in the formation of catalytic materials.
	Convective Phenomena Investigations	A family of experiments, dealing with convection due to surface tension, vibration, and electrical fields.
Ceramics and Glasses	Immiscible Glasses for Advanced Applications in Optics	Investigates the role of gravity in the inability to mix glasses having dissimilar densities.
Biological	Continuous Electrophoresis for Biological Separations	Provide continuously flowing separation of biological materials by *electrophoresis*, applying an electric field across the solution.
	Human Cell and Antibody Research	Determines the difference in cell behavior, for use in cancer research and investigation of aging processes.

student experiment program for microgravity experiments onboard the space shuttle. Several industrial concerns have since donated space accommodations in scientific equipment on the shuttle and on rockets, as well as engineering and scientific manpower during integration of the experiments in the spacecraft.

The Role of the International Space Station

The use of the International Space Station (ISS) during the first decade of the twenty-first century is an important milestone in the growth and maturing of the research phase of the materials processing in space program. The International Space Station provides continuing, long-duration microgravity capability to conduct experiments with the participation of astronauts and cosmonauts. The list of experiments delivered to the ISS on the last shuttle flight, STS-135, in July 2011 is indicative of the kind of research that can be conducted on the station. One example was a joint U.S.-Canada experiment designed to test the components of a possible future re-fueling station in space at which satellites would be able to recharge their batteries and propulsion systems in order to extend their lifetime in space. An example of the medical experiments delivered by STS-135 was Ultrasound-2, an upgraded and improved model of a cardiac ultrasound machine that had been in use on the station for ten years. The machine provides important information about the effects of humans living in low-gravity environments over extended periods of time. In another ongoing medical experiment, the Commercial Biomedical Test Module (CBTM-3) will be used to test the effects of low gravity environments on bone loss in mice. Agricultural experiments on the ISS are of increasing interest also. One such experiment was designed to test the effect of low gravity on the infectiousness of bacteria on important crop plants, while another experiment was designed to study the effect of microgravity on rate of germination and growth pattern in seeds. Finally, two new types of smart phones were delivered to the station to test their effectiveness in the special conditions of low gravity.

 See also **Crystal Growth (Volume 3) • International Space Station (Volumes 1 and 3) • Microgravity (Volume 2) • Zero Gravity (Volume 3)**

Resources

Books and Articles

Cassanto, John M. "A University among the Stars." *International Space Business Review* 1, no. 2 (July/August 1986): 77–84

Catchpole, John. *The International Space Station: Building for the Future.* Berlin; New York: Springer, 2008.

Furniss, Tim, David Shayler, and Michael D. Shayler. *Praxis Manned Spaceflight Log, 1961-2006.* New York: Springer, 2007.

Gabriel, Kamiel S. *Microgravity Two-phase Flow and Heat Transfer.* Dordrecht: Springer, 2007.

Russomano, Thais, Gustavo Dalmarco, and Felipe Prehn Falcao. *The Effects of Hypergravity and Microgravity on Biomedical Experiments.* San Rafael, CA: Morgan & Claypool, 2008.

Sharpe, R. J., and M. D. Wright. *Analysis of Microgravity Experiments Conducted on the Apollo Spacecraft.* Huntsville, AL: National Aeronautics and Space Administration, Marshall Space Flight Center, 2009.

Websites

Externally Triggered Microcapsules - Patent 7968117. <http://www.docstoc.com/docs/83135803/Externally-Triggered-Microcapsules---Patent-7968117> (accessed October 24, 2011).

Microencapsulation Electrostatic Processing (MEPS). <http://www.nasa.gov/centers/marshall/news/background/facts/MEPS.html> (accessed October 24, 2011).

Space Station Research and Technology. <http://www.nasa.gov/mission_pages/station/research/index.html> (accessed October 24, 2011).

STS-135: The Final Mission. <http://www.shuttlepresskit.com/STS-135/STS135.pdf> (accessed October 24, 2011).

Werner, Debra. "Firm Seeks $8 Million For Experiment That Survived Columbia Accident." *Space.com.* <http://www.space.com/news/090202-sn-columbia-experiment-money.html> (accessed October 24, 2011).

Made with Space Technology

To meet the many goals of space exploration and aeronautical development, the National Aeronautics and Space Administration (NASA) and the aerospace industry seek many innovations in a number of science and technology fields. This storehouse of knowledge continues to provide a broad technical foundation for stimulating secondary applications of these different developments. Each application is a result of "spinoffs" of both astronautics and aeronautical research.

A spinoff is a technology that has been transferred to uses other than the purpose for which it was developed. In the early twenty-first century, it is difficult to find an area in everyday life into which a space-based spinoff has not penetrated, yet many people are unaware of the existence of these breakthroughs.

Spinoffs in Medical Applications

Walking through the emergency ward of modern hospitals reveals many changes in equipment stemming from early U.S. manned space programs like Apollo*.

* **Apollo** American program to land men on the Moon; *Apollo 11, Apollo 12, Apollo 14, Apollo 15, Apollo 16,* and *Apollo 17* delivered twelve men to the lunar surface (two per mission) between 1969 and 1972 and returned them safely back to Earth

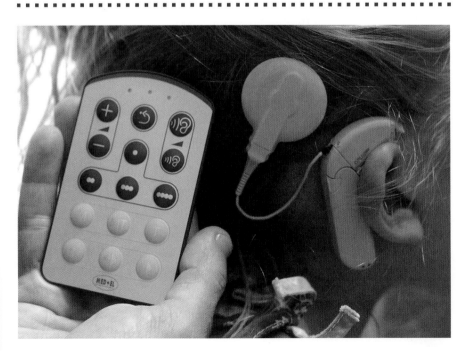

A young woman displays her cochlear implant and its remote control. Cochlear implants were invented by NASA engineer Adam Kissiah. These devices detect sound waves with a microphone, interpret them, and then transmit information about them directly to a person's auditory nerve. They can help even profoundly deaf people to hear. © *Harry Lynch/Raleigh News & Observer/MCT via Getty Images.*

Materials in Wheelchairs The spacecraft and rockets used to take humans to the Moon were developed from new materials that were lightweight yet very strong. Engineers developed new methods of construction and new alloys and composite materials for these missions. Many of these new developments found use in everyday life here on Earth.

An advanced wheelchair is one example. To address the needs of the wheelchair user, researchers at the NASA Langley Research Center in Virginia and the University of Virginia's Rehabilitation Engineering Center developed a wheelchair made from aerospace composite materials much lighter but stronger than common metals.

This 11.3-kilogram (25 pounds of equivalent weight on Earth) wheelchair offers the strength and weight-bearing capability of a normal 22.6-kilogram (50-equivalent-pound) wheelchair, which can also be collapsed for storage and transport. Robotic and tele-operator technologies for space-related programs have also been adapted to develop a voice-controlled wheelchair and manipulator as an aid to paralyzed and severely handicapped people. At the heart of this system is a voice-controlled analyzer that uses a minicomputer. The patient speaks a command into a microphone connected to a computer that translates the commands into electrical signals, which then activates appropriate motors to cause the desired motion of the wheelchair or manipulator. The manipulator can pick up objects, open doors, turn knobs, and perform a variety of other functions.

The Unistick Another breakthrough for the handicapped from the space program is called Unistick. For the later Apollo Moon landings in the early 1970s, NASA developed a Lunar Rover that allowed astronauts to drive around on the lunar surface, greatly enhancing their ability to explore more of the Moon around their landing site. The rover was designed to allow an astronaut to drive one-handed, using an aircraft-like joystick to steer, accelerate, and decelerate (brake) the vehicle. On Earth, this technology is being applied to a system that allows people who have no lower limbs to drive with the use of a joystick, which combines the functions of a steering wheel, brake pedal, and accelerator.

MRI Technology Another spinoff into the medical field is magnetic resonance imaging (MRI), which enables magnetic field and radio waves to peer inside the body. Unlike x rays, MRI scans are able to see into bones. By applying computerized image enhancement technology developed to read Earth-resources satellite photographs, experts have been able to provide thermatic maps of the human body, using color to indicate different types of tissue, making tumors or blood clots easy to find.

Nitinol in Dentistry In dentistry, straightening teeth requires months or even years of applying corrective pressure by means of arch wires, or braces. A new type of arch-wire material called Nitinol (or, nickel titanium, NiTi) now helps reduce the number of brace changes because

The suits worn by these firefighters at the French Guiana Space Center, near Kourou in French Guiana, were adapted from space suit technology. The varying protective gear has been specifically designed to be able to handle different types of emergency situations. © *Roger Ressmeyer/Corbis.*

of its elasticity. This new material, an alloy of titanium and nickel, has an ability to return to its original shape after bending. Many satellite antennas or other hardware could be compacted inside a satellite during launch, then later expanded to full size when in space. This same property allows braces made of Nitinol to exert continuous pull on teeth, reducing the number of dentist visits and changes in braces.

Spinoffs in Other Applications

The field of firefighting and fire prevention has benefited greatly from aerospace spinoffs. Spinoff applications include protective outer garments for workers in hazardous environments; a broad range of fire-retardant paints, foams, and ablative coatings for outdoor structures; and different types of flame-resistant fabrics for use in the home, office, and public transportation vehicles. Many new flame-resistant materials, primarily developed to minimize fire hazards in the space shuttle, have resulted in new, lightweight substances that resist ignition. When exposed to open flame, the material decomposes. This same material is now used in the production of seat cushions and panels for doors, walls, floors, and ceilings. This new fire-resistant material has particular application to commercial aircraft, ships, buses, and rapid-transit trains, where toxic smoke is the major cause of fire fatalities.

One of the biggest fire-related technology transfers is the breathing apparatus worn by firefighters for protection against smoke inhalation. Until the 1970s, the breathing apparatus used by firefighters was large, heavy, and restrictive. The NASA Johnson Space Center in Houston, Texas, developed a breathing system weighing one-third less than conventional systems. The system included a facemask, frame and harness, warning device, and an air bottle. In the first decade of the twenty-first century, many breathing systems incorporate space technology in some form.

Anticorrosion paints, developed for many structures at the NASA Kennedy Space Center in Florida, have found a market in an easily applied paint that incorporates a high ratio of potassium silicate, but which is water-based, nontoxic and nonflammable. With these properties, a hard ceramic finish with superior adhesion and abrasion resistance is formed within an hour of application. Once applied to many structures that are exposed to salt spray and fog (such as bridges, pipelines, and ships), the lifespan of these structures can be dramatically increased.

As of the first decade of the twenty-first century, more than thirty thousand applications of space technology have been brought down to Earth to enhance everyday life. These spinoffs are used in such areas as weather forecasting, consumer and recreation, health and medicine, global positioning system (GPS) networks, remote sensing and telemetry, communication systems, environmental safety and protection, transportation, and computer technology. So many spinoffs have resulted from space technology that NASA maintains a Spinoff Database and publishes *Spinoff*, an annual publication featuring NASA technologies that have been commercialized over forty years of research, development, and transfer to the private sector of the United States and, ultimately, to the world.

 See also **Made in Space (Volume 1)**

Resources

Books and Articles

Bijlefeld, Marjolijn. *It Came from Outer Space: Everyday Products and Ideas from the Space Program.* Westport, CT: Greenwood Press, 2003.

European Space Agency Technology Transfer Programme Office. *Down to Earth: How Space Technology Improves Our Lives.* Noordwijk, Netherlands: European Space Agency, 2009.

Websites

European Space Agency Technology Transfer Programme Office. <http://www.esa.int/SPECIALS/TTP2/> (accessed October 11, 2011).

NASA's Forays Yield Earthly Spin-offs. CNN. <http://edition.cnn.com/2009/TECH/space/07/20/space.nasa.spinoffs/index.html> (accessed October 11, 2011).

NASA Spinoff. National Aeronautics and Space Administration. <http://www.sti.nasa.gov/tto/> (accessed October 11, 2011).

NASA Spinoff Database. National Aeronautics and Space Administration. <http://www.sti.nasa.gov/spinoff/database> (accessed October 11, 2011).

Market Share

The modern space age can be said to have begun in the late 1950s with the launch of experimental satellites by the Soviet Union (now Russia; *Sputnik* in 1957) and the United States (*Explorer I* in 1958). For the next three decades, space programs were operated almost entirely by governmental agencies or private companies contracted by such agencies. The age of space commercialization did not actually begin, then, until the early 1980s with the rise of companies such as Arianespace in France and RKK Energia (S.P. Korolev Rocket and Space Corporation Energia) in Russia and the first purely commercial efforts of American companies such as Lockheed Martin, Boeing, and Northrup Grumman. Since that time, competition for leadership in the space industry has been intense among these companies and their younger rivals, as well as among the United States, Russia, the European Space Agency, China, Japan, and a number of other nations.

One way of measuring that competition is to follow the market share of commercial satellite revenues enjoyed by various countries of the world. Satellite manufacture, launch, and servicing provide by far the largest segment of income to space industries around the world. Throughout the decade of the 1990s, the United States was by far the world's leader in this measure, accounting for 60–65 percent of the world's satellite revenues. According to the Satellite Industry Association (SIA), the total value of the commercial satellite industry worldwide rose from about $8 billion to more than $12 billion between 1996 and 1998.

Since that time, those revenues have oscillated between a low of $8 billion in 2005 and a high of $12 billion in 2006. Over that period of time, the U.S. market share fell from its peak of 78 percent in 1998 to about 40 percent in 2001 to less than 30 percent in 2010. A critical factor in this change was the decision by the U.S. Congress in the late 1990s to include the U.S. aerospace industry under the government's International Traffic in Arms Regulations (ITAR) provisions. The Congress took this action because of concerns that essential technology involved in the manufacture and use of satellites might find its way to nations whose agendas are not necessarily compatible with those of the United States, China being perhaps the best example.

As a result of that action, some nations discontinued ordering satellites, satellite parts, and other components of the satellite industry from U.S. suppliers and turned, instead, to suppliers in other nations or, in some cases, began producing those components in their own nations. As an example, China now has a so-called "ITAR-free" agreement with Europe that allows the transfer of satellite technology from Europe to China of the type that is no longer permitted in the United States. Some experts suggest that as long as the satellite industry is covered by ITAR provisions, the United States will continue to lose market share in the space industry to other parts of the world.

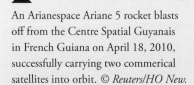

An Arianespace Ariane 5 rocket blasts off from the Centre Spatial Guyanais in French Guiana on April 18, 2010, successfully carrying two commerical satellites into orbit. © *Reuters/HO New.*

A somewhat different measure of worldwide competitiveness in the space industry is the number and commercial value of satellite launches in various countries of the world. In its most recent survey of these trends, the U.S. Federal Aviation Administration (FAA) found that Russia had dominated the field of satellite launches in the five year period between 2006 and 2010, with 56 out of 121 commercial launches occurring from its bases, for a market share of 46 percent of that industry. In second position was Europe, with 27 launches (22 percent), followed by the United States (19 launches; 16 percent); China (two launches; 2 percent), and India (one launch; 1 percent). Sixteen launches were conducted by multinational collaborations. Although Russia dominated the field in terms of sheer launch numbers, the leader in the actual commercial value of launches was Europe, which, throughout the five year period, consistently ranked first in the world in this category. In 2006, Europe recorded a total value of $560 million in its satellite launches compared to $374 million for Russia, $140 million for U.S. launches, and $350 million for multinational launches. No other nation realized significant profits from commercial satellite launches.

That pattern remained consistent throughout the 2006–2010 period. In the last of these years, for example, European launches accounted for almost 54 percent of the world total of $2,453 million value of all satellite launches, followed by Russia (34 percent), and the United States (12 percent). No other nation or combination of nations accounted for any appreciable commercial launch value in 2010, although both China and Japan have launched commerical satellites (some only partially successful) in the 2006–2010 period.

Worldwide satellite launch services are provided by a handful of companies, including International Launch Services (ILS), Boeing, Lockheed Martin, and Arianespace. According to data provided by the Office of Transportation and Machinery of the International Trade Administration of the U.S. Department of Commerce on an annual basis, the dominant launch provider in the period between 2008 and 2010 was ILS, with 26 of 63 commercial launches (41 percent), followed by Arianespace (16 launches; 25 percent), and Boeing Launch Services (13; 21 percent). Minor players in the field included Space Explorations Technologies (SpaceX; five launches), Lockheed Martin (two launches), and Orbital Sciences Corporation (one launch). (Note that the number of satellites launched in a given year is different from the number of launches because many launches carry more than one satellite into orbit.)

 See also **Legislative Environment (Volume 1) • Marketplace (Volume 1) • Regulation (Volume 1) • Space Industries (Volume 4)**

Resources

Books and Articles

Encyclopedia of Emerging Industries, 6th ed. Detroit: Gale, 2011.

Global Apace Industry Stakeholder Mapping. London: Frost & Sullivan, 2009.

Plunkett, Jack W. *Plunkett's Infotech Industry Almanac 2010: The Only Comprehensive Guide to Infotech Companies and Trends.* Houston, TX: Plunkett Research, Ltd., 2010.

Websites

2011 Commercial Space Transportation Forecasts. <http://www.faa.gov/about/office_org/headquarters_offices/ast/media/2011%20Forecast%20Report.pdf> (accessed October 20, 2011).

The Satellite Market. European Space Agency. <http://telecom.esa.int/telecom/www/object/index.cfm?fobjectid=456> (accessed October 22, 2011).

Satellite Markets and Research. <http://www.satellitemarkets.com/> (accessed October 21, 2011).

Space Launch Report. <http://www.spacelaunchreport.com/log2010.html> (accessed October 21, 2011).

State of the Satellite Industry Report 2011. Satellite Industry Association. <http://www.sia.org/PDF/2011%20State%20of%20Satellite%20Industry%20Report%20(June%202011).pdf> (accessed October 22, 2011).

* **remote sensing** the act of observing from orbit what may be seen or sensed below on Earth or another planetary body

Marketplace

Since the 1960s, the market for commercial space operations has been limited almost entirely to communications satellites and commercial rocket launchers, with some tentative ventures, beginning in the 1990s, in the areas of remote sensing* and weather observation. Because of the value of the information they carry—television signals, telephone links, and various digital data—communication satellites have been able to support a fleet of costly launch vehicles. Remote sensing satellite systems were first developed in the 1960s and 1970s. Since then, many satellites have been launched into space for the global observation of Earth's land surface, atmosphere, and oceans and other water surfaces.

For instance, the National Aeronautics and Space Administration's (NASA's) Earth Observing System (EOS) began launching a series of artificial satellites and instruments in 1997. The first one in the series is called Sea-viewing Wide Field-of-view Sensor, which is shortened to *SeaWiFS*. The U.S. satellite was launched on August 1, 1997. Since then, numerous other EOS satellites have been launched. For instance, the French/U.S. *Cloud-Aerosol Lidar and Infrared Pathfinder Satellite Observations*, or *CALIPSO*, was launched on April 28, 2006. As of 2011, *CALIPSO* is continuing its mission to measure aerosols and clouds in Earth's atmosphere. More missions have followed. For example, NASA's *Aquarius* EOS instrument was launched on June 10, 2011, as part of the Argentina spacecraft *SAC-D* (in Spanish, an acronym for *Satelite de Aplicaciones Cientificas-D*, meaning *Satellite for Scientific Applications-D*). The instrument was designed to measure worldwide sea surface salinity to help better predict global climate conditions.

Remote sensing technology has come quite a way between the 1990s and 2010s, with private companies such as Google providing access to remote sensing data to the public through its virtual mapping program called Google Earth (beginning in 2005). The program maps the Earth's surface with the use of satellite images, aerial photography, and other such imaging techniques. In the 2010s, Google Earth is available in various formats, including as an app (short for application) for many mobile devices.

Over the next few decades, the most lucrative industry in space will probably be tourism. On Earth in the early twenty-first century, tourism is the second largest export industry. (The largest is energy, in the form of petroleum.) A tourist industry in space will reduce the cost of getting into orbit because of the sheer volume of launches required. This industry will require launch vehicles not only for transporting people but also for transporting the space-borne facilities tourists will be visiting and for resupplying those facilities. Demand for low-cost launchers will increase by orders of magnitude, promoting competition and driving down costs. As of 2011, Virgin Galactic is expecting to begin suborbital flights into space for the paying public in or around 2013. Its *WhiteKnightTwo* launching vehicles will take *SpaceShipTwo* spacecraft (both made by Scaled Composites) to about 18 kilometers (11 miles) above Earth. Then, *SpaceShipTwo* will separate from its carrier aircraft and complete its manned trip to space, about 100 kilometers (62 miles) above Earth. Other companies pursuing similar space tourism activities include Armadillo Aerospace, Blue Origin, Masten Space Systems, Sierra Nevada, and XCOR Aerospace.

Less costly launchers promise new markets for industrial processes in space. Many industrial processes may benefit dramatically from operating in the weightless environment. So far, no such venture has been cost-effective; the cost of getting the machines and materials into space and the finished products back exceeds the potential sales of the materials produced.

Electrophoresis is a process that uses electric fields to separate fluids; it is used especially in the pharmaceuticals industry to make very valuable (and very expensive) drugs. A team made up of Johnson & Johnson and McDonnell Douglas flew a prototype electrophoresis system on four NASA space shuttle flights, with an eye to making it a commercial venture. Initially, it looked as if producing pharmaceuticals in orbit would make sense from a business standpoint, but the companies ultimately determined that it would be less costly to make their products on the ground. However, dramatically lower launch costs would turn the business equations around.

Lower launch costs open space to a host of other industries. Most of the potential markets identified to date are in esoteric high-tech fields, such as super-strength drawn fibers, single-crystal metals, and protein crystals. Others are more familiar, such as movie and television production, for which space would provide an excellent shooting location. These markets are only forerunners of new markets that will open as commercial business moves into space. Current research in space processing methods might lead to some surprising markets for both industrial and consumer products.

New Markets

Transgenic plants, which are made by crossing the genes from diverse species, promise to create whole new species with new flavors and dramatically increased crop yields. In an experiment flown on the space shuttle, a rose plant produced some new, very desirable fragrances in its microgravity environment while orbiting about Earth. The production of other bio-engineered agriculture products is possible from the confines of outer space. Moreover, several prototype systems have been developed that may one day lead to the deployment of huge electrical power plants in orbit, or even on the Moon, that collect energy directly from the Sun and transmit the power to Earth on microwave beams. Zeolite crystals are yet another product that one day may be produced in space. These crystals command high prices in the chemical processing industry because of their ability to selectively filter out specific chemicals. Though they are scarce on Earth, they can be manufactured efficiently in space.

Opportunities for new markets in space extend to the medical industry as well, which will benefit from improved efficiency in the production of pharmaceuticals and entire new technologies, such as components for bone replacement. Protein synthesis, compound doping, composite material manufacturing, medical procedure enhancements, and biomedical monitoring are only a few of a vast number of ways to possibly expand the medical industry in space.

Finally, developing industries in space create new markets to meet the demands of the space-borne industries. People working in space need places to live, work, and play; they need food to eat, clothes to wear, and transportation systems to get around. In short, they need everything that people need on Earth, and each of these needs is a new market for the space entrepreneur.

Of the many ways to use outer space as a commercial market, some of the ways yet to be mentioned include such topics as centrifuge facilities for product development, closed-loop (completely recycled) life support systems, nanotechnology products and processes, and robotics. For instance, NASA has been testing a robot on the International Space Station (ISS) called "Robonaut 2" (or R2) since February 2011, when the STS-133 crew delivered it to the station. R2 was designed to help the ISS Expedition crew with various tasks inside the structure. Based on its performance so far, General Motors (GM) is working with NASA to develop the next-generation Robonaut. In the future, other such robots may replace humans on extravehicular activities, such as repairs to the outside of spacecraft, satellites, and other such space structures.

 See also **Launch Industry (Volume 1) • Launch Services (Volume 1) • Made in Space (Volume 1) • Space Tourism, Evolution of (Volume 4) • Tourism (Volume 1)**

Resources

Books and Articles

Catchpole, John E. *The International Space Station: Building for the Future.* Berlin: Springer, 2008.

Dubbs, Chris, and Emeline Paat-Dahlstrom. *Realizing Tomorrow: The Path to Private Spaceflight.* Lincoln: University of Nebraska Press, 2011.

Handberg, Roger. *International Space Commerce: Building From Scratch.* Gainesville: University Press of Florida, 2006.

Jamison, David E., editor. *The International Space Station.* New York: Nova Science, 2010.

Lutes, Charles D, and Peter L. Hays, editors. *Toward a Theory of Spacepower: Selected Essays.* Washington, D.C.: National Defense University Press, 2011.

McCurdy, Howard E. *Space and the American Imagination.* Baltimore: Johns Hopkins University Press, 2011.

Reference Guide to the International Space Station. Washington, D.C.: National Aeronautics and Space Administration, 2010.

Tumlinson, Rick, and Erin Medlicott, editors. *Return to the Moon.* Burlington, Ontario, Canada: Apogee, 2005

Websites

Center for Space Resources. Colorado School of Mines Research Center. <http://spaceresources.mines.edu/> (accessed November 3, 2011).

The Earth Observing System. National Aeronautics and Space Administration. <http://eospso.gsfc.nasa.gov/> (accessed November 3, 2011).

Google Earth. Google. <http://www.google.com/earth/index.html> (accessed November 3, 2011).

NASA Spinoff. Office of Technology Transfer, National Aeronautics and Space Administration. <http://www.sti.nasa.gov/tto/> (accessed October 21, 2011).

Office of the Chief Technologist. National Aeronautics and Space Administration. <http://www.nasa.gov/offices/ipp/home/index.html> (accessed November 3, 2011).

Robonaut 2. National Aeronautics and Space Administration. <http://www.nasa.gov/pdf/469616main_Robonaut2_factsheet.pdf> (accessed November 3, 2011).

U.S. National Spectrum Requirements: Projections and Trends. National Telecommunications and Information Administration. <http://www.ntia.doc.gov/osmhome/EPS/openness/sp_rqmnts/contents.html> (accessed November 3, 2011).

Virgin Galactic Unveils Suborbital Spaceliner Design. Space.com. <http://www.space.com/news/080123-virgingalactic-ss2-design.html> (accessed October 24, 2011).

Virgin Galactic. <http://www.virgingalactic.com/> (accessed November 3, 2011).

Whalen, David J. *Communications Satellites: Making the Global Village Possible.* National Aeronautics and Space Administration. <http://www.hq.nasa.gov/office/pao/History/satcomhistory.html> (accessed November 3, 2011).

Wisconsin Center for Space Automation and Robotics. <http://wcsar.engr.wisc.edu/> (accessed November 3, 2011).

McCall, Robert
American Illustrator
1919–2010

Robert T. (Bob) McCall, one of the world's leading illustrators of space themes, was named to the Society of Illustrators Hall of Fame in 1988. His bold, colorful canvases depict the visions of America's space program since its beginnings.

McCall was born in 1919 in Columbus, Ohio. He grew up on magazines describing airplanes and futuristic spacecraft. When he went to the Ohio State Fair, he saw his first airplane, a World War I (1914–1918) bomber. By, this time, McCall had found his talent for drawing, and he was drawing everything that interested him, especially things that peaked his adventuresome spirit. After graduation from high school, McCall won a scholarship to the Columbus Fine Art School. During World War II (1939–1945), McCall enlisted in the Army Air Corps and became a bombardier instructor. After the war, he and his wife, artist Louise McCall, moved to Chicago and later New York City, where he worked as an advertising illustrator, working for such magazines as *Life, Saturday Evening Post,* and *Popular Science.* He especially enjoyed illustrating in the aviation field. Through the Society of Illustrators, McCall was invited to produce paintings for the U.S. Air Force.

In the early 1960s, McCall was one of the first artists selected for the National Aeronautics and Space Administration's (NASA) fine arts program after producing future space concepts for *Life* magazine. This connection led to a number of patch designs for space missions, a U.S. Postal Service commemorative stamp set, and murals at NASA field centers. Several astronauts include his artwork in their collections.

McCall's most visible work is a six-story mural in the Smithsonian's National Air and Space Museum in Washington, D.C., which is seen by over six million visitors annually. His most widely recognized work is the painting of a massive double-ringed space station for the Stanley Kubrick (director) and Arthur C. Clarke (writer) science fiction film *2001: A Space Odyssey* (1968).

McCall's training included studies at the Columbus School of Art and Design and the Art Institute in Chicago in the late 1930s. He donated some of his paintings to the Air Force, the Pentagon, Air Force Academy, and many air bases around the world. McCall helped to found the American Society of Aviation Artists. Among his many honors are being a recipient of the Uri Gagarin Medal (1988), receiving an honorary doctor of visual arts degree from Columbus College of Art and Design (1998), being inducted into the Arizona Aviation Hall of Fame (2001), receiving the Elder Statesman of Aviation Award from the National Aeronautic Association (2002), and receiving the Douglas S. Morrow Public Outreach Award by the Space Foundation (2003). McCall passed away, from heart failure, in 2010 at the age of 90.

 See also **Artwork (Volume 1)** • **Bonestell, Chesley (Volume 4)** • **Rawlings, Pat (Volume 4)**

Resources

Books and Articles

Asimov, Isaac. *Our World in Space,* art by Bob McCall. Greenwich, CT: New York Graphic Society, 1974.

Bova, Ben. *Vision of the Future: The Art of Robert McCall.* New York: Harry N. Abrams, 1982.

Bradbury, Ray. "Introduction." In *The Art of Robert McCall: A Celebration of Our Future in Space,* captions by Tappan King. New York: Bantam Books, 1992.

Websites

McCall Studios. <http://www.mccallstudios.com/> (accessed October 13, 2011).

Robert McCall (1919–2010). Smithsonian National Air and Space Museum. <http://blog.nasm.si.edu/2010/03/05/robert-mccall-1919-2010/> (accessed October 13, 2011).

Military Customers

The military space program is a significant but largely unseen aspect of space operations. Nearly two dozen countries currently have, or had in the past, some kind of military space program, but the U.S. program dwarfs the efforts of all these other countries combined.

Military space operations are divided into five main areas: reconnaissance* and surveillance, signals intelligence, communications, navigation, and meteorology*. Only the United States and Russia operate spacecraft in all five areas. Several other countries have long used communications satellites for military purposes. Beginning in the 1990s, several countries in addition to Russia and the United States began developing reconnaissance satellites.

Reconnaissance and Surveillance

Reconnaissance and surveillance involve the observation of Earth for various purposes. Dedicated reconnaissance satellites, like the United States' Improved Crystal (also called "Advanced KENNAN" and "Ikon") and the Russian Terilen, take photographs of targets on the ground and relay them to receiving stations in nearly real time. These satellites, however, cannot take continuous images like a television camera. Instead, they take a black-and-white photograph of a target every few seconds. Because they are in low orbits and are constantly moving, they can photograph a target for only a little over a minute before they move out of range. The best American satellites, which are similar in appearance to the Hubble Space Telescope, can see objects about the size of a softball from hundreds of miles up in space but they still cannot read license plates, as of the latter part of the 2000s. The Russians also occasionally use a system that takes photographs on film and then returns the film to Earth for processing. This provides them with higher-quality photos. The United States abandoned this technology in the 1980s after developing superior electronic imaging technology.

Other nations, such as France and Japan, operate reconnaissance satellites that can see images on the ground about one to three feet in length. France operates its *Helios* satellite while Japan directs the operations of its *Information Gathering Satellite.* From the late 1970s until the mid–1990s, China had a film-based system, which is no longer operational. China now operates its *Fanhui Shi Weixing* satellites. In addition, India (such as CARTOSAT and RISAT), Israel (Ofeq and TecSAR), and Egypt (EgyptSat) operate satellites capable of making visual observations of the ground. Some private companies, like Google, operate commercial imagery satellites and sell images on the World Wide Web. These satellites are much less capable than the larger military satellites but their products have improved significantly and are in demand.

Other surveillance satellites, such as the American *DSP* (Defense Support Program), which is part of the Satellite Early Warning System, and *Space-Based Infrared System* (SBIRS, pronounced "sibirs") and the Russian *Oko* (or "eye"), are equipped with infrared* telescopes and scan the ground for the heat produced by a missile's exhaust. They can be used to warn of missile attack and can predict the targets of missiles fired hundreds or thousands of miles away. There are also satellites that look at the ground in different wavelengths* to peer through camouflage, try to determine what objects are made of, and analyze smokestack emissions.

* **infrared** portion of the electromagnetic spectrum with waves slightly longer than visible light

* **wavelength** the distance from crest to crest on a wave at an instant in time

Satellites such as this artist's rendering of an U.S. Air Force NAVSTAR Global Positioning System (GPS) BlockIIF, are used increasingly in military operations by countries with global interests. © *AP Images.*

▶

* **low Earth orbit** an orbit between 300 and 800 kilometers (185 and 500 miles) above Earth's surface

* **geostationary orbit** a specific altitude of an equatorial orbit where the time required to circle the planet matches the time it takes the planet to rotate on its axis. An object in geostationary orbit will always remain over the same geographic location on the equator of the planet it orbits

* **radar** a technique for detecting distant objects by emitting a pulse of radio-wavelength radiation and then recording echoes of the pulse off the distant objects

Signals Intelligence

Signals intelligence satellites can operate either in low Earth orbit* or in extremely high, geostationary* orbit, where they appear to stay in one spot in the sky. These satellites listen for communications from cellular telephones, walkie-talkies, microwave transmissions, radios, and radar*. They relay this information to the ground, where it is processed for various purposes. Contrary to popular myth, these satellites do not collect every conversation around the world. There is far more information being transmitted every day over the Internet than can be collected by even the best spy agency.

Communications

Communications satellites operate in several different orbits for various purposes. The most common communications satellites operate in geosynchronous orbit. Some, like the U.S. Navy's *UHF Follow-On* (UFO) satellite fleet (where UHF stands for ultra-high frequency), are used to communicate with ships at sea, along with various fixed and mobile terminals. Others, like the Air Force's massive *Milstar* satellite, are used to communicate with troops on the ground and submarines equipped with small dish antennas. Still other communications satellites are used to relay reconnaissance pictures to ground stations or to troops in the field. Some satellites are used to relay data and commands to and from other satellites.

Russia operates a number of military communications satellites, including some that store messages for a brief period before relaying them to the ground. Several other countries, such as the United Kingdom, Spain, and

France, have either military communications satellites or a military communications package installed on a commercial satellite. But few countries have the global military communications requirements of the United States.

Navigation and Meteorology

Navigation satellites are also vital to military forces. Sailors have used the stars to navigate for centuries. Beginning in the early 1960s, the U.S. Navy developed a satellite system to help it navigate at sea. This was particularly important for ballistic* missile submarines that stayed submerged for most of their patrols and could only occasionally raise an antenna above the waves to determine their position.

In the 1980s, the U.S. Air Force started operating the Global Positioning System (GPS), which allowed anyone equipped with a receiver to locate his or her position on Earth to within about nine meters (30 feet) or less. Initially, the system was set up so that civilian users received less-precise information than military users, but this feature was disabled in 2000. The U.S. GPS system is now widely used throughout the world by military and civilian users alike. The U.S. military continues to operate the system, and reportedly has the ability to disrupt its use by enemies of the United States within a given region. Russia operates a system called GLONASS that is similar to GPS, but that provides better coverage in northern latitudes. It was also originally a military program.

Accurate weather information is critical to military operations. The United States and Russia operate meteorology satellites* for military use. However, since the end of the Cold War, separate military and civilian meteorology satellites have been viewed as an unnecessary expense, and the military systems have gradually been merged with their similar civilian counterparts.

Antisatellite Defense ("Star Wars")

Antisatellite (ASAT) and missile defense (Strategic Defense Initiative [SDI] or "Star Wars") satellites are not currently part of any nation's arsenal. However, the United States, the Soviet Union (now Russia), and China have all developed such weapons at times in the past. ASAT weapons are difficult to develop and operate and they have limited usefulness. It is extremely difficult to use a satellite to shoot down ballistic missiles. In the future, satellites may be used to intercept missiles, but it is unlikely that this will happen for a long time.

During the Cold War, both superpowers (the United States and the Soviet Union) studied the possibility of placing nuclear weapons in orbit, but neither country did so. A bomb in orbit will spend most of its time nowhere near the target it needs to hit, unlike a missile on the ground, which will always be in range of its target. In addition, controlling a system of orbiting bombs would be difficult.

* **ballistic** the path of an object in unpowered flight; the path of a spacecraft after the engines have shut down

* **meteorology satellites** satellites designed to take measurements of the atmosphere for determining weather and climate change

Military Role of Humans in Space

There has never been a clear military role for humans in space, despite decades of study by both superpowers. During the 1960s, the United States explored several piloted military space systems. One of these was the *DynaSoar* spaceplane, which was canceled in 1963 after the Air Force could not find a clear mission for it. Another of these was the *Manned Orbiting Laboratory* (MOL). MOL was to carry a large reconnaissance camera, and two astronauts were to spend up to a month in orbit, photographing objects on the ground. The United States canceled MOL in 1969 after it became clear that humans were not needed for the job and robotic systems could perform the task reliably and in many cases better than humans. The Soviet Union briefly operated crewed space stations similar to MOL but abandoned these for the same reason as the United States.

Summary

Around the world, military operations are increasingly using commercial satellites to accomplish their missions. Commercial communications satellites are particularly useful and relatively cheap. In addition, commercial reconnaissance satellites are finding many military uses, enabling countries that cannot afford their own satellites to buy photos of their adversaries.

Satellites are not required for many local military operations. However, if a country is operating far from its borders or has global interests, they are a necessity. Only a few countries are willing to pay the expense of operating military space systems, but that number is growing.

In January 2007, China destroyed one of its nonfunctioning weather satellites, not with a military satellite from space, but with a ground-based multistage missile that contained an anti-satellite payload. The missile was thought by the U.S. military to be the first anti-satellite missile to have destroyed a satellite in space since the United States destroyed one in 1985. In February 2008 the United States destroyed one of its own spy satellites, which was malfunctioning, with a missile launched from a ship stationed in the Pacific Ocean. Both actions show that many countries prefer to use ground-based interceptor missiles to destroy artificial objects in space, rather than attempt the more complicated method of positioning anti-satellite devices in space.

These two anti-satellite missile launches demonstrated that China and the United States could potentially shoot down the satellites of countries that they come into conflict with. The United States, Russia, and China are key participants in the potential use of space for defense—what is sometimes termed an "arms race" in space. However, many other countries are attempting to prevent such an arms race in space with international agreements.

 See also **Global Positioning System (Volume 1)** • **Military Exploration (Volume 2)** • **Military Uses of Space (Volume 4)** • **Navigation From Space (Volume 1)** • **Reconnaissance (Volume 1)** • **Satellites, Types of (Volume 1)**

Resources

Books and Articles

Arbatov, Alexei, and Vladimir Dvorkin, eds. *Outer Space: Weapons, Diplomacy, and Security.* Washington, DC: Carnegie Endowment for International Peace, 2010.

Coletta, Damon, and Frances T. Pilch, eds. *Space and Defense Policy.* London: Routledge, 2009.

Johnson-Freese, Joan. *Space as a Strategic Asset.* New York: Columbia University Press, 2007.

MacDonald, Bruce W. *China, Space Weapons, and U.S. Security.* New York: Council on Foreign Relations, 2008.

Norris, Pat. *Spies in the Sky: Surveillance Satellites in War and Peace.* Berlin: Springer; Chichester, UK: Praxis Publishing, 2008.

Richelson, Jeffrey T. *America's Space Sentinels: DSP Satellites and National Security.* Lawrence: University Press of Kansas, 1999.

Robertson, Ann E. *Militarization of Space.* New York: Facts on File, 2010.

Spires, David. *Beyond Horizons: A Half Century of Air Force Space Leadership.* Rev. ed. Washington, DC: U.S. Government Printing Office, 1998.

Websites

Kestenbaum, Kevin. *Chinese Missile Destroys Satellite in 500-Mile Orbit.* NPR. <http://www.npr.org/templates/story/story.php?storyId=6923805> (accessed October 27, 2011).

U.S. Missile Hits Spy Satellite. New Scientist. <http://www.newscientist.com/article/dn13359-us-missile-hits-spy-satellite.html> (accessed October 27, 2011).

Mining   *See Asteroid Mining (Volume 4); Natural Resources (Volume 4); Resource Utilization (Volume 4); Space Resources (Volume 4)*

Mueller, George

American Engineer and Corporate Leader
1918–

George Edwin Mueller is an American engineer and corporate leader whose work and career span the development of the U.S. space program. Born July 16, 1918, Mueller holds a master's degree in electrical engineering from Ohio State University, and worked at Bell Laboratories

George Mueller headed NASA's Apollo Manned Space Flight Program from 1963 to 1969. He is known as the "Father of the Space Shuttle." © *Bettmann/Corbis.*

* **Gemini** the second series of American-piloted spacecraft, crewed by two astronauts; the Gemini missions were rehearsals of the spaceflight techniques needed to go to the Moon

* **Apollo** American program to land men on the Moon; *Apollo 11, Apollo 12, Apollo 14, Apollo 15, Apollo 16, Apollo 17* delivered twelve men to the lunar surface (two per mission) between 1969 and 1972 and returned them safely back to Earth

before subsequently earning his doctorate degree in physics from Purdue University (Indiana). His career has focused on the development and success of the U.S. space program.

As head of the National Aeronautics and Space Administration's (NASA's) Apollo Manned Space Flight Program from 1963 to 1969, Mueller led the program that put Americans on the Moon. He was in charge of the Gemini*, Apollo*, and Saturn programs. In addition, he coordinated the activities of about 20,000 industrial firms, 200 universities and colleges, and hundreds of thousands of individuals into a concerted effort. His leadership made it possible to meet the challenge set in 1961 of not only landing men on the Moon before the end of the decade, but also their safe return to Earth. Mueller's position within NASA also involved him developing the Space Transportation System (STS), what is commonly called the Space Shuttle program, before he left NASA.

After the successful completion of the second landing on the Moon by the Apollo 12 mission, Mueller returned to industry where he was senior vice president of General Dynamics Corporation (1969–1971) and chairman and chief executive officer (1981–1983) and president (1971–1980) of Systems Development Corporation. He was the senior vice-president of Burroughs Corporation from 1982 to 1983, president of Jojoba Propagation Laboratories from 1983 to 1995 and chair of Desert King Jojoba Corporation from 1983 to 1995.

Then, from 1995 to 2004, Mueller was the chief executive officer of Kistler Aerospace Corporation, where he was leading the development and operations of the two-stage-to-orbit launch vehicle Kistler K-1, intended to become the world's first fully reusable aerospace vehicle. From 2004 to 2006, Mueller was the company's chairperson and chief vehicle architect. Mueller left the company when it was acquired by Rocketplane Limited, Inc., in 2006.

Mueller is the recipient of many prestigious awards, including the Rotary National Award for Space Achievement, which was awarded to him in 2002. He has been awarded six honorary doctorates. Among his many other awards are three NASA Distinguished Service Medals, the Apollo Achievement Award, the Medal of Pairs, the American Institute of Aeronautics and Astronautics Goddard Medal, the National Award for Space Achievement, and the American Astronautical Society Lloyd V. Berkner Award.

 See also **Apollo (Volume 3)** • **Apollo Lunar Landing Sites (Volume 3)** • **Gemini (Volume 3)** • **NASA (Volume 3)**

Resources

Books and Articles

Murray, C., and C. B. Cox. *Apollo.* Burkittsville, MD: South Mountain Books, 2004.

Orloff, Richard W. *Apollo: The Definitive Sourcebook.* Berlin: Springer, 2006.

Websites

George E. Mueller, Father of the Space Shuttle Program, to Receive 2002 Rotary National Award for Space Achievement. Rocketplane Kistler, Inc. <http://www.rocketplanekistler.com/newsinfo/pressreleases/012402.html> (accessed October 13, 2011).

Mueller, George Edwin. Encyclopedia Astronautica. <http://www.astronautix.com/astros/mueller.htm> (accessed October 13, 2011).

NASM Oral History Project: Dr. George Mueller. Smithsonian National Air and Space Museum. <http://www.nasm.si.edu/research/dsh/TRANSCPT/MUELLER1.HTM> (accessed October 13, 2011).

Two Commercial Space Companies Join Forces. Rocketplane Global, Inc. <http://www.rocketplane.com/press/20060307a.html> (accessed October 13, 2011).

▲
Elon Musk. © *Brendan Smialowski/Getty Images.*

Musk, Elon

Businessman and Space Industry Leader
1971–

Elon Musk is an entrepreneur and space visionary. He founded a company for transporting cargo and people into outer space called Space Exploration Technologies Corporation, better known as SpaceX. By 2011, Musk had led SpaceX in developing an entirely new line of rockets, as well as the Dragon capsule for launching people and cargo into orbit. One intended use for the Dragon capsule is to ferry astronauts to the International Space Station (ISS), launched atop SpaceX's Falcon 9 rocket.

In addition to SpaceX, Musk has founded or co-founded a number of other companies, including two Internet companies, a solar energy company, and the Tesla electric automobile company. Musk resides in Bel Air, California, with wife Talulah Riley; he has five sons from a previous marriage to Justine Musk.

Addressing Major Challenges

Elon Musk was born on June 28, 1971, in South Africa. Musk had his eye on ultimately living in the United States when he emigrated from South Africa to Canada, the country of his mother's birth, at the age of seventeen. After living for a short period in Canada, Musk moved to the United States to attend the University of Pennsylvania, where he obtained bachelor's degrees in business and physics. It was around this time Musk decided that he wanted to tackle some of the major challenges in the fields of energy, the Internet, and space transportation.

* **payload** any cargo launched aboard a rocket that is destined for space, including communications satellites or modules, supplies, equipment, and astronauts; does not include the vehicle used to move the cargo or the propellant that powers the vehicle

Musk began a graduate program in physics, but almost immediately dropped out to join his brother in starting an online content company called Zip2. Elon Musk sold his interest in Zip2 in 1999. He then started another Internet company that, after a merger, ultimately became PayPal, the highly successful online payment service. Musk sold his interest in PayPal in 2002. By this time, Musk was a very wealthy man.

Making Spaceflight More Affordable

Turing his attention to the problem of affordable space transportation, Musk founded SpaceX in 2002. The overarching motivation for his interest in developing an aerospace company was to help bring about the colonization of outer space. As Mr. Musk sees it, there are very important reasons to see humanity colonize other worlds. The most important reason is to ensure the survival of the human species in case of a cataclysm here on Earth (for instance, a devastating collision with a large comet or asteroid). Such a cataclysmic event could wipe out civilization on Earth. A second reason Musk sees for colonizing space is that it will profoundly enrich human culture.

As Musk sees it, the high cost of launching payloads* into space has been the limiting factor to a greater human presence there. Therefore, a radical reduction in the cost of launching payloads into orbit is essential. Musk is not alone in these views—several other commercial space enterprises were started in the 1980s and 1990s with the goal of lowering spaceflight costs. But with few exceptions those new space ventures failed. Musk has cited three main reasons for the failure of many previous space start-ups: collectively, their personnel lacked the necessary technical skills; there was insufficient funding necessary to see the company through the development stage; and an over-reliance on breakthroughs in the technology of spaceflight—breakthroughs that usually failed to materialize.

As both chief executive officer (CEO) and chief technology officer (CTO) of SpaceX, Musk has been in position to address the kinds of problems that have plagued other space start-ups. From the beginning of SpaceX, Musk brought in talented people to perform the necessary tasks of building a new aerospace company, while at the same time being careful not to bring in too many people, too quickly. The result has been a steady and balanced growth in personnel with complimentary sets of skills.

In the area of technology, Musk has not tied SpaceX's success to over-hyped technological breakthroughs in rocket propulsion, but instead has largely relied on proven rocket systems and technologies, with an emphasis on safety and reliability. For instance, the first two rocket models successfully launched by SpaceX—the Falcon 1 and Falcon 9—use liquid oxygen and rocket-grade kerosene (RP-1), which is much less expensive, safer, and easier to store and handle than some other types of liquid rocket fuels, such as the liquid hydrogen that was used in the space shuttles.

Funding for SpaceX has consisted of a combination of capital investment and revenue from various customers for its launch services. By 2006,

* **low Earth orbit** an orbit between 300 and 800 kilometers (185 and 500 miles) above Earth's surface

Musk himself had invested $100 million of his own money in the company. Investment has come from other investors as well. This continuous infusion of capital into SpaceX while it was developing its own in-house rockets and rocket engines helped the company get from startup enterprise to viable launch company.

The first rocket launch attempt by SpaceX was with the Falcon 1 model in 2006. This initial attempt failed, as did two subsequent attempts using Falcon 1's. In September 2008, a Falcon 1 successfully launched into space, followed by another Falcon 1 launch success in 2009. The much larger Falcon 9 rocket was first successfully launched in mid 2010, with a second launch near the end of 2010 that placed the company's Dragon capsule into low Earth orbit*. After a short time orbiting the Earth, the Dragon capsule parachuted into the Pacific Ocean. This marked the first time ever that a private company had launched, orbited, and recovered its own spacecraft.

From the very beginning of rocket and capsule development at SpaceX, Musk has kept in mind the principles of safe, reliable, and affordable spaceflight. Though not yet demonstrated (as of late 2011), Musk has insisted that he will strive for reusability of the SpaceX line of rockets, meaning that at least one rocket stage in each of the Falcon-series line of rockets will be recovered and reused. This reusability forms a key part of attaining lower costs for SpaceX launches as compared to its competitors.

In 2008, SpaceX was awarded a contract to deliver crew and supplies to the International Space Station utilizing funding from the Commercial Orbital Transportation Services (COTS) program, which is directed through the National Aeronautics and Space Administration (NASA). By late 2011, SpaceX announced it had more than thirty orders for launches using its rockets scheduled through the year 2017. Future plans at SpaceX include development of a heavy-lift rocket launcher (Falcon X), and developing a completely reusable rocket system.

An Influential Space Industry Leader

Elon Musk has received multiple awards in recognition of his efforts in furtherance of the commercial space industry. In 2010, the Kitty Hawk Foundation proclaimed Elon Musk a "living legend in aviation" for his efforts towards creating SpaceX's Falcon 9 rocket and Dragon space capsule. Musk is also well known for his involvement in Tesla Motors, the electric vehicle company, which he helped establish in 2003.

 See also **Aerospace Corporations (Volume 1)** • **Launch Industry (Volume 1)** • **Launch Services (Volume 1)** • **Private Space Companies (Volume 1)**

Resources

Books and Articles

Belfiore, Michael P. *Rocketeers: How a Visionary Band of Business Leaders, Engineers, and Pilots is Boldly Privatizing Space.* New York: Smithsonian Books, 2007.

Herman, Richard T. and Robert L. Smith. *Immigrant, Inc.: Why Immigrant Entrepreneurs Are Driving the New Economy (and how they will save the American worker)*. Hoboken, New Jersey: John Wiley & Sons, 2010.

Websites

About Tesla: Elon Musk. Tesla Motors. <http://www.teslamotors.com/about/executives/elonmusk> (accessed October 14, 2011).

Company Overview. Space Exploration Technologies. <http://www.spacex.com/company.php> (accessed October 14, 2011).

NOP

Navigation from Space

For hundreds of years, travelers have looked to the sky to help navigate their way across oceans, deserts, and land. Whether using the angle of the Sun above the horizon or the night stars, celestial bodies guided explorers to their destinations. In the twenty-first century, people still look to the sky for direction, but now they are using satellites that orbit Earth to determine their location. In fact, it is quite common to see people using what is called the Global Positioning System (GPS), which is a particular type of global navigation satellite system (GNSS)—system of global coverage of positions on Earth provided by orbiting satellites—to answer the age-old question: Where am I?

Evolution of Satellite Navigation

The idea of using satellites for navigation was conceived when the satellite *Sputnik 1* was launched by the Soviet Union in 1957. At that time, U.S. scientists developed a way to track Sputnik's orbit using the time delay or Doppler shift of the radio signal being broadcast by the satellite. The scientists proposed that this process could be used in the opposite way for navigation. Specifically, using a satellite with a known orbit, one's position could be determined by observing the time delay or Doppler shift of a radio signal coming from that satellite.

The concept of being able to determine a position from satellites appealed to the U.S. Navy. To test the idea, they developed the TRANSIT (Transit) satellite navigation system (also called NAVSAT, for Navy Navigation Satellite System). By 1964, Polaris submarines were using Transit to update the inertial navigation systems onboard the submarines. During roughly the same period, the U.S. Air Force also had a satellite navigation program under development. In the early 1970s, the Navy and Air Force programs merged into one program called the Navigation Technology Program. This program evolved into the NAVSTAR (Navigation System with Timing and Ranging) GPS—the space navigation system used today. Commonly called GPS, it became fully operational on June 26, 1993, when the twenty-fourth Navstar satellite was placed into orbit.

How the Global Positioning System Works

In the early 2010s, the GPS system was utilizing twenty-four to thirty-one active satellites (with three to four retired spares that can be reactivated, if needed) that circle Earth in an approximate 20,000-kilometer (12,400-mile) orbit above Earth. The satellites are in orbits that are

The position that this soldier obtains through his Global Positioning System receiver is communicated through the triangulation of the signals of at least four orbiting satellites. © *Leif Skoogfors/ Corbis.*

inclined at about 55° with respect to the equator. The satellites are in six orbital planes, each of which has four operational satellites. The other satellites are redundant satellites that help to improve the accuracy of the system. At any given point on Earth, at least six satellites are above that local horizon. In March 1994, the full twenty-four-satellite constellation was in place in orbit and the network became fully functional the following year. Users of this navigational system need a GPS receiver. There are many commercial manufacturers of these devices. They are sold in most stores that sell electronic equipment, and come in a range of prices depending upon the functions performed by the unit (such as presenting the user's location on a display), its accuracy, and so forth. By 2011, some brands of "personal GPS trackers," which can be worn on the wrist or around the neck, could be purchased for under $40.

Each satellite in the GPS transmits a signal with information about its location and the current time. Signals from all of the satellites are transmitted at the same time. A GPS receiver receives these signals at different times because some satellites are closer than others to the receiver. The distance to the satellite is determined by calculating the amount of time it takes the signal to reach the receiver. The position of the receiver is determined by triangulation, except that in this case, the distance to four GPS satellites is used to determine the receiver's position in three dimensions.

Alternatives to the Global Positioning System

The United States allows anyone around the world to use the GPS system as a free resource. For many years, however, there has been a concern in other countries that the United States could deny access to the GPS system at any time. This has led to attempts by other nations at developing alternative satellite navigation systems. The most notable of these emerging systems is a European Space Agency venture called Galileo.

The Galileo system will be under the control of the nations of the European Union (EU), and will provide major economic benefits to Europe. When the Galileo navigation system is completed, it will consist of 30 satellites providing worldwide navigation coverage. As in the U.S. GPS system, the optimal number of satellites in Galileo will be 24, with the additional satellites used as spares in case of equipment failures. The first two Galileo navigation satellites were launched atop a Russian Soyuz rocket at the spaceport located in Kourou, French Guiana on October 21, 2011. Galileo is designed to be compatible with the U.S. GPS navigation system. However, there are differences between the EU Galileo system and the U.S. GPS system. Some of the satellites in Galileo's constellation will be in orbits with greater inclination to the equatorial plane than the GPS satellites. This will give northern Europe better coverage than that provided by GPS today. Another difference between the two systems is that Galileo will have a higher accuracy than the U.S. system—within a meter of the actual position. Navigational services using the Galileo system are scheduled to begin by 2014.

Russia's satellite navigation system is called GLONASS (short for Global Navigation Satellite System). Much like its American counterpart, the GLONASS system was initially conceived and funded to meet the navigational needs of the military. Due to economic problems, the GLONASS system fell into disrepair in the late 1990s and throughout much of the 2000s. But the Russian government made new investments in order to resurrect GLONASS, and has projected that the system would have its full complement of 24 primary navigation satellites, along with several spares, by the end of 2011.

China is working to develop its own satellite navigation system called Beidou, named after the Big Dipper constellation. As of late 2011, the Beidou-2 navigation system consists of eight satellites. More satellites will be added, so that the full Beidou-2 system (also called the COMPASS system) is anticipated to consist of at least 30 satellites offering worldwide navigational services by 2020.

India and Japan are also developing satellite navigation systems, but their systems are expected to be only regional in nature—covering only their country and surrounding territories. India's system is called Indian Regional Navigational Satellite System (IRNSS), and is designed to operate with a total of seven satellites. India expects its system to be ready for full operations by 2014.

Japan's system is named the Quasi-Zenith Satellite System (QZSS), which is designed to use three satellites in orbits around Earth so that at least one of the three is at or near its zenith (highest altitude above Earth) at any one time. The QZSS system is designed to work with America's GPS system in order to provide much higher positioning data for users in Japan and the Pacific region than can GPS alone. As of late 2011, the QZSS system is budgeted to operate using four satellites. Japan hopes to have its navigation system operational by 2013.

 See also **Global Positioning System (Volume 1) • Military Customers (Volume1) • Navigation (Volume 3) • Reconnaissance (Volume 1) • Satellites, Types of (Volume 1)**

Resources

Books and Articles

Clarke, Bill. *Aviator's Guide to GPS.* New York: McGraw-Hill, 1998.

Hinch, Stephen W. *Outdoor Navigation with GPS.* 3rd ed. Birmingham, AL: Wilderness Press, 2010.

McNamara, Joel. *GPS For Dummies.* Hoboken, NJ: Wiley, 2008.

Misra, Pratap and Per Enge. *"Global Positioning System: Signals, Measurements, and Performance.* Lincoln, Mass.: Ganga-Jamuna Press, 2010.

Stearns, Edward V. B. *Navigation and Guidance in Space.* Englewood Cliffs, NJ: Prentice-Hall, 1963.

Hermann Oberth, in his office at the Army's Redstone Arsenal in Alabama, working on a mathematical formula in 1958. Oberth came to the United States in 1956 from Germany as a consultant on space travel. © *AP Images.*

Websites

Compass Satellite Navigation System (Beidou). SinoDefence.com. <http://www..sinodefence.com/space/satellite/compass-beidou2.asp> (accessed October 20, 2011).

Galileo Navigation. European Space Agency. <http://www.esa.int/esaNA/galileo.html> (accessed October 20, 2011).

Global Positioning System: History. National Park Service, U.S. Department of the Interior. <http://www.nps.gov/gis/gps/history.html> (accessed October 20, 2011).

GLONASS. Information Analytical Centre, Russian Federal Space Agency. <http://www.glonass-ianc.rsa.ru/en/GLONASS/> (accessed October 20, 2011).

GPS.gov. United States Government. <http://www.gps.gov/> (accessed October 20, 2011).

Oberth, Hermann
Austro-Hungarian Physicist
1894–1989

Hermann Julius Oberth, who was born on June 25, 1894, in the Transylvanian town of Hermannstadt, is considered a founding father of rocketry and astronautics. In the 1920s, Oberth, whose childhood fantasies had been inspired by the novels of Jules Verne, wrote an influential publication *The Rocket into Planetary Space,* which discussed many aspects of rocket travel. Later, he expanded that work into a larger volume, *The Road to Space Travel,* which won wide recognition.

After World War I (1914–1918), Oberth studied physics at the University of Munich (Germany), where he realized that the key to space travel was the development of multistage rockets. Despite this important insight, Oberth's doctoral thesis on rocketry was rejected in 1922. However, in 1923, he published *The Rocket into Planetary Space,* which was followed by a longer version in 1929. In the final chapter, Oberth foresaw "rockets … so [powerful] that they could be capable of carrying a man aloft."

In the 1930s, Oberth proposed to the German War Department the development of liquid-fueled, long-range rockets. Oberth worked with the rocket pioneer Wernher von Braun during World War II (1939–1945) to develop the V-2 rocket for the German army. During this period, Robert Goddard was launching liquid-fueled rockets in the United States. After the war, Oberth and von Braun collaborated again at the U.S. Army's Ballistic Missile Agency in Huntsville, Alabama. Oberth contributed many important ideas regarding spaceflight, including the advantages of an orbiting telescope.

Oberth died on December 29, 1989, while in a Nuremberg, Germany hospital, at the age of 95 years. After his death, the Hermann Oberth Society built the Hermann Oberth Space Travel Museum (Hermann-Oberth-Raumfahrt-Museum) in the city of Feucht, Germany. In astronautics, the Oberth effect was named after Oberth. It states that larger changes in final speeds are possible the closer a rocket is to a gravitational body, when equal burns are assumed. In addition, a crater on the Moon was named after him.

 See also **Goddard, Robert Hutchings (Volume 1) • Verne, Jules (Volume 1) • von Braun, Wernher (Volume 3)**

Resources

Books and Articles

Friedman, Herbert. *The Amazing Universe.* Washington, DC: National Geographic Society, 1975.

Heppenheimer, T. A. *Countdown: A History of Space Flight.* New York: Wiley, 1997.

McDonough, Thomas R. *Space: The Next Twenty-Five Years.* New York: Wiley, 1987.

Ordway, Frederick I., and Mitchell R. Sharpe. *The Rocket Team.* Burlington, ON: Collector's Guide Publishing, 2008.

Websites

Hermann Oberth. U.S. Centennial of Space Commission. <http://www.centennialofflight.gov/essay/SPACEFLIGHT/oberth/SP2.htm> (accessed October 13, 2011).

Hermann Oberth: Father of Space Travel. Kiosek.com. <http://www.kiosek.com/oberth/> (accessed October 13, 2011).

The Hermann Oberth Space Travel Museum. <http://www.oberth-museum.org/index_e.htm> (accessed October 13, 2011).

Payloads and Payload Processing

The machines, equipment, hardware, and even people that are carried into space atop rockets are often called payloads. The term originated in World War I (1914–1918) during efforts to determine the amount of cargoes and people that could be carried by land tanks. The term is also often applied to the amount of useful weight that can be lifted by airplanes and inside trucks. Without a useful amount of payload—the

Payloads carried on rockets can include experiment packages, satellites for deployment, or space station hardware or supplies. © *NASA/Roger Ressmeyer/ Corbis.*

"pay" carrying load—any transportation system would be of minimum value since the objective of a transport is to carry cargoes from destination to destination. This is true whether the transport in question is a rocket or a car and the payload consists of satellites or groceries. Payloads can consist of nearly anything that researchers, government, or industry seek to place into space. Satellites, robotic probes, or instrument packages can act as payloads. In human spaceflight programs, payloads can be the astronauts themselves, along with their life-sustaining equipment and supplies.

In space transportation, unmanned payloads arrive in space with minimum activity involving people. If the transport is an expendable, throwaway rocket, people are not present when the robotic craft arrives in space. Even when a space shuttle was used for launching a payload, astronaut interaction with the payload during a flight is kept at a minimum except under unusual circumstances. Thus, all of the payloads sent into space are carefully prepared before the launching and their checkouts and activation automated to the maximum extent possible. Because people will not be present when these payloads arrive in space, payload processing and pre-launching preparation is an important part of the flight itself.

Payload Design and Storage

Payload preparation actually begins when the payload is under design. Space engineers often design a satellite to absorb the types of effects that the launching system—a rocket—places upon the machine. These can include the effects of the thrust of the rocket and the amount of gravity that its thrust into space generates on the payload and everything else aboard the rocket. Depending on the flight path chosen, the type of rocket, and the final destination planned for the payload, this can be many times the pull of gravity experienced on Earth's surface. Other effects, such as friction, heat, vibration, and vacuum*, also affect the payloads as they rise through the atmosphere and move out into the space environment.

Once the craft reaches its planned destination in space, designers must factor in the final environmental conditions, such as radiation and the surface conditions of a planet if a landing is planned. If the planetary destination is far away, engineers must build the craft to sustain the long flight. If the spacecraft is flying toward the Sun, it must be shielded from the harsh and continuous heat and other radiation streaming out from the Sun. If the craft is flying in the opposite direction, then the craft and its electronics must be heated to keep warm during its long cruise in the coldness of space.

Once a payload has been designed and manufactured, it must be kept in storage until the time draws near for its launch. Usually the manufacturer prepares a storage container and location that maintains the payload in environmentally friendly conditions as the launch is awaited. This period could last months or even years. For example, when the space shuttle *Challenger* exploded in 1986 all shuttle missions were placed on hold. Their payloads had to be stored for several years because of this unexpected delay. Large satellites such as the Hubble Space Telescope and military spacecraft bound for a shuttle ride had to be specially stored during the delay.

* **vacuum** an idealized region wherein air and all other molecules and atoms of matter have been removed, though such a complete absence of particles is never actually observed; within interstellar space density estimates range from about a hundred to a thousand atoms per cubic meter

Preparation for Launch

As the date of a planned launch draws nearer, payloads are shipped to the launching base where the flight will take place. Following its arrival from the manufacturer, the payload is re-checked to assure that it has not been damaged or affected in transit. Sometimes this includes partially dismantling the payload and conducting extensive re-checkouts. Complicated payloads such as Russian or Japanese modules attached to the International Space Station were shipped only partially built, with construction completed at the launching site itself. Once engineers have assured themselves that the payload has arrived at the launch site without damage, the next phase of preparation usually consists of readying the payload for mating with its rocket transport.

Shuttle Launches When the launching vehicle was a space shuttle (the shuttle fleet was active from 1981 to 2011), much of the preparation process served to ensure that the payload posed no risk to astronauts on the shuttle. Careful review was conducted of the payload's fuels, its electrical systems, and any rocket engines that might be part of its design. Once that step was completed, the craft was then checked for the method by which it would to be attached to the shuttle's cargo bay. Attachments, release mechanisms, and other devices that would act to deploy the payload away from the shuttle or allowed it to be operated while still attached inside the bay were tested and verified ready for flight.

At a certain stage in the final launch preparations the payload was moved from its preparation facility to the launching pad and installed inside the shuttle. Once in place, many of the tests and verifications were repeated to assure workers that the payload and its shuttle interfaces were working together. Unlike cargoes that fly inside commercial airliners, cargoes that were launched aboard the space shuttles were partially integrated with the shuttle itself. This even included the selection of the location where the payload was attached to the shuttle bay.

When all of these steps had been completed, a complete dress rehearsal of the final days of the countdown and liftoff was conducted. Called the Terminal Countdown Demonstration Test, this simulated launching even included suiting up the astronaut crew and having them board the shuttle just as they would have done on the day of the actual flight. The payload was activated at the same level it would have been on launch day, and the test went all the way up to the point where the rocket engines would be ignited to start the actual mission into space. If all went well with this test, the payload and the shuttle were deemed ready for their space mission.

Such preparations will again be critical to payload processing when the United States manned space program returns to space with its new generation of manned space vehicles and rockets. These space shuttle methods will prove invaluable as NASA and private companies within the United States continue the exploration of space in the twenty-first century.

▲

Mission specialists assemble a structure in the payload bay of space shuttle Endeavour. *NASA. Reproduced by permission.*

Expendable Rocket Launches When the launching vehicle is an expendable rocket, the process is somewhat less complex. Once at the launching site, the checkout and testing is conducted and the craft made ready for installation atop the rocket. In the United States, France, and China, for example, the test and integration procedure is done with the rocket and payloads stacked vertically. Russian space launch vehicles use a horizontal integration technique. Whichever method is used, the payload is attached to the final propulsive stage of the rocket or to its own rocket stage, and the completed assembly is carried to the rocket pad or final assembly building and becomes part of the overall launch vehicle.

As was the case with the shuttle, tests are conducted to verify that the attachments have been correctly made and that the rocket's computers are "talking," or exchanging data, with the payload computers. A dress rehearsal of the launch is also conducted, although it is usually less extensive than what was done for the shuttles. A successful completion of this test clears the way for the final countdown. Rocket fuels and explosive devices to separate the rocket's stages in flight or to destroy the craft if it veers off course are loaded into the rocket. Checks of the weather along the vehicle's flight path are also conducted.

When liftoff occurs, information on the health of the payload is sent by radio transmissions to tracking stations along the path that the rocket takes towards space. When the point in the flight is reached where the payload becomes active, it comes alive through radio commands, and begins its own role in achieving its space mission goal. If a malfunction occurs, radio data give mission controllers and engineers information on the cause, so that future versions of the rocket and payload can be redesigned to avoid the trouble.

Present-day launching rockets have an average of one chance in ninety-five to ninety-seven of experiencing a launching disaster. The most reliable rockets thus far designed have been the Apollo-Saturn rockets and the space shuttle boosters. The Saturn rockets had a perfect flight record in their missions from 1961 through 1973. The boosters for the space shuttle failed once (during launch phase) in 135 missions (as of August 31, 2011).

NASA maintains its Space Station Processing Facility (SSPF) at the Kennedy Space Center (KSC), in Florida. The 42,500 square meter (457,000 square foot) building was built specifically to process flight hardware for the International Space Station. The SSPF consists of two processing bays, an airlock, control rooms, laboratories, and logistical areas, along with office space and a cafeteria.

Many other payload processing facilities are located around the world. For instance, Astrotech maintains payload processing facilities at KSC, along with facilities at Vandenburg Air Force base in California and port facilities at Long Beach, California, specifically for sea launches. As another example of payload processing facilities around the world, the S5 satellite preparation building for Arianespace is located at the Guiana Space Center in French Guiana, on the northern Atlantic coast of South America.

 See also **Launch Services (Volume 1) • Launch Vehicles, Expandable (Volume 1) • Launch Vehicles, Reusable (Volume 1) • Payload Specialists (Volume 3) • Payloads (Volume 3) • Satellites, Types of (Volume 1) • Spaceports (Volume 1) • Space Shuttle (Volume 3)**

Resources

Books and Articles

Darrin, Ann Garrison, and Beth Laura O'Leary, editors. *Handbook of Space Engineering, Archaeology, and Heritage.* Boca Raton, FL: CRC, 2009.

Hale, Wayne, editor. *Wings in Orbit: Scientific and Engineering Legacies of the Space Shuttle.* Washington, D.C.: National Aeronautics and Space Administration, 2010.

Jedicke, Peter. *Great Moments in Space Exploration.* New York: Chelsea House, 2007.

Lewis, Richard. S., and Alcestic R. Oberg. *The Voyages of Columbia: The First True Spaceship.* New York: Columbia University Press, 1984.

Ordway, Frederick, III, and Mitchell R. Sharpe. *The Rocket Team.* Burlington, Ontario, Canada: Collectors Guide, 2008.

Skurzynski, Gloria. *This is Rocket Science: True Stories of the Risk-taking Scientists Who Figure Out Ways to Explore Beyond Earth.* Washington, D.C.: National Geographic, 2010.

Tewari, Ashish. *Advanced Control of Aircraft, Rockets, and Spacecraft.* Hoboken, NJ: Wiley, 2011.

Treadwell, Terry C. *Stepping Stones to the Stars: The Story of Manned Spaceflight.* Stroud: History Press, 2010.

Websites

Checkout, Assembly, and Payload Processing Services (CAPPS). The Boeing Company. <http://www.boeing.com/defense-space/space/capps/index.html> (accessed September 5, 2011).

Facilities. Astrotech Corporation. <http://www.astrotechcorp.com/about-us/facilities> (accessed September 7, 2011).

Inside the Multi-Payload Processing Facility. NASA. <http://www.nasa.gov/centers/kennedy/news/facts/ksc/MPPF-06.html> (accessed September 5, 2011).

Spaceport Introduction. Arianespace. <http://www.arianespace.com/spaceport-intro/overview.asp> (accessed September 7, 2011).

Space Shuttle and Payload Processing Tour. Kennedy Space Center, National Aeronautics and Space Administration. <http://science.ksc.nasa.gov/shuttle/countdown/tour.html> (accessed September 5, 2011).

Space Station Processing Facility (SSPF). NASA. <http://science.ksc.nasa.gov/facilities/sspf.html> (accessed September 5, 2011).

Privatization *See Commercialization (Volume 1)*

R

Reconnaissance

The first U.S. military space mission was directed towards reconnaissance (the act of observing or spying), and that remains the most important mission of the military's reconnaissance satellites, offering capabilities that cannot be obtained by any other means. As of 2011, the following countries were known to possess military satellite reconnaissance systems: the United States, Russia, France, India, Germany, Italy, Spain, Japan, Israel, China, South Korea, and the United Kingdom. Other countries with satellites in orbit, such as Iran, Turkey, and Brazil, have reconnaissance capabilities, but their military use is disputed by the countries in question. Since the late 1990s, several private companies have offered to sell satellite imagery of increasingly high quality. In 2004, Google Inc. began offering such images through its Google Earth Web site. The government of virtually any country can now buy detailed pictures of any place it wants to see.

At the most basic level, reconnaissance involves looking at an area of Earth to determine what is there. Among the ways this can be done from space, the primary methods are visual and radar* reconnaissance. Visual reconnaissance can be conducted in black and white or in color, although black-and-white images provide more detail. The major problem is that visual reconnaissance is impossible when the target is covered by clouds. Radar can penetrate cloud cover, but the images it returns are of lower quality. Radar reconnaissance is more challenging, and fewer countries operate dedicated radar satellites.

The United States was the first country to consider the use of satellites for reconnaissance. In 1946, the RAND Corporation conducted a study for the U.S. Air Force of the potential military uses of satellites, and reconnaissance was high on the list. However, the high cost of launching a satellite into orbit was prohibitive. In 1954, RAND conducted a much more extensive study of reconnaissance satellites and their capabilities. RAND proposed an atomic-powered satellite carrying a television camera. When the U.S. Air Force began a reconnaissance satellite program, the television camera proved impractical, and a "film-scanning" system was chosen instead. That system would take a photograph, develop the film aboard the satellite, and then scan the image and transmit it to Earth. Solar panels were substituted for the atomic power supply. The Atlas ICBM (intercontinental ballistic missile) was to be used to launch the satellite into orbit, but the U.S. Air Force was unwilling to fund the program until after *Sputnik 1* was launched in October 1957.

* **radar** a technique for detecting distant objects by emitting a pulse of radio-wavelength radiation and then recording echoes of the pulse off the distant objects

The U.S. Air Force's TacSat-3 tactical reconnaissance satellite in Earth orbit. © *Erik Simonsen/Photographer's Choice/ Getty Images.*

* **ballistic** the path of an object in unpowered flight; the path of a spacecraft after the engines have shut down

After the advent of the space age, satellite reconnaissance received much more attention in the United States, which feared that the Soviet Union had large numbers of ballistic* missiles at sites deep within that country. The U.S. Air Force funded a series of film-scanning satellites called SAMOS (or, Samos)—which was a series of signals intelligence (SIGINT) reconnaissance satellites during the 1960s and early 1970s. The Central Intelligence Agency (CIA) was placed in charge of an "interim" program called CORONA (or, Corona), which is considered the first U.S. reconnaissance satellite system. Its satellites operated from June 1959 to 1972, taking photographs of perceived hostile areas such as within the Soviet Union, China, and other countries. Unlike Samos, the Corona satellites returned their film to Earth in small capsules that were caught in midair by an airplane trailing a cable.

After a string of failures, the first Corona satellite returned its film to Earth in 1960. The pictures were grainy and showed relatively little detail but provided a wealth of information on the Soviet Union, including the fact that the Soviet Union did not have more missiles than did the United States. Corona became more successful, and its images improved in quality, whereas Samos experienced numerous failures. At its best Corona could photograph objects on the ground that were a minimum of 1.8 to 2.7 meters (six to nine feet) long. That length did not allow for acceptable observations and measurements, and so the U.S. Air Force canceled Samos and began a satellite program called GAMBIT (or, Gambit). Like Corona, Gambit returned its film to Earth, but it could photograph much smaller objects. Corona was discontinued in 1972, but Gambit kept flying until 1985, and late models of the satellite could photograph objects as small as a baseball—with some reports stating clear images down to the size of a golf ball.

Developments in Reconnaissance Technology

The next major leap in reconnaissance satellite technology occurred in 1976, when the United States launched a satellite known as KENNAN (Kennan), later renamed CRYSTAL (Crystal). Kennan could transmit electro-optical digital images directly to the ground, using a camera similar to a common digital camera. The images were black and white and took several minutes to transmit, but this was far faster than the days or weeks required with the film-return system. These satellites could see objects no smaller than a softball. With the increase in speed came a change in the ways the satellites were employed. Instead of being used to prepare long-range plans and studies, they could now be used in crisis-type situations, and the president could make instant decisions based on satellite photographs. The series of satellite, also known as KH, for Key Hole, were launched from late 1976 to 1990. Later versions of Kennan are probably still in use, but these satellites are limited by their inability to see through clouds.

In the late 1980s the United States launched a radar satellite called LACROSSE (Lacrosse) (later renamed ONYX [Onyx]) that could look through clouds and smoke. The major drawback of Lacrosse was that it could see objects no less than 0.9 meter (three feet) long. In 1999, the U.S. government initiated a Future Imagery Architecture (FIA) program for the design and development of a high-tech series of reconnaissance satellites. The program initiated a shift from using a few large reconnaissance satellites to employing numerous smaller satellites. The purpose of this shift is to decrease the amount of time it takes to photograph any spot on the ground. However, after over five years of work, the government finally cancelled the program after the Boeing Company failed to produce the satellites and continued to have cost overruns and delays in meeting deadlines.

The United States and Germany have supposedly been developing a joint military reconnaissance satellite program known as HiROS. The claim was made in early 2011, and was associated with leaked diplomatic cables associated with the controversial WikiLeaks website, which made public some 250,000 classified documents. However, German government officials have strongly denied the claims that the HiROS observation satellite program has any military purposes.

Soviet/Russian Reconnaissance Satellites

The Soviet Union developed similar systems, usually trailing about three to seven years behind the United States. Its first reconnaissance satellite, Zenit, was similar to the first Soviet spacecraft to launch a man into orbit, *Vostok 1.* Unlike Corona, Zenit returned both the film and the camera to the ground in a large capsule. Over five hundred of the Zenit series were launched by the Soviets (and later by the Russians) between 1961 and 1994. The Russians replaced Zenit with a higher-resolution system called Yantar, which later turned into series called Orlets, Resurs, and Persona. In the 1980s, the Terilen series was the first Soviet reconnaissance satellite program that was able to transmit "real-time" digital images from space to Earth without having to return part or all of the spacecraft. As of late 2011, the latest launch of a Yantar reconnaissance satellite had occurred in April 2010.

Chinese Reconnaissance Satellites

Throughout the 2000s and early 2010s, China was rapidly building up its reconnaissance satellite capability. Its series of Yaogan reconnaissance satellites includes both optical imaging, as well as synthetic aperture radar (SAR), which is for all-weather viewing of ground targets. The *Yaogan 9A, Yaogan 9B,* and *Yaogan 9C* were launched upon a single rocket in March 2010. These three spacecraft form a reconnaissance satellite constellation used to track warships. *Yaogan 7* and *Yaogan 11* are thought by military analysts to capture visible images. The *Yaogan 8* and *Yaogan 10* spacecraft are reportedly SAR-equipped radar reconnaissance satellites. As of late 2011, the latest of the Yaogan-series of satellites, *Yaogan 10* and *Yaogan 11,* were launched in August and September 2010, respectively.

Other Countries with Military Reconnaissance Satellites

In mid 2010, Israel launched the *Ofeq-9* reconnaissance satellite. The successful launch gave Israel six spy satellites with which to watch regional powers, such as Iran and Syria. With a constellation of six spy satellites, the Israelis were assured of continuous coverage of the Middle East from space.

Japan has developed an extensive reconnaissance satellite capability called Information Gathering Satellite (IGS). A series of IGS satellites was launched during the 2000s, with launches continuing into the 2010s. On September 23, 2011, Japan launched its latest reconnaissance satellite, the *IGS 6A.*

The Indian Space Research Organization (ISRO), India's counterpart to the U.S. space agency NASA (National Aeronautics and Space Administration), has developed the Indian radar imaging satellite (RISAT) system. *RISAT 2* was successfully launched on April 20, 2009, the first Indian RISAT to make it into orbit about Earth. *RISAT 2* uses synthetic aperture radar to peer through clouds and darkness to image surface features. As of late 2011, another RISAT satellite was scheduled for launch in early 2012.

The government of Iran launched its second satellite into orbit with its own domestic rocket in June 2011 (previous satellite launches were preformed by Russia). Though the satellite—named *Rasad-1,* meaning *observation* weighed only 34 pounds (equivalent to 15.5 kilograms), neighbors such as Israel were concerned about Iran's growing rocketry and satellite prowess.

Commercial Reconnaissance Technology

In the early twenty-first century, commercial satellites such as *IKONOS* (launched on September 24, 1999, and still operational as of late 2011), operated by American companies Lockheed Martin Corporation and Space Imaging (later renamed GeoEye), provides satellite imagery of virtually any place on Earth for a fee. In addition, Google, Inc., using technology largely developed by Keyhole, Inc., provides images of Earth taken from images obtained off satellites, aircraft, and devices and software using the geographic information system (GIS). Commercial satellites generally show objects on the ground that are as small as 0.9 meter (three feet) long, and this is useful for many civilian and military purposes. As with the military satellites, these images are still pictures, not the moving images shown in spy movies. They sometimes are referred to as "the poor man's reconnaissance satellite," but they can dramatically increase the power of a military force by allowing the users to know what their adversaries are doing from a vantage point that the vast majority of the world cannot reach.

 See also **Global Positioning System (Volume 1) • Military Customers (Volume 1) • Military Exploration (Volume 2) • Military Uses of Space (Volume 4) • Remote Sensing Systems (Volume 1) • Satellites, Types of (Volume 1)**

Resources

Books and Articles

Day, Dwayne A., Brian Latell, and John M. Logsdon, eds. *Eye in the Sky: The Story of the Corona Spy Satellites.* Washington, DC: Smithsonian Institution Press, 1998.

Graham, Thomas and Keith A. Hansen. *Spy Satellites and Other Intelligence Technologies That Changed History.* Seattle: University of Washington Press, 2007.

Norris, Pat. *Spies in the Sky: Surveillance Satellites in War and Peace.* Berlin: Springer, 2008.

Websites

Day, Dwayne A. "Ike's gambit: The KH-8 reconnaissance satellite." *The Space Review.* <http://www.thespacereview.com/article/1283/1> (accessed October 25, 2011).

Short, Nicholas M. "Remote Sensing Tutorial. Military Intelligence Satellites." *Goddard Space Flight Center, National Aeronautics and Space Administration.* <http://rst.gsfc.nasa.gov/Intro/Part2_26e.html> (accessed October 25, 2011).

Taubman, Philip. "In Death of Spy Satellite Program, Lofty Plans and Unrealistic Bids." *The New York Times.* <http://www.nytimes.com/2007/11/11/washington/11satellite.html?pagewanted=1&_r=1&adxnnl=1&adxnnlx=1250528582-78j5VMefBtWG3ieQCJk5hg#step1> (accessed October 25, 2011).

Regulation

Commercial space activities conducted by U.S. companies are regulated by the federal government in four major areas: communications, space launches, remote sensing*, and limitation of the transfer of technology for reasons of national security and industrial policy.

Communications are regulated by the Federal Communications Commission (FCC). The FCC was established in the 1930s to regulate radio (and later television) broadcasts and the use of the radio portion of the electromagnetic spectrum, assuring that the signals from one station would not interfere with those from another station. When commercial communications satellites arrived in the mid–1960s, the FCC had three decades of regulatory experience. The office within the FCC that issues licenses for satellites is the Satellite Division (SD) (http://www.fcc.gov/ib/sd/) of the International Bureau (IB). Licensing assures that any proposed

* **remote sensing** the act of observing from orbit what may be seen or sensed below on Earth or another planetary body

new satellite will not interfere with other satellites or with any other operating radio applications, on Earth or in space.

The Commercial Space Launch Act (CSLA) of 1984, along with its 2004 amendment, regulates all commercial launches, reentries, or landings conducted by U.S. companies. Under the CSLA, each launch or reentry must have a license. The Office of Commercial Space Transportation is part of the Federal Aviation Administration (FAA) and is the federal government agency that issues these licenses. Its Web site (http://www.faa.gov/about/office_org/headquarters_offices/ast/) contains all the relevant rules, laws, regulations, and documents needed to obtain a launch license. The Office of Commercial Space Transportation conducts a policy review, a payload* review, a safety evaluation, an environmental review, and a financial responsibility determination based on the data in the license application before issuing or refusing a license. The purpose of a launch license is to assure that "the public health and safety, safety of property, and the national security and foreign policy interests of the United States" are properly considered.

Commercial remote sensing from space is regulated under the 1992 Remote Sensing Policy Act and its associated regulations and administration policies. The act directs the secretary of commerce to administer its provisions, and those duties have been delegated to the National Environmental Satellite, Data, and Information Service (NESDIS) of the National Oceanic and Atmospheric Administration (NOAA), an agency of the Department of Commerce. NESDIS (http://www.nesdis.noaa.gov/) runs the nation's weather satellites, and the International and Interagency Affairs Office (IIAO; http://www.nesdisia.noaa.gov/) within NESDIS issues the licenses needed to operate private space-based remote sensing systems.

NESDIS/IIAO reviews these applications in consultation with the Department of Defense (for national security), the Department of State (for foreign policy), and the Department of the Interior (which has an interest in archiving remote sensing data). Once an application has been determined by NESDIS/IIAO to be complete (all the required documents and data have been submitted), by law NOAA has to issue an up-or-down license determination within one hundred and twenty days. Documents, background data, instructions, and examples are available at NESDIS/IIAO's Web site to aid license seekers.

Under the law, a licensee must operate its space-based remote sensing system(s) so that the national security interests of the United States are respected and the international obligations of the nation are observed. A licensee must maintain positive control of its system(s) and maintain clear records of the sensing those systems have done. A U.S. licensee also must agree to "limit imaging during periods when national security or international obligations and/or foreign policies may be compromised." This is called "shutter control." The federal government can, in time of international stress (war or conflict) tell licensees what they can and cannot take images of while operating their business.

The major law in the area of trade control is the Arms Export Control Act (AECA) and its associated regulations, the International Traffic in Arms Regulations (ITAR). Virtually any subject involving space falls under ITAR. Equipment for ground stations for satellite control; transmitters; rocket engines; computer software for controlling a rocket, a satellite, or a ground station; rockets; and satellites are all subject to control and licensure under ITAR.

Licenses and regulation under the AECA and ITAR are administered by the U.S. Department of State and its Directorate of Defense Trade Controls (DDTC), which is part of the Bureau of Political-Military Affairs. These organizations are aided in their work by the Defense Threat Reduction Agency (DTRA) of the Department of Defense (DoD). DTRA's Web site (http://www.dtra.mil/) contains documents, background data, and instructions to aid license seekers, including electronic means for the filing and tracking of license applications.

The United States, one of its founding members, is a party to the Missile Technology Control Regime (MTCR), to which thirty-four other countries, including the founding countries of Canada, France, Germany, Italy, Japan, the United Kingdom belong. Equipment and technology are controlled under this regime to limit the proliferation of weapons of mass destruction through efforts to control the availability of delivery systems (rockets). The State Department and the Department of Defense attempt to assure that space companies that export services or products adhere to the goals of the MTCR.

During most of the 1990s, space-related trade control was the responsibility of the Department of Commerce, specifically the Bureau of Industry and Security (http://www.bis.doc.gov/) and the International Trade Administration (http://www.trade.gov/). Both of these agencies now play a reduced role in regulating the export of space-related trade products and services, but their main role at present is primarily to support the activities of the Department of State.

The shift of the regulation and licensing of space-related trade from the Department of Commerce to the Department of State resulted from a law passed by Congress, which wanted to eliminate what it felt was a looseness in U.S. trade control that had led to the transfer of sensitive space technology. This statutory change had unintended consequences, making it extremely difficult for companies, such as Orbital Sciences Corporation, to communicate with a division of its own company based in a foreign country. Under this regime, satellite engineers cannot talk to their counterparts in another country unless it possesses a license. In the case of Orbital Sciences, it could not communicate to its people in the United Kingdom. These restrictions became so stringent that Orbital sold its Canadian-based division because of the difficulties presented by these mandated trade restrictions. Congress has since passed new legislation to address this problem.

 See also **Law (Volume 4)** • **Law of Space (Volume 1)** • **Legislative Environment (Volume 1)** • **Licensing (Volume 1)**

Resources

Books and Articles

Jakhu, Ram S., ed. *National Regulation of Space Activities.* Dordrecht, Netherlands: Springer, 2010.

Websites

The Arms Export Control Act. U.S. Department of State, Directorate of Defense Trade Controls. <http://www.pmddtc.state.gov/regulations_laws/aeca.html> (accessed October 27, 2011).

Bureau of Industry and Security. U.S. Department of Commerce. <http://www.bis.doc.gov/> (accessed October 27, 2011).

Commercial Space Launch Amendments Act of 2004. Federal Aviation Administration. <http://www.faa.gov/about/office_org/headquarters_offices/ast/media/PL108-492.pdf> (accessed October 27, 2011).

Consolidated ITAR 2011. U.S. Department of State, Directorate of Defense Trade Controls. <http://www.pmddtc.state.gov/regulations_laws/itar_consolidated.html> (accessed October 27, 2011).

Defense Threat Reduction Agency. U.S. Department of Defense. <http://www.dtra.mil/> (accessed October 27, 2011).

International and Interagency Affairs Office. National Environmental Satellite, Data, and Information Service. <http://www.nesdisia.noaa.gov/> (accessed October 27, 2011).

International Trade Administration. U.S. Department of Commerce. <http://www.trade.gov/> (accessed October 27, 2011).

Land Remote Sensing Policy. National Aeronautics and Space Administration, Ames Research Center. <http://geo.arc.nasa.gov/sge/landsat/15USCch82.html> (accessed October 27, 2011).

The Missile Technology Control Regime. <http://www.mtcr.info/english/index.html> (accessed October 27, 2011).

National Environmental Satellite, Data, and Information Service. National Oceanic and Atmospheric Administration. <http://www.nesdis.noaa.gov/> (accessed October 27, 2011).

Office of Commercial Space Transportation. Federal Aviation Administration. <http://www.faa.gov/about/office_org/headquarters_offices/ast/> (accessed October 27, 2011).

Satellite Division of the International Bureau. Federal Communications Commission. <http://www.fcc.gov/ib/sd/> (accessed October 27, 2011).

Remote Sensing Systems

International research efforts have been undertaken to study the complex and interconnected processes that affect Earth's atmosphere, oceans, and land. Essential information for this research is provided by fleets of satellites and aircraft equipped with sensors that collect enormous amounts of Earth data. These systems are called remote sensing systems.

The variety of data that can be obtained through remote sensing systems is vast. The world scientific community uses remote sensing systems to obtain information about ocean temperature, water levels and currents, wind speed, vegetation density, ice sheet size, the extent of snow cover, rainfall amounts, aerosol concentrations in the atmosphere, ozone levels in the stratosphere*, and many other important variables to better understand how natural phenomena and human activities impact global climate. Decision makers such as environmental resource managers, city planners, farmers, foresters and many others often use remote sensing systems to better run their businesses and to improve people's quality of life. This article presents a number of applications of remote sensing systems to farming, water quality analysis, water resources in arid regions, noxious weed detection, urban sprawl, urban heat islands, and many others. This field is in rapid evolution and many new applications may appear in the years to come.

Satellite Remote Sensing

Space is an excellent vantage point from which to study air, sea, and land processes both locally and globally. It provides the bird's-eye view that captures all the information in a single image. Satellite observations have definite advantages over ground or aircraft observations. Ground observations are labor intensive, time consuming, and costly. Aircraft observations require less labor and time but are still costly. In spite of their high initial cost, satellites are a cheaper way to do observations as they may take data continuously during their lifetime over the whole globe. Satellites can also observe areas difficult to access on the ground and provide regular revisits of the same areas showing surface feature changes over time.

Satellite observations are made with sensors that measure the brightness of electromagnetic radiation either reflected or emitted by ground features. Electromagnetic radiation includes not only wavelengths of visible light with its various colors but also many wavelengths invisible to the human eye such as ultraviolet* and infrared*, as well as microwaves, radio waves, and gamma rays.

Resolution refers to the smallest size object that can be identified. A one-kilometer (0.6-mile) resolution satellite will produce images made of small squares with uniform brightness representing one-kilometer by one-kilometer squares on the ground. In general, objects smaller than one kilometer cannot be distinguished in such an image.

Whereas visible and near-infrared radiation observed by sensors is actually solar radiation* *reflected* by ground features, thermal infrared

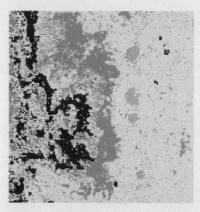

Data obtained through remote sensing systems show some of New Mexico's variegated terrain. Remote sensing systems can also be used to collect data varying from vegetation density to aerosol concentrations in the atmosphere. *NASA.*

* **stratosphere** a middle portion of a planet's atmosphere above the tropopause (the highest place where convection and "weather" occurs)

* **ultraviolet** portion of the electromagnetic spectrum with shorter wavelengths than visible light but longer wavelengths than x rays

* **infrared** portion of the electromagnetic spectrum with waves slightly longer than visible light

* **solar radiation** total energy of any wavelength and all charged particles emitted by the Sun

radiation is radiation *emitted* by ground features. Thermal infrared radiation provides information about the temperature of the emitting objects. Other sensors actively illuminate Earth and measure the reflected signal. Radar is an active remote sensing system that is very useful in areas that are often covered with clouds. Whereas visible and infrared radiation is blocked by clouds, radar waves (those using microwaves or radio waves) penetrate clouds, thus enabling observations of Earth from space in almost all weather conditions.

Remote sensing data can be compared to an ore that contains gold (information) from which a piece of jewelry can be made (knowledge). Remote sensing is at its best when it is used to answer specific and well-posed questions. The end result of the processing of data, information extraction, and analysis is the answer to these questions.

Many remote sensors are placed onboard aircraft. Satellites may take several days, even weeks before revisiting a specific area on Earth, whereas aircraft can be commissioned to take remote sensing data over that area on a moment's notice. They also operate at significantly lower altitudes and produce higher resolution data than satellites when fitted with the same type of sensor. Finally, many new National Aeronautics and Space Administration (NASA) sensors are tested on aircraft before being put on satellites. Aircraft remote sensing has an important role to play both for global climate change studies and for more immediate applications such as the ones described below.

Global Environmental Observations

People live in a rapidly changing world facing major global challenges. A rapidly increasing world population demanding accelerated economic development strains Earth's resources. Remote sensing systems are being used to investigate a number of areas related to the global environment, including global climate change, rain forest deforestation, the health of the oceans, the size of polar ice covers, and coastal ecosystem health.

Global Climate Change One of the most ambitious and far-reaching programs of environmental investigation is the U.S. Global Change Research Program (USGCRP). This effort is part of a worldwide program to study global climate change, which involves changes in the global environment that could affect Earth's ability to support life.

A strongly debated climate change issue is global warming, which results from increased atmospheric levels of greenhouse gases—such as carbon dioxide—which trap heat in the lower atmosphere, preventing it from escaping into space.

In 2011, the Carbon Dioxide Information Analysis Center estimated that fossil fuel use and other industrial activities had resulted in the release of 337 billion tons of carbon into the atmosphere since 1751, with more than half of the total occurring since the mid–1970s. The concentration of carbon dioxide in the atmosphere was approximately 280 parts per million (ppm) before the Industrial Revolution (late-eighteenth and early-nineteenth centuries). However, by 1958, the carbon dioxide concentration climbed to about 315 ppm in the atmosphere and, in 2011, climbed even higher to about 392 ppm. Experts believe that almost all of the increase in carbon dioxide in the atmosphere results from human activity—primarily the burning of fossil fuels (coal, oil, and natural gas) and secondarily from the clearing of forests and other land uses and various industrial processes. The first satellite dedicated entirely to the observation of greenhouse gases in Earth's atmosphere is the *Japanese Greenhouse Gases Observing Satellite* (GOSAT), also known as *Ibuki* (which means "breath" in Japanese). *Ibuki* was launched into orbit on January 23, 2009, and almost immediately began returning information about greenhouse gases and a variety of other atmospheric phenomena. By 2010, data obtained from *Ibuki* was being released to the public, including vertical maps of carbon dioxide and methane concentrations in the atmosphere.

The average global surface temperature of Earth has also increased during this timeframe between the early nineteenth century and the early twenty-first century. Specifically, according to the Intergovernmental Panel on Climate Change, the average global surface temperature increased 0.74°C (with an uncertainty of 0.32°C) or 1.33°F (with an uncertainty of 0.32°F) during the twentieth century. The year 1998 was considered the warmest of the twentieth century and, so far, 2005 and 2010 have tied as the warmest years ever recorded on Earth. The first decade of the twenty first century set almost all records for high temperatures, with nine of Earth's ten warmest years having occurred since 2001, and all 12 of the warmest years in recorded history having occurred since 1997. The question that some people debate is whether this warming is directly related to the human production of carbon dioxide, due to natural processes, or is a combination of the two. Whatever is the answer to this question, the trend is clear: the consequences may be severe for the human species and all organisms on Earth.

Specifically, plants grow by absorbing atmospheric carbon dioxide, storing the carbon in their tissue. Rain forests and ocean phytoplankton are great carbon dioxide absorbers. So are corals and shellfish, which make calcium carbonates* that end up in the bottoms of oceans.

Trees and plankton may grow faster—left to themselves—if the level of carbon dioxide in the atmosphere increases. This could provide a mechanism limiting atmospheric carbon dioxide concentrations. Unfortunately, people pollute the oceans, which kills phytoplankton and coral reefs, and destroy tropical rain forests.

Scientists worldwide inventory and monitor rain forests, phytoplankton, and coral reefs in an effort to estimate their impact on the concentration of carbon dioxide in the atmosphere. Their main sources of information are from satellite remote sensing data. The warming of Earth's lower atmosphere results in the melting of glaciers and polar ice sheets. The extra liquid water produced raises ocean water levels. Indeed, sea level rose ten to 25 centimeters (four to ten inches) during the twentieth century and glaciers continue to melt in the twenty-first century and

* **carbonate** a class of minerals, such as chalk and limestone, formed by carbon dioxide reacting in water

are melting, in many cases, faster than predicted by scientists previously. Data from a number of satellites are used by the National Oceanic and Atmospheric Administration (NOAA) to measure the rate of ice melting in Antarctica and Greenland, two major causes of sea level rising.

Rain Forests The Global Observations of Forest Cover is an international effort to inventory worldwide forest cover and to measure its change over time. From these observations, which are based on high-resolution satellite remote sensing, scientists produce digital deforestation maps.

Deforestation is a politically sensitive topic. Developed nations pressure developing countries such as Brazil and Indonesia to stop the deforestation process, arguing that the rain forests in these countries are virtual lungs for the world's atmosphere. Developing countries with tropical rain forests argue that the deforested areas are important and necessary sources of revenue and food as they are used for agricultural activities. The debate prompted an international meeting, the United Nations Conference on Environment and Development in Rio de Janeiro, Brazil, in June 1992. This conference resulted in the Rio Declaration on Environment and Development (often abbreviated as Rio Declaration) that sets the basis for a worldwide sustainable development—an economic development that does not deplete natural resources and that minimizes negative impact on the environment.

A great deal of data is now available about rain forest deforestation in most parts of the world. According to the Food and Agriculture Organization (FAO) of the United Nations, the total amount of forest cover in 62 nations of the world fell by 10,240,133 hectares, or 8.3 percent between 1990 and 2005. The rate of deforestation was, however, much greater in some parts of the world than in others. The worst hit were African nations, such as Burundi and Togo, which lost nearly half their forest cover between 1990 and 2005 (47.4 and 43.6 percent, respectively), Nigeria (35.7 percent), and Benin (29.2 percent). Some nations elsewhere experienced similar devastation, including the Philippines (32.3 percent loss), Honduras (37.1 percent loss), and Indonesia (24.1 percent loss).

Oceans Phytoplankton and coral reefs in the oceans significantly contribute to the removal of atmospheric carbon dioxide. Acid rain and other pollutants adversely affect coral reefs and phytoplankton. Satellite remote sensing is used to inventory coral reefs and phytoplankton worldwide. Indeed, because of their wide distribution and remote locations, coral reefs can practically be inventoried and monitored only from space. Phytoplankton is easy to recognize from space because it has a distinctive green color.

Satellite sensors are also used to measure other ocean characteristics such as topography and ocean temperature. For example, a partnership between the United States and France developed *TOPEX-Poseidon,* a satellite that monitored global ocean circulation and global sea levels in an effort to better understand global climate change, specifically the links between the oceans and the state of the atmosphere. The satellite operated from 1992 to 2006. The *Jason-1* continued the mission of *TOPEX-Poseidon* when it was launched into space on December 7, 2001. In addition, the *Ocean Surface Topography Mission* (OSTM) satellite was launched onboard the *Jason-2* mission on June 20, 2008. *Jason-1* and *OSTM* (sometimes called Jason-2) are located on opposite orbits around Earth, so are able to cover in tandem much more of Earth's surface. As of late 2011, the two satellites are producing a host of data with numerous practical applications, including research on coral reefs, data on fishery resources, surveys of marine mammal populations and migration routes, maps of ocean currents for use by shipping fleets, hurricane and El Nino forecasting, and climate change studies.

Radar remote sensing has several important uses over oceans. Reflected signals from radar are sensitive to water surface roughness. The rougher areas reflect the radar signal better and appear brighter. Smooth areas are dark as they barely reflect radar signal. This feature helps locate and monitor oil spills on the ocean surface because oil makes the ocean surface smooth and thus appears dark on radar* images.

Polar Ice Covers The U.S. Landsat satellites, the Canadian RADARSAT spacecraft, and the European radar satellites ERS (European Remote-sensing Satellite) have been actively used to monitor the ice sheets in Antarctica and Greenland. The Landsat program is the longest operating satellite remote sensing program in the world. The first Landsat was launched on July 23, 1972, and it operated until January 6, 1978. Six other Landsat satellites have also been launched; two remain functional, Landsat 5, launched on March 1, 1984; and Landsat 7, launched on April 15, 1999. Both satellites continue to return very large volumes of data with applications in agriculture, cartography, geology, forestry, regional planning, surveillance, education, and national security. One example of the practical use to which Landsat data can be put was the mapping in April to July 2011 of the Wallow North forest fire in Arizona. Photographs taken by the satellite were used by Earthbound workers to begin restoration efforts almost as soon as the fire had been extinguished.

The Land-Sea Interface Beaches provide a lively, productive habitat for wildlife and a buffer against coastal storms. Salt marshes produce nutrient-rich "sludge" as a basis for the food chain while providing nurseries for juvenile fishes and habitat for shrimp, crabs, shellfish, turtles, and waterfowl. Coastal habitats are essential to the feeding, reproduction, and migration of fish and birds. However, development, and the sand pumping, jetties, and seawalls that come with it, is overwhelming beaches. Salt marshes are under constant threat from short-sighted development schemes that require they be drained and filled.

* radar images images made with radar illumination instead of visible light that show differences in radar brightness of the surface material or differences in brightness associated with surface slopes

NOAA has a Coastal Remote Sensing program that is using remote sensing, along with other technologies, to help coastal resource managers improve their management of aquatic and coastal ecosystems. The data sets and products provided by this program include ones dealing with ocean color, coastal topography and erosion, water quality, and the monitoring and tracking of harmful algal bloom.

Satellite remote sensing can thus play a central role in monitoring the health of coastal waters. The challenge is to provide decision makers with the knowledge derived from the remotely sensed data and to educate them about the mechanisms at work in coastal waters using satellite images.

Several commercial companies also provide remote sensing images and data from satellites for littoral* water and ocean monitoring. The DigitalGlobe company, of Longmont, Colorado, launched its *QuickBird* satellite on October 18, 2001, which contains a 60-centimeter (23-inch) resolution panchromatic camera and a 2.4-meter (7.8-foot) resolution multispectral camera. On September 18, 2007, the company launched its *WorldView-1* satellite, which possesses a 50-centimeter (20-inch) resolution panchromatic camera. Both satellites are intended to show detailed coastal features, including beach structure, sandbars, and wave patterns. Its *WorldView-2* satellite, with similar hardware and objectives, was launched on October 8, 2009.

GeoEye launched its commercial satellite, *OrbView-2,* in 1997, to measure phytoplankton and sediment concentration in oceans and inland lakes, data that are useful for environmental applications such as coastal pollution monitoring and "red tide" tracking. The company followed in 2003 with the launch of *OrbView-3.* However, it became inoperable in 2007, and its orbit decayed on March 3, 2011. Its *GeoEye-1* satellite was launched in 2008, and it is operating in orbit, as of late 2011. Red tides are the result of dying algae* producing a rapid multiplication of the bacteria that feed on them. These bacteria in turn deplete the water of its oxygen, killing marine life. Red tides can make mussels and oysters dangerous to eat as they produce toxins that can be life threatening to consumers. GeoEye's latest satellite, *GeoEye-2* is scheduled for launch in early 2013.

These examples are by no means exhaustive of the many applications of satellite remote sensing, the numerous satellites in orbit, or the large number of new satellites planned. Satellite remote sensing is a business in rapid expansion, particularly on the commercial side. Mainly government-sponsored satellites have provided Earth data until recently, but commercial satellite providers have entered the scene and they will play an increasingly important role in the future. This in turn has spurred the geographic information business.

Land Features

Satellite remote sensing was first used by the intelligence communities of the United States and the Soviet Union to spy on each other's military targets, starting in the early 1960s. In the 1970s, the United States initiated the Landsat program—a civilian program monitoring Earth's land resources—and in the 1980s NASA launched the Mission to Planet Earth program with an emphasis on understanding the global climate and monitoring the human impact on it.

In the late 1970s and in the 1980s, several other countries—such as India and France—launched remote sensing satellites to gather land surface data in an effort to monitor their agriculture and land use processes. Remote sensing information helps these countries establish national policies and monitor compliance. Since these satellites orbit* over the entire Earth, they can provide data about many other locations. Data from both the Indian Remote Sensing (IRS) satellites (by the Indian Space Research Organization) and the French Satellites Pour l'Observation de la Terre (SPOT) are sold in the United States. In addition, Canada, the European Union, and Japan have put radar remote sensors into space. Recently, several U.S. companies (GeoEye and DigitalGlobe, being the leading companies) have obtained permission from the U.S. government to launch very-high-resolution satellites capable of seeing objects on the ground as small as one meter (about a yard). This activity may spur another information revolution in the 2010s similar to the personal computer explosion of the 1980s and the burgeoning of the Internet in the 1990s.

The enabling factors in this so-called spatial information revolution include higher resolution, more reliable sensors, more powerful personal computers, the Internet, the civilian use of the U.S. Global Positioning System (GPS), and significantly improved geographic information system (GIS) software. In 2011, very-high resolution, color images of any part of the world are available on the Internet in almost real time for a modest fee. For instance, a farmer with a computer connected to the Internet is able to monitor crops in a field, while police officers are able to observe traffic jams in a big city in real time.

Environmental Observations   Land observations from space have an endless list of applications. A few examples are watershed analysis (including water resources inventory and water quality analysis), noxious weed detection, monitoring land use change over time, tracking urban sprawl and the loss of agricultural land, erosion monitoring, observing desertification in semiarid lands, tracking natural hazards such as floods and fires, agricultural land inventory, and crop yield prediction. Many of these observations have a significant economic impact and enhance the quality of life of citizens and user communities.

In the Middle East and in Africa water scarcity has become a serious geopolitical issue. Ecosystems rarely recognize political boundaries and several countries share common water resources. The Nile River, for example, flows through eight countries before reaching Egypt, a country that experiences very little rainfall and relies almost entirely on the waters of the Nile for its agriculture and drinking water resources. Actions upstream by other governments can severely affect Egypt. Similar problems exist in the Middle East between Syria, Jordan, and

* **orbit** the circular or elliptical path of an object around a much larger object, governed by the gravitational field of the larger object

Remote sensing can be used to aid farmers in making decisions about irrigation and fertilization. Color variations show crop density in the top image, whereas the middle and bottom images indicate water deficits. Fields 120 and 119 indicate a severe need to be irrigated. *Landsat 7 Science Team/USDA Agricultural Research Service/NASA.*

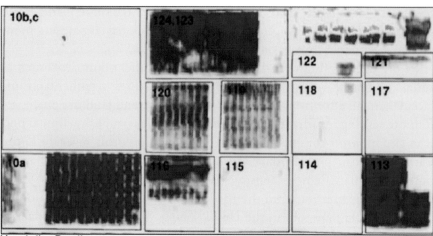

Vegetation Density

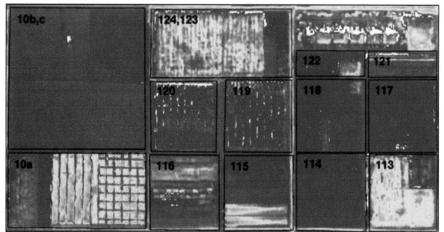

Water Deficit

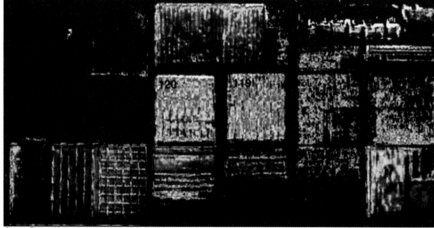

Crop Stress

Israel, countries that share common aquifers (natural underground water reservoirs) and other water resources.

Satellite remote sensing may play an essential role for peace by providing information about new water resources as well as accurate maps of existing known resources and a means to monitor use. Radar remote sensing in particular can be helpful in discovering new water resources in arid areas. For instance, the Canadian RADARSAT system has discovered new underground water flows in the African desert.

In the American West, noxious weeds spread at an alarming rate, overcoming other species of vegetation, destroying ecological balance, and even killing livestock. It is very impractical and costly to locate these weeds from the ground in the semiarid expanses of Arizona, New Mexico, and Utah. Satellite remote sensing helps locate these weeds either through their spectral reflectance pattern or by observing their blooming at specific times of the year when no other vegetation blooms. This information can then be used in efforts to eradicate the weeds.

Agricultural Applications The agricultural applications of remote sensing are particularly useful. French SPOT satellites are used to determine what crops are planted where and how many acres of a given crop are planted in a region. Crop health is monitored over time, and claims of crop loss to drought or other natural disaster can be verified using satellite images.

MDA Federal Inc. (formerly Earth Satellite Corporation or EarthSat) uses remote sensing data to provide weekly information about worldwide crop conditions on the Internet. The company, for example, claims to make 95% correct yield predictions for cocoa, sugar, and coffee crops two months ahead of harvest. Information such as this is extremely useful to growers needing to decide what crops to plant. If a wheat glut is predicted in South America in winter, informed farmers in the Northern Hemisphere will not plant wheat in early spring. Spatial technologies also give farmers new tools to better manage crops. A yield map over a field shows areas of higher and lower productivity.

In precision farming, a field is not treated as a homogeneous whole. Rather, as conditions—such as soil composition and soil fertility—vary across fields, the farmer's treatment of the field also varies. Thus, irrigation, liming, fertilizers, and pesticides are not applied uniformly across a field but are varied according to need using a variable spreader with a GPS antenna and a computer program that has in its memory information about the local needs of the field. High-resolution satellite images can provide information about crop health. This information is put on a GIS used by the farmer to divide his fields into zones, each zone being treated differently. Precision farming has several advantages over traditional farming. Because of differential treatment of field zones, there is less fertilizer, water, and/or pesticides used because applications are made in response to local needs only. There is, thus, economic benefit to the farmer. There is also less impact on the environment because fertilizer is not squandered in areas where it is not needed, reducing leaching into runoff water.

In the late 1990s, thermal remote sensing data of fields in Alabama and Georgia showed a strong correlation between temperature maps of corn fields in June and yield maps of the same fields at harvest, at the end of August. These data were obtained using the NASA Atlas sensor onboard

The Power of Radar

Radar can penetrate one to four meters (three to thirteen feet) of sand, revealing covered-up structures invisible to the eye and to other remote sensing bands. Radar remote sensing from the U.S. space shuttle has uncovered lost cities under the desert sands of Egypt and Arabia.

a Stennis Space Center Lear Jet. Results indicate that thermal infrared remote sensing may predict crop yields with high accuracy several months before harvest. Thermal infrared emission from plants is a measure of their temperature. A healthy plant pumps water from the ground, vaporizes it (perspires), and stays cool by doing so. Less-healthy plants exposed to the hot summer Sun cannot keep cool and show a "fever."

Urban Observations Whereas in 1950 less than one-third of the world's population lived in urban areas, almost half the population lives in cities in the first decade of the twenty-first century. Projections indicate that in 2025 two-thirds of the growing world population will be city dwellers. Most of the city population increase will occur in developing countries where serious challenges are expected. In rich countries such as the United States, city development is characterized by urban sprawl using up an ever-increasing proportion of available land. A 1997 U.S. Department of Agriculture study reported that nearly 6.5 million hectares (16 million acres) of U.S. forestland, cropland, and open spaces were converted to urban use between 1992 and 1997. This unfortunate trend continues into the twenty-first century. For instance, the California Farmland Trust performed a study in the mid-2000s that showed the state of California lost 538,273 acres of farmland to urban uses from 1990 to 2004. The worst part of using farmland for urban development, the study found, is that the best farmland is often being converted to urbanization. California is the top food producer in the United States, with nearly 27 million acres used for agriculture, and nine million of those acres considered as high-quality farmland.

Rapid growth and changes of urban geography require detailed, accurate, and frequently updated maps. Such maps can be produced faster, cheaper, and with considerably less human labor by using very-high-resolution satellites such as GeoEye's *IKONOS* than by using

REMOTELY SENSING THE HEAT ISLAND EFFECT

The replacement of vegetated areas with concrete cover in high-density urban areas has a number of environmental drawbacks. One of these problems is known as the heat island effect. Pavement and concrete-covered areas can raise the temperature of cities 10° (based on the Celsius scale), or 50° (based on the Fahrenheit scale) above the temperature in surrounding areas that have kept their vegetation. Such microclimates in cities are not only uncomfortable for the people who live there but also significantly increase power consumption for cooling purposes and increase levels of harmful ground-level ozone. The heat island effect has been studied extensively with remote sensing data obtained from airborne sensors over cities such as Atlanta (Georgia), Sacramento (California), Salt Lake City (Utah), and Baton Rouge (Louisiana).

ground-based data acquisition. A number of satellite remote sensing companies, such as the French company Spot Image, provide services and products for land and urban planners and for businesses such as real estate and insurance companies.

This information may be used to decide in which region to expand urbanization, where to build roads, and how to develop transportation infrastructure. Frequently updated and accurate maps from very-high-resolution satellites will also be useful for infrastructure designs—power cables, water lines, sewer lines, urban transportation systems, and so on.

Businesses can use very-high-resolution urban satellite observations in conjunction with other data—such as demographics—to choose the appropriate location for a franchise or a new store by extrapolating information about urban growth trends. Construction companies can use images taken by satellites, such as GeoEye's *IKONOS* or DigitalGlobe's *QuickBird,* to plan large-scale construction projects. These very-high-resolution satellites are able to identify and locate, with a great deal of accuracy, such surface features as buildings, parking lots, and their elevation.

Urban expansion and loss of farmland can also be monitored using radar remote sensing, such as that provided by the Canadian RADARSAT system. The advantage of radar is that it "sees" through clouds and at night. Thus, regions that are often covered with clouds and do not lend themselves to visible light and near-infrared remote sensing can be imaged using radar illumination.

Wireless communications in cities require a judicious distribution of relays atop tall buildings to avoid blind spots. A three-dimensional model of the cityscape is thus essential. Currently, such models are produced from radar and stereoscopic remote sensing from aircraft. Since cityscapes change rather quickly as new skyscrapers or other tall buildings are built, there is a need for updates. High-resolution radar or stereoscopic visible data from space-based satellites may in the future prove cheaper than aircraft for such applications. As of 2004, Google Inc. provides space-based images to the public through its Google Earth. Many of the images come from satellites owned by DigitalGlobe and GeoEye. In addition, many map images used by Microsoft and Yahoo! in their search engines use these satellites.

Conclusion

This rapid tour of satellite and airborne remote sensing applications shows how useful this technology can be to resolve global, regional, or very local challenges when combined with GIS. It also gives a flavor of a future where geospatial* information will permeate all activities on Earth and create tremendous business opportunities.

 See also **Global Positioning System (Volume 1) • Military Customers (Volume 1) • Military Uses of Space (Volume 4) • Natural Resources (Volume 4) • Reconnaissance (Volume 1) • Satellites, Types of (Volume 1)**

* **geospatial** relating to measurement of Earth's surface as well as positions on its surface

Resources

Books and Articles

Campbell, James B., and Randolph H. Wynne. *Introduction to Remote Sensing*, 5th ed. New York: Guilford Press, 2011.

Chuvieco, Emilio, and Alfredo Huete. *Fundamentals of Satellite Remote Sensing*. Boca Raton, FL: CRC Press, 2010.

Cracknell, Arthur Philip, and Ladson Hayes. *Introduction to Remote Sensing*, 2nd ed. Boca Raton, FL: CRC Press, 2008.

Horning, Ned, et al. *Remote Sensing for Ecology and Conservation: A Handbook of Techniques*. Oxford: Oxford University Press, 2010.

Purkis, Samuel J., and Victor Klemas. *Remote Sensing and Global Environmental Change*. Chichester, UK: Wiley-Blackwell, 2011.

Websites

Carbon Dioxide Information Analysis Center. <http://cdiac.esd.ornl.gov/home.html> (accessed October 25, 2011).

Climate Change. Environmental Protection Agency. <http://www.epa.gov/climatechange/index.html> (accessed October 25, 2011).

Coastal-Change and Glaciological Maps of Antarctica. U.S. Geological Survey. <http://pubs.usgs.gov/fs/2005/3055/> (accessed October 25, 2011).

Coastal Remote Sensing (CRS) Program. Coastal Services Center, National Oceanic and Atmospheric Administration. <http://csc.noaa.gov/> (accessed October 25, 2011).

DigitalGlobe. <http://www.digitalglobe.com/> (accessed October 25, 2011).

GeoEye. <http://www.geoeye.com/CorpSite/> (accessed October 25, 2011).

GOSAT Project. <http://www.gosat.nies.go.jp/index_e.html> (accessed October 25, 2011).

The Landsat Program. <http://landsat.gsfc.nasa.gov/> (accessed October 25, 2011).

MDA Federal Inc. <http://www.mdainformationsystems.com/> (accessed October 25, 2011).

NOAA: 2010 Tied for Warmest Year on Record. <http://www.noaanews.noaa.gov/stories2011/20110112_globalstats.html> (accessed October 25, 2011).

A World Imperiled: Forces behind Forest Loss. <http://rainforests.mongabay.com/0801.htm> (accessed October 25, 2011).

* **velocity** speed and direction of a moving object; a vector quantity

RLV *See Launch Vehicles, Reusable (Volume 1)*

Rocket Engines

From the first rockets built by the Chinese over a millennium ago to the precision engines used by modern missiles, rocket engines all work in accordance with Isaac Newton's Third Law of Motion: For every action there is an opposite and equal reaction. In a rocket engine, hot gas expelled at high velocity* generates thrust in the opposite direction. The most common means of doing this uses chemical reactions to produce the hot gas. The first rockets used solid propellants, such as black powder, but they were very inefficient. Liquid-propellant rocket engines, first developed in 1926 by Robert H. Goddard (1882–1945), are generally much more efficient and opened the way to spaceflight.

The Origins of Modern Engines

Atlas and Delta launch vehicles were originally U.S. Air Force (USAF) rockets developed in the 1950s. To power these missiles, Rocketdyne developed a family of rocket engines that burned kerosene and liquid oxygen (LOX) based on German V-2 rocket technology obtained after World War II (1939–1945). As these rockets were adapted to their new role as launch vehicles in the 1960s, still larger versions of their engines (such as Rocketdyne's 1.5 million-pound thrust F-1) were built for the Saturn rockets that sent Apollo missions to the Moon.

Delta II and III rockets have used the 200,000-pound thrust RS-27A rocket engine, which is an updated descendant of the MB-3 used in the original Delta rocket. The Delta IV rocket system uses Rocketdyne's RS-68 (Rocket System 68) rocket engine. The RS-68 engine produces 663,000-pounds of thrust (at sea level atmospheric pressure). The RS-68 is the largest hydrogen-fueled rocket engine in the world. The engine is expendable, meaning it is used only once. The RS-68 was designed for simplicity and lower cost as compared to the multi-use Space Shuttle Main Engine (SSME).

Besides the Delta IV, the Atlas V rocket is the only other American expendable rocket capable of lifting large payloads into orbit. The Atlas V uses the RD-180 engine, which utilizes kerosene and LOX, to power its first stage, and an RL10 engine burning liquid hydrogen and LOX to power its Centaur upper stage. Some configurations of the Atlas V use strap-on booster rockets to increase thrust. Energomash, the corporate descendant of the Soviet design bureau that developed many Russian rocket engines, worked with the American firm Pratt & Whitney to build the 860,568-pound thrust (measured at sea level pressure) RD-180 used on the American Atlas V.

A close-up view of a space shuttle's main engine test firing shows how hot gas is expelled. At a high velocity the expelled gas generates thrust in the opposite direction, enabling liftoff. *NASA.*

The maiden flight of the Atlas V took place on August 21, 2002. Since that time, and as of November 1, 2011, there have been a total of 27 Atlas V launches. All of these launches have been successful, with only one anomalous flight, on June 15, 2007. During that launch, the Centaur upper stage engine shut down prematurely resulting in the payload's being successfully placed into orbit, but at an altitude lower than originally intended. The anomaly was subsequently traced to a faulty valve in the Centaur upper stage.

Boosting Performance

While modern solid rockets are less efficient than liquid-propellant engines, their simplicity and relatively low cost make them ideal for certain roles. For decades many American launch vehicles used solid rocket upper stages. The Delta II, with its Star 48B motor, continued this series' use of solid rocket third stages. Some small launch vehicles, such as the American Pegasus, Taurus, and Athena, as well as the Japanese J-1 and M-5, use solid rockets in all their stages to reduce costs.

Solid rocket motors strapped to the first stage of a launch vehicle have also proved to be an economical means of increasing a rocket's payload* capability. Since 1964, the U.S. Delta rocket line used increasingly larger clusters of Castor solid rocket motors to help enhance the design's performance. The Delta II rocket, as well as the (discontinued) Delta III rocket, utilize as many as nine GEM-40 motors built by Alliant Techsystems.

The use of solid rocket boosters was most apparent in the venerable Titan family of launch vehicles. A pair of 300-centimeter (120-inch) diameter solid rocket motors was attached to the USAF (United States Air Force) Titan II missile core in 1965 to produce the Titan IIIC, which had over four times the payload capability. The Titan used the Aerojet-General LR87 and LR91 engines burning liquid hypergolic* propellants that ignite spontaneously on contact. Successive Titans have used more powerful solid boosters attached to upgraded cores to further increase the payload. The Titan IVB used a pair of solid rocket motor units built by Alliant Techsystems to produce 3.4 million pounds of thrust at liftoff. Largely due to cost, safety, and environmental issues associated with its fuel, the Titan line of rocket vehicles was terminated in 2005.

A means of boosting rocket performance is by using cryogenic* propellants, such as liquid hydrogen and LOX, which have twice the efficiency of most other propellants. The first engine to use these cryogenic propellants was the 15,000-pound thrust RL-10 engine built by Pratt & Whitney and used in the high-performance Centaur upper stage since 1960. The Centaur, with improved versions of the RL-10, has been used in combination with the Atlas and Titan, while the RL-10B-2 (a variant of the RL-10 engine) has been used in the second stage of the Delta III and IV (Delta III and the Titan family of rockets are no longer in service).

Modern Rockets That Use Both Cryogenic and Solid Fuels

The U.S. space shuttle was designed to use a combination of solid rocket motor technology and high-efficiency cryogenic rocket engines. The space shuttle program ended in July 2011 with the launch of mission STS-135 on the orbiter *Atlantis*. The space shuttle used a pair of solid rocket motors (officially designated Solid Rocket Boosters [SRBs]) built by ATK Launch Systems Group (originally named Thiokol) to generate 5.3 million pounds of thrust for liftoff, while a trio of Rocketdyne-built SSMEs, generating 375,000 pounds of thrust each, supplied most of the energy needed to reach orbit. Unlike most other large rocket engines in use today, the space shuttle's SSME liquid rocket engines were designed from the start to be reused many times.

Other launch vehicles use similar arrangements of solid-fuel boosters and cryogenic engines, such as the European Ariane 5 and Japanese

* **payload** any cargo launched aboard a rocket that is destined for space, including communications satellites or modules, supplies, equipment, and astronauts; it does not include the vehicle used to move the cargo nor the propellant that powers the vehicle

* **hypergolic** fuels and oxidizers that ignite on contact with each other and need no ignition source

* **cryogenic** related to extremely low temperatures; the temperature of liquid nitrogen or lower

* **Apollo** American program to
land men on the Moon; *Apollo 11,
Apollo 12, Apollo 14, Apollo 15,
Apollo 16,* and *Apollo 17* delivered twelve men to the lunar
surface (two per mission)
between 1969 and 1972 and
returned them safely back to
Earth

H-IIA (or H2A). The Delta IV uses the cryogenic RS-68 engine with various booster rockets. Built by Rocketdyne, the RS-68 was the first totally
new American rocket engine design since the SSME was designed in
1971. The booster rockets are often smaller solid-fuel rockets attached to,
and arranged around, the first stage of the rocket. The smaller, but fast-
burning, booster rockets initially give a big boost in upward thrust, but
exhaust their fuel long before the liquid-fueled rocket engines, at which
time they are jettisoned from the main rocket.

Nuclear Rockets

A very efficient type of rocket engine utilizes nuclear energy. The vast
majority of rocket engines in use today utilize chemical reactions to
produce a heated gas that in turn produces a rocket's thrust. In contrast, nuclear engines make use of a compact nuclear reactor to heat
liquid hydrogen or other fluid thereby generating a thrust with more
than twice the efficiency of conventional chemical rocket engines.
During the 1960s, the National Aeronautics and Space Administration
(NASA) developed the Nuclear Engine for Rocket Vehicle Applications
(NERVA) with a reactor built by Westinghouse Electric and the engine
built by Aerojet-General. Before work stopped in 1972, in part due to
post-Apollo* budget cuts, NERVA was intended for use in advanced lunar
and interplanetary missions. The potential use of nuclear rocket engines
is somewhat controversial since they utilize a radioactive material that is
inherently dangerous to most life, including humans, for very long periods
of time if it were to leak or otherwise be released into the environment.

Advanced Engine Concepts

One of the more novel rocket engine designs is the aerospike engine.
A chief advantage of the aerospike engine is that it maintains its aerodynamic efficiency as the rocket it powers flies through the atmosphere
on its way to outer space. The XRS-2200 aerospike rocket engine was
developed by Rocketdyne for the now-defunct X-33 rocket program.
Aerospike engines are members of the class of rocket engines known as
"altitude compensating nozzle" engines. A launch vehicle with an aerospike engine is projected to use 25 to 30% less fuel at low altitudes (where
rockets typically have the greatest need for additional thrust) as compared to conventional chemical rocket engines. Aerospike engines have
been studied for a number of years and appear in many Single-Stage-To-
Orbit (SSTO) rocket designs. However, as of November 2011, aerospike
engines were not yet in commercial production.

Space Tourist Industry

In the first decade of the twenty-first century, many companies were
formed with the intention of sending people into space on a paying basis.
The first successful step in that process occurred in 2004, when the Tier
One program of the Scaled Composites corporation won the $10 million
Ansari X Prize for launching humans on two sub-orbital flights within a

period of two weeks. The program used a spacecraft called SpaceShipOne, which was carried into space by a jet-powered aircraft called White Knight One. SpaceShipOne was retired on October 4, 2004, immediately after winning the Ansari X Prize. Its successor, SpaceShipTwo was already under development, along with the modification of the original White Knight One, called White Knight Two. Scaled Composites has entered into an agreement with Richard Branson's Virgin Galactic company to begin offering recreational trips into space, probably sometime in or shortly after 2012. Flights will cost $200,000 per person, and more than 430 tickets had been sold as of late 2011, with a downpayment of 10 percent on the purchase price required for a reservation. SpaceShipOne was powered by a hybrid rocket engine* burning HTPB (a type of solid polymer) along with nitrous oxide. Though not as efficient as a liquid-fueled rocket engine, the hybrid engine that will be used for Virgin Galactic's tourist rocket is very reliable and safe.

▶ *See also* **Launch Industry (Volume 1)** • **Launch Services (Volume 1)** • **Launch Vehicles, Expendable (Volume 1)** • **Launch Vehicles, Reusable (Volume 1)** • **Reusable Launch Vehicles (Volume 4)** • **Rockets (Volume 3)**

* **hybrid rocket engine** a rocket engine utilizing propellants that are in two different states of matter; one state is solid, while the other state is either gas or liquid

Resources

Books and Articles

Baker, David. *Jane's Space Systems and Industry 2009-2010.* Surrey, UK: Jane's Information Group, 2009.

Butrica, Andrew J. *Single Stage to Orbit: Politics, Space Technology, and the Quest for Reusable Rocketry.* Baltimore: The Johns Hopkins University Press, 2003.

Hujsak, Edward. *All about Rocket Engines.* La Jolla, CA: Mina-Helwig, 2009.

Kitsche, Wolfgang. *Operation of a Cryogenic Rocket Engine: An Outline with Down-to-earth and Up-to-space Remarks.* Berlin: Springer, 2011.

Sutton, George Paul, and Oscar Biblarz. *Rocket Propulsion Elements*, 8th ed. Hoboken, NJ: Wiley, 2010.

Turner, Martin J. L. *Rocket and Spacecraft Propulsion: Principles, Practice and New Developments.* Chichester, UK: Springer, 2010.

Websites

Beginner's Guide to Rockets. Glenn Research Center, National Aeronautics and Space Administration. <http://exploration.grc.nasa.gov/education/rocket/> (accessed October 24, 2011).

Cryogenic Rocket Engine. <http://www.slideshare.net/guestfadacb/cryogenic-rocket-engine> (accessed October 24, 2011).

Riding Nuclear Rockets to Mars Will Get Us There Incredibly Fast, but Is It the Craziest, Most Irresponsible Idea Ever? <http://www.motherboard. tv/2011/7/18/riding-nuclear-rockets-to-mars-will-get-us-there-incredibly-fast-but-is-it-the-craziest-most-irresponsible-idea-ever> (accessed October 24, 2011).

Rocket Engines. Aircraft Engine Historical Society. <http://www.engine history.org/rocket_engines.htm> (accessed October 24, 2011).

Space Tourism Comes Closer to Liftoff. <http://travel.usatoday.com/ flights/story/2011/04/Space-tourism-travel-comes-closer-to-fruition/46549950/1> (accessed October 24, 2011).

Gene Roddenberry. © *Ron Sachs/ Consolidated News Pictures/Getty Images.*

Roddenberry, Gene
American Television Producer and Writer and Futurist
1921–1991

Eugene Wesley Roddenberry (better known as Gene Roddenberry), creator of the television series *Star Trek,* saw space as a place for learning new ideas and ways of thinking. Born in El Paso, Texas, on August 19, 1921, Roddenberry was a pre-law student in college for three years before becoming interested in aeronautical engineering. In 1941, he trained as a flying cadet in the U.S. Army Air Corps. During World War II (1939–1945), he took part in eighty-nine missions and sorties and was decorated with the Distinguished Flying Cross and the Air Medal.

During the war, Roddenberry began to write, selling stories to flight magazines. Back in the United States, he first became a commercial pilot for Pan American World Airways. Later, Roddenberry went to Hollywood intending to write for television. He joined the Los Angeles Police Department to support his family but also to gain life experiences and soon sold scripts to such shows as *Goodyear Theatre, Dragnet,* and *Have Gun Will Travel.*

Roddenberry's creation, the series *Star Trek,* was described by him as "Wagon Train to the Stars," referring to the popular western television show *Wagon Train.* Rodenberry's new series, produced by Desilu Studios, debuted in 1966. Although it had low ratings, the series developed a loyal following. However, it was canceled after only three seasons. It rebounded later when many movies and televisions series were spun off from this original concept. In fact, it was the first television series to have an episode preserved in the Smithsonian Institution, where a 3.3-meter (11-foot) model of the U.S.S. *Enterprise* is also exhibited on the same floor as the Wright brothers' original airplane. The first U.S. space shuttle was named *Enterprise* in honor of this fictional spacecraft.

While making *Star Trek,* Roddenberry gained a reputation as a futurist, speaking on the subject at universities, the Smithsonian, meetings of the National Aeronautics and Space Administration (NASA), and Library of Congress gatherings.

Roddenberry died from heart failure on October 24, 1991, in Santa Monica, California. A year later, a canister of his ashes was taken into space aboard the space shuttle *Columbia*, during NASA mission STS-52. In 1997, another capsule carrying some of his ashes was launched into space from a Pegasus XL rocket.

Roddenberry received a star on Hollywood's Walk of Fame in 1986. The impact crater Roddenberry on the planet Mars is named after Gene Roddenberry, as is asteroid 4659 Roddenberry, which was discovered in 1981. In 2007, Roddenberry was inducted into the Science Fiction Hall of Fame, held within the Science Fiction Museum in Seattle, Washington. The *Star Trek* franchise he created remains popular; a 2009 feature film of the same name was a financial and critical success.

 See also **Burial (Volume 1) • Careers in Writing, Photography, and Film (Volume 1) • Entertainment (Volume 1) •** *Star Trek* **(Volume 4)**

Resources

Books and Articles

Alexander, David. *Star Trek Creator: The Authorized Biography of Gene Roddenberry.* New York: Roc, 1994.

Engel, Joel. *Gene Roddenberry: The Myth and the Man Behind Star Trek.* New York: Hyperion, 1994.

Van Hise, James. *The Man Who Created Star Trek, Gene Roddenberry.* Las Vegas, NV: Pioneer Books, 1992.

Whitfield, Stephen E., and Gene Roddenberry. *The Making of Star Trek.* New York: Ballantine Books, 1968.

Websites

Gene Roddenberry. Star Trek. <http://www.startrek.com/database_article/roddenberry> (accessed October 13, 2011).

Roddenberry.com: Where SciFi Begins. <http://www.roddenberry.com/> (accessed October 13, 2011).

S

Satellite Industry

When people watch the Olympics do they think of satellites? Maybe they should. For many years companies have been televising sporting events such as the Olympic games by way of satellites, which popularized the phrase "live via satellite" and helped create a common impression of what commercial satellites can do for humanity. It was, in fact, a boxing match pitting Muhammad Ali ("The Greatest") against ("Smokin'") Joe Frazier in 1975 when satellites were first used to broadcast a single sporting event to the entire world. While satellites still bring the public sports, news, and entertainment programming from around the world each day, the commercial satellite industry can, and is, doing much, much more—from delivering high-speed Internet content to taking pictures from space of objects on Earth that are as small as a soccer ball.

* **elliptical** having an oval shape

* **geosynchronous** remaining fixed (as seen from Earth) in an orbit 35,786 kilometers (22,300 miles) above Earth's surface

* **orbit** the circular or elliptical path of an object around a much larger object, governed by the gravitational field of the larger object

Historical Development of the Industry

Most people are familiar with movies such as *Apollo 13* (1995) and *The Right Stuff* (1983)—the latter movie depicted the start of the U.S. civilian space program within the National Aeronautics and Space Administration (NASA)—but many do not realize that the commercial satellite industry actually developed right alongside the government space program in the early 1960s. The satellite industry was kick-started back in July 1962, when scientists at AT&T Bell Laboratories decided to build the world's first commercial satellite, dubbed *Telstar,* after losing a competition for NASA's active satellite program. NASA later offered to launch the *Telstar* satellite into an elliptical* orbit during which it transmitted brief live television transmissions across the Atlantic Ocean for the first time. *Telstar* had a tremendous worldwide impact by showing the amazing potential of satellite communications.

A few years later, in April 1965, the International Telecommunications Satellite Organization (INTELSAT) launched *Intelsat 1* (nicknamed Early Bird), the world's first commercial geosynchronous* satellite. So-called "GEO" (Geosynchronous Earth Orbit) satellites orbit* Earth at 35,800 kilometers (22,300 miles) in a belt directly above Earth. At that point in space, the satellite orbits Earth at the same speed as Earth's rotation—making the satellite appear to be fixed in the same location in the sky. It was futurist Arthur C. Clarke, author of the film *2001: A Space Odyssey* (1968), who first predicted back in 1945 that the entire world could be connected wirelessly by placing three communications satellites in geosynchronous orbit.

The Early Bird satellite was built by Hughes Aircraft, the company founded by eccentric billionaire Howard Hughes. The satellite had the capacity to carry 480 telephone channels (240 simultaneous calls) and had the power of an ordinary household light bulb. For that era, Early Bird's capabilities were impressive considering that the largest under-sea transoceanic telephone cable at the time carried only 256 channels. However, today's largest telecommunications satellites are more than 500 times as efficient and generate more than 15 kilowatts of power, allowing a single spacecraft to carry tens of thousands of simultaneous telephone calls or hundreds of channels of television programming to satellite dishes on the ground as small as 46 centimeters (18 inches).

For the first two decades of its existence, the satellite industry worked to connect large companies to other large companies across oceans, deserts, and other vast, remote regions. Telephone companies first employed satellites to connect calls where there were no undersea cables. Later it was television networks, such as ABC, NBC, and CBS in the United States, which used satellites to transmit programs to their local affiliate stations—so-called point-to-multipoint distribution—and helped the satellite industry grow. In the late 1970s, cable television companies began to get into the act, using satellites to downlink channels such as CNN and MTV, and then retransmit the signals over coaxial cables to homes.

The satellite industry can mark 1976 as the year it began to evolve from a purely business-to-business model to one that also included business-to-consumer or so-called retail services. That year a Stanford University (California) professor named Taylor Howard designed and built the first backyard satellite dish. Howard used his 4.9-meter (16-foot) home satellite dish to receive HBO and other television programs that previously were carried only by cable television companies. By 1980, Howard had sold the blueprints on how to build his dish to over five thousand people, and the direct-to-home (DTH) satellite industry was born. By 1985, several DTH equipment companies were shipping more than five hundred thousand home satellite systems to consumers across the United States.

It was really the 1990s that witnessed the greatest changes in satellite technology, which in turn brought fundamental changes to the commercial satellite industry. The introduction in 1994 of high-powered Direct Broadcast Satellite (DBS) services, which used new digital compression technology and more powerful spacecraft, allowed consumers to receive hundreds of channels of digital-quality programming on a dish about the size of a pizza pan. The DTH industry has since continued to grow with more than eighty million subscribers in the United States and a total of more than 745 million subscribers around the world subscribing to DTH satellite television services in 2011.

Key Segments of Today's Satellite Industry

History aside, the best way to understand the satellite industry today is to divide it into four key segments: satellite services—transmitting voice, data, and television signals to businesses and consumers; ground equipment—designing and manufacturing satellite dishes, large Earth stations, software, and consumer electronics; satellite manufacturing—building spacecraft, components, and electronics; and launch—building space launch vehicles and carrying satellites into orbit.

Each year, the Satellite Industry Association (SIA) surveys over 70 members and various key companies around the world to determine the state of the industry. The SIA reports worldwide employment and revenue in each of the segments. The SIA reported in 2011 that the commercial satellite industry generated $168.1 billion in revenue worldwide in 2010, which was a 4.5% increase over the previous year.

Satellite Services The largest segment of the industry is satellite services, which worldwide generated $101.3 billion in revenue in 2010, a 9% increase over 2009. About $11.1 billion in revenue in this sector was generated by companies that lease transponder* capacity to programming companies such as the Discovery Channel and ESPN, as well as to long-distance telephone companies such as MCI and AT&T. The bulk of the services revenue, at $83.1 billion, came from satellite television and radio broadcasting.

While traditional satellite service providers such as IntelSat, Eutelsat, GE American Communications, and SES Astra continue to lease capacity

* **transponder** bandwidth-specific transmitter-receiver units

* **point of presence** an access point to the Internet with a unique Internet Protocol (IP) address; Internet service providers like AOL generally have multiple POPs on the Internet

Trucks equipped with satellite uplinks like these can transmit signals to orbiting communication satellites, which can retransmit the signal around the world. This technology has enabled live television broadcasts from almost anywhere a truck can reach. © *Paul J. Richards/AFP/Getty Images.*

to television and telephone companies, a growing portion of their business now comes from data services. Internet Service Providers (ISPs) are now using satellites in countries throughout the developing world to link directly to the Internet backbone in the United States. Those providers set up a small 1-to-3-meter (3-to-10-foot) dish and in one hop can link directly to a point of presence* (POP) on a fiber-optic backbone. Another application being pursued by satellite operators is to broadcast common content—such as Stephen King's latest short story or streaming audio/video clips—over the World Wide Web (Web) to local ISPs around the world where the content can be retrieved by nearby Web surfers. In using satellites for the same kind of point-to-multipoint distribution service used by television broadcasters, popular web sites can manage the flow of traffic on the Web and avoid crashing their servers or networks.

* **low Earth orbit** an orbit between 300 and 800 kilometers (185 and 500 miles) above Earth's surface

Mobile satellite services such as those offered by Globalstar, Inmarsat, Thuraya, and Terrestar are yet another emerging part of the services business. These companies provide voice and data service to thousands of ships, planes, cars, and people in parts of the world that are not served by cellular or traditional wired telephone networks. These systems often use constellations of satellites in either GEO or low Earth orbit* to serve laptop-and handset-sized mobile terminals. Such systems allow pipeline workers, merchant ships, and other mobile users to communicate even in the most remote places on Earth.

The year 2001 witnessed the inception of Satellite Digital Audio Radio Services (SDARS) for much of North America. Originally, two companies, XM Satellite Radio and Sirius Satellite Radio both broadcast audio programming to the United States and Canada. Meanwhile, a company called WorldSpace (which later changed its name to 1worldspace) broadcast to parts of Africa, the Middle East, and Asia from stations in India. Among them, the SDARS broadcast hundreds of channels of digital music, news, sports, and entertainment programming to cars, homes, and boom boxes. On July 25, 2008, the U.S. Federal Communications Commission (FCC) approved the merger of the XM and Sirius satellite radio services, resulting in a single satellite radio network for the United States and Canada under the name Sirius XM Radio, Inc. At about the same time, in October 2008, 1worldspace filed for Chapter 11 bankruptcy protection. As of late 2011, the company is no longer in active operation.

Ground Equipment Growth in the ground equipment sector of the commercial satellite industry has begun to slow significantly from its peak of 34% between 2007 and 2008. In 2010, total income from the sale of satellite dishes, mobile satellite phones, and Earth stations that communicate with satellites in orbit reached $51.6 billion, an increase of only three percent over the preceding year. About 85% of that total came from consumer sales, with the remainder coming from network sales. An increasing portion of the ground equipment market is made up of DTH systems. Since its introduction to American consumers in 1994, DBS dishes and set-top boxes have been the fastest-selling consumer electronics product of all time—outselling video cassette recorders (VCRs), personal computers (PCs), and color televisions during their first year on the market.

Companies such as Hughes, RCA, Sony, and Gilat manufacture these dishes for both consumers and large corporations that own private satellite networks called Very Small Aperture Terminals (VSATs). VSATs are a little-known but important part of the overall telecommunications network. They allow retail companies, such as Target and Blockbuster, as well as gas stations such as those run by ExxonMobil, to verify credit cards and control their inventories. Prices for VSAT dishes dropped from $10,000 to $20,000 per terminal in 1980, to less than $1,000 per terminal in 2011, helping fuel sales of VSATs to many large and small corporations.

* **spot beam technology** narrow, pencil-like satellite beam that focuses highly radiated energy on a limited area of Earth's surface (about 160 to 800 kilometers [100 to 500 miles] in diameter) using steerable or directed antennas

* **ion propulsion** a propulsion system that uses charged particles accelerated by electric fields to provide thrust

* **expendable launch vehicles** launch vehicles, such as a rocket, not intended to be reused

Satellite Manufacturing Consumer satellite services such as DBS and SDARS would not be possible without advances in satellite manufacturing. In terms of power, capacity (bandwidth), and lifetime in orbit, large telecommunications satellites at the turn of the millennium were twenty times more capable than satellites manufactured only a decade previous. In the first decade of the twenty-first century, satellite capabilities continued to increase dramatically, thanks to features such as spot beam technology*.

Both the number of transponders and the overall power of satellites have increased over the years. Each transponder that used to be able to carry a single analog channel can now carry several simultaneous digital channels. Increases in power are tied to more efficient solar panel and battery technology. By increasing the power of the satellite in space, satellite operators can dramatically reduce the size of receiving dishes on Earth. Another major achievement is the use of ion propulsion* technology for satellite station-keeping. By using ion propulsion to generate the thrust that keeps the satellite oriented towards Earth, satellite manufacturers have been able to increase the number of years that a satellite is able to provide service before it runs out of fuel. Altogether, these technologies have had a major impact on the ability of satellites to compete with terrestrial telecommunications technologies.

Satellite manufacturing, including payments to prime contractors and their subcontractors, accounted for $10.8 billion worldwide in 2010—a decline of 20% from 2009. The decline could be attributed primarily to a reduction in the number of large satellites launched by governmental agencies. Leading satellite manufacturers include Boeing Satellite Systems, Lockheed Martin, Orbital Sciences, and Space Systems/Loral in the United States and Thales Alenia Space and EADS Astrium in Europe. This segment of the industry experienced rapid consolidation in the past few years as several European companies merged in order to compete with U.S. companies.

In 2010, there were twenty-eight geosynchronous-orbit (GEO) satellites ordered worldwide, a decrease of 31% from the forty-one ordered in 2009. Of this number, sixteen went to U.S. manufacturers, six went to European manufacturers, four went to Russian manufacturers, and two went to Chinese companies.

Launch Of course, the satellite industry would not exist if it were not for the expendable launch vehicles* (ELVs) that launch commercial spacecraft into orbit. The worldwide launch industry generated revenues of $4.3 billion in 2010, a decrease of 4% from 2009. Companies such as Arianespace, International Launch Services (ILS), United Launch Alliance, Sea Launch, Orbital Sciences, and China Great Wall Industry Corporation (CGWIC) sell rockets that launch satellites into outer space.

The launch segment of the industry has also changed dramatically since the 1970s when U.S. Air Force rockets were used to launch commercial satellites. The U.S. decision to shift all satellite launches from ELVs to

the space shuttle helped spur the Europeans to develop their own ELV—
the Ariane rocket. In the wake of the space shuttle *Challenger* tragedy
in 1986, the United States decided to fly satellites aboard ELVs once
again, and U.S. companies got back into the launch services market. In
the 1990s, Chinese, Russian, and Ukrainian rockets began to be used to
launch commercial satellites. The market in the early twenty-first century
is more competitive than ever, resulting in lower launch costs for satellite
operators.

Space Sciences, 2ⁿᵈ Edition

* **geostationary orbit** a specific
altitude of an equatorial orbit
where the time required to circle
the planet matches the time it
takes the planet to rotate on its
axis. An object in geostationary orbit will always appear
to remain over the same geographic location on the equator
of the planet it orbits

Meanwhile, launch service companies have worked steadily to increase the lift capabilities of their rockets to accommodate the heavier, more powerful satellites. Vehicles such as the Ariane 5 rocket are capable of delivering 6.5 metric tons (equivalent to 7.2 tons in the U.S. customary system) to geostationary* orbit, and their ability is expected to only increase in the coming years. Sea Launch—an international cooperative venture including Boeing (United States), RSC Energia (Russia), Yushnoye (Ukraine), and Kvaerner (Norway)—launches satellites from a converted offshore oil-drilling platform and command ship that motor to a site on the equator in the middle of the Pacific Ocean in an effort to increase lift capability. International Launch Services now uses powerful Russian-built RD-180 rocket engines to increase the lift capability of the workhorse Atlas ELV.

Emerging Technologies

Outside of the four major industry segments—communication services, ground equipment, satellite manufacturing, and launch—there are a host of other emerging technologies that are generating increasing revenue and interest. Commercial remote sensing satellites, such as GeoEye's IKONOS spacecraft, are now capable of taking pictures from space clear enough to resolve objects on the ground less than one meter (40 inches) in size. Such images are used by farmers, geologists, and urban planners to assist them in their jobs. Software that links these images with maps generated using coordinates from the U.S. Air Force's Global Positioning System (GPS) provides an important new source of information to businesses that use scarce natural resources here on Earth.

The satellite industry has come a long way since AT&T's *Telstar 1* satellite first proved that space could be used for moneymaking commercial ventures. The continued growth in Internet data and new information technologies are expected to drive the commercial satellite industry in the coming decades. The growth in these services markets will fuel demand for more satellites, satellite dishes, and rocket launches. Arthur C. Clarke's vision of a world connected via satellite has become a reality. Today's visionaries see viable markets for solar power generation and routine space tourism in the relatively near future. Do not count them out—many people thought President John F. Kennedy's pledge to put a man on the Moon would never be fulfilled.

 See also **Communication Satellite Industry (Volume 1) • Navigation from Space (Volume 1) • Reconnaissance (Volume 1) • Remote Sensing Systems (Volume 1) • Satellites, Types of (Volume 1) • Small Satellite Technology (Volume 1)**

Resources

Books and Articles

Clarke, Arthur C. *2001: A Space Odyssey.* New York: New American Library, 1968.

Labrador, Virgil S., and Peter I. Galace. *Heavens Fill with Commerce: A Brief History of the Communications Satellite Industry.* West Covina, CA: SatNews Publishers, 2005.

Payne, Silvano, and Hartley Lesser. *2010 International Satellite Directory: The Complete Guide to the Satellite Communications Industry.* Sonoma, CA: SatNews Publishers, 2010.

Pelton, Joseph N. *Future Trends in Satellite Communications: Markets and Services.* Chicago: International Engineering Consortium, 2005.

Roddy, Dennis. *Satellite Communications.* 4th ed. New York: McGraw-Hill Professional, 2006.

Via Satellite's Satellite Industry Directory 2011. Rockville, MD: Access Intelligence Llc, 2010.

Websites

Satellite Broadcasting & Communications Association. <http://www.sbca.com/> (accessed October 14, 2011).

Satellite Industry Links. <http://www.satellite-links.co.uk/> (accessed October 14, 2011).

State of the Satellite Industry Report, August 2011. Satellite Industry Association. <http://www.sia.org/PDF/2011_State_of_Satellite_Industry_Report_(August%202011).pdf> (accessed October 14, 2011).

Whalen, David J. "Communications Satellites: Making the Global Village Possible." National Aeronautics and Space Administration (NASA) History Division. <http://www.hq.nasa.gov/office/pao/History/satcomhistory.html> (accessed October 14, 2011).

Satellites, Types of

Not long after the Soviet Union launched the first satellite in 1957, satellites began to play an increasingly important role in the lives of people around the world. The first satellites were small because of the lack of powerful launch vehicles, and almost all had scientific or military missions. However, as larger rockets became available and engineers used new technologies to build more efficient payloads*, the first prototypes of many of the satellites still in use in the early 2010s appeared, and consequently changed the world.

Observing Earth

One of the earliest classes of satellites was designed to observe Earth from orbit. Among the first were military reconnaissance* satellites, such as the U.S. Corona and the Soviet Zenit, which took photographs with film

* **payload** any cargo launched aboard a rocket that is destined for space, including communications satellites or modules, supplies, equipment, and astronauts; it does not include the vehicle used to move the cargo nor the propellant that powers the vehicle

* **reconnaissance** a survey or preliminary exploration of a region of interest

* **polar orbits** orbits that carry a satellite over the poles of a planet

* **geosynchronous orbit** the altitude above the surface of the Earth (22,300 miles up) where an orbiting object will be traveling around the Earth at the same speed that the Earth spins. This results in the orbiting object remaining over the same region of the Earth at all times, although it will progress north and south along a straight line (as seen from Earth) over the course of a day, unless its orbit is directly over the equator, which results in a geostationary orbit

* **infrared** portion of the electro-magnetic spectrum with waves slightly longer than visible light

* **remote sensing** the act of observing from orbit what may be seen or sensed below on Earth or another planetary body

* **cartographic** relating to the making of maps

that had to be returned to Earth to be developed. Over the years, sophisticated electronic imaging technology made it possible for spy satellites to obtain very high-resolution images and transmit them almost immediately to analysts. This technology has been useful in gauging a potential adversary's intentions, for verifying compliance with treaties, and in other important ways.

In 1960, early television surveillance technology was used for the first weather satellites. Today, those satellites (flying in polar orbits* and geosynchronous* orbits) are equipped not only with cameras but with a range of sensors that employ the latest infrared* technology. In addition to providing the weather pictures that people see every day on television, these satellites supply meteorologists with the highly detailed information they need to track storms and predict the weather. This application of satellite technology alone has saved countless lives.

Starting in the 1970s some of this technology was applied to remote sensing * satellites such as Landsat. Instead of monitoring military targets at high resolution, these satellites monitor Earth's natural resources on a more moderate scale. These data provide the information needed to locate new sources of raw materials and determine the effects of natural disasters and pollution on the environment. Because this information is so valuable, many commercial remote sensing satellites, such as the French SPOT (Satellite Pour l'Observation de la Terre), have been launched and their data have been sold to a wide range of government and private users.

Recently declassified reconnaissance and cartographic* photographs from U.S. and former Soviet spy satellites are now available, giving researchers more varied long-term data on the environment. Radar mapping technology originally used by the military to make observations through clouds has found numerous civilian applications.

▶

NASA and NOAA's *GOES-N* (Geostationary Operational Environmental Satellite) weather satellite is prepared for launch inside a clean room near Kennedy Space Center. © *Ben Cooper/Science Faction/Getty Images.*

314

Voices in the Sky

By far the most common satellite type launched, and perhaps the one that has had the greatest impact on people's lives, is the communication satellite, abbreviated "SATCOM" or "comsat." Beginning in 1958 early experimental comsats operated as relays or repeaters in relatively low orbits, in part because of the lack of powerful launch vehicles and the crude nature of their electronics. Although such low orbiting comsats still have a place, a large number of them are required to provide continuous communications coverage around the globe.

As early as 1946 the space visionary and author Arthur C. Clarke (1917–2008) recognized the value of placing comsats in geosynchronous orbits. From an altitude of 35,786 kilometers (22,300 miles) above the equator, satellites match Earth's spin and appear to hang motionless in the sky. From this great height, over one-third of the planet's surface can be seen, allowing a satellite to relay signals over long distances. After several successful experiments the first commercial geosynchronous comsat, *Early Bird* (*Intelsat I*), was launched in 1965. In the succeeding decades, improved rockets allowed larger comsats to be launched. Combined with major advances in microelectronics, each of the hundreds of active comsats presently in orbit have thousands of times the capacity of their earliest ancestors.

Although geosynchronous comsats are useful at low latitudes, they appear too close to the horizon at high or polar latitudes. To overcome this problem, since 1965 the Soviet Union (and later Russia) has launched Molniya satellites into highly elliptical* 12-hour orbits (now called Molniya orbits) inclined to the equator at 63.4°. This type of orbit allows them to be seen high above the horizon over most of Russia's territory for long periods. From this vantage point, Molniya satellites can relay television and telephone signals across that nation's vast expanses. During the Cold War (1945–1991), such orbits were used by some signal intelligence, or sigint, satellites to intercept radio signals. Sigint satellites also are used to track ships at sea, locate radar installations, and monitor other activities such as various types of radio transmissions.

A type of comsat known as a navigation satellite, or navsat, has become important to military and civilian users. Operating in precisely known orbits thousands of miles above Earth, these satellites broadcast a precise timing signal. Signals from three or more navsats can be used to determine a position on or above the Earth's surface to within a few meters (yards) or less. The first experimental navsats were built by the U.S. Navy in the 1960s and were used by ships to determine their exact positions at sea. Today a U.S. constellation of satellites forming the Global Positioning System (GPS) allows military and civilian users to accurately determine their locations anywhere in the world. Elsewhere in the world, there are several global or regional navigation satellite systems that are partially operational or expected to become fully operational in the near future.

* **elliptical** having an oval shape

Voyager 2 is an example of a science satellite, designed to collect data that enabled scientists to discover more about the outer solar system. *Illustration by Pat Rawlings. NASA.*

The European Union's Galileo navigation system will provide European countries a navigation system under their mutual control, and is projected to be a major economic benefit to the participating countries. As of October 1, 2011, the launch of the first two navsats in the Galileo system was scheduled for later that same month (October 20). When completed, the Galileo navigation system will consist of 30 navsats providing worldwide navigation coverage (the optimal system consists of 24 navsats, just like the U.S. GPS system—the additional satellites are spares in case of equipment failures in the normal constellation of 24 navsats). Galileo is designed to give better navigation coverage with its satellites in the high latitude (polar) regions than is available with the GPS system of the United States. Moreover, Galileo is designed to be compatible with the U.S. GPS navigation system. Navigational services using the Galileo system are scheduled to begin by 2014.

The Chinese regional system called Beidou-1 consisted of three navigational satellites and commenced serving Chinese and some other regional customers in the year 2000. Throughout the 2000s the Chinese added more navsats for increased navigational capabilities, and in April 2011 an eighth Beidou satellite was launched into orbit, completing the regional phase of the Beidou-2 navsat system. The full Beidou-2 system (also called the COMPASS system) is anticipated to consist of at least 30 navsats and offer worldwide navigational services by 2020.

Russia's navigational system is called GLONASS (Global Navigation Satellite System). Like its American counterpart, GPS, the GLONASS system was conceived in order to serve military navigational needs. The former Soviet Union began GLONASS in the mid 1970s. Russia inherited the GLONASS system after the fall of the Soviet Union, and it was fully operational by 1995. However, economic problems in the 1990s meant that replacement satellites were not supplied in a timely manner, and by 2001 only six of the originally-intended 24 navsats were functional within the GLONASS system. The Russian government has made the necessary expenditures to revitalize GLONASS, and announced in mid 2011 that GLONASS was expected to have its full complement of 24 primary navsats, along with several spare satellites, by the end of that year.

As of late 2011, the government of India was proceeding with plans to have a regional system called IRNSS (Indian Regional Navigational Satellite System) available for full operations sometime in 2014. The IRNSS is designed to operate with a total of seven navsats. Finally, there is a Japanese regional system called QZSS (Quasi-Zenith Satellite System), which is designed to use three satellites in an orbital configuration above Earth so that at least one of the three satellites is at its zenith (highest altitude above the Earth) at any one time. The QZSS system is designed to work with America's GPS system to provide navigational data; by doing so, the combined system will provide much higher positioning data for users in Japan and the Pacific region than GPS alone. At one point, the government of Japan increased the number of satellites for the QZSS system from three to seven, but after the 2011 earthquake and tsunami, followed by the nuclear disaster at Fukushima, budgetary constraints lowered the expected number of satellites for the QZSS system down to four.

Science in Space

Whereas a large number of satellites with practical applications have been launched, science satellites still provide important information about the space environment and the universe beyond. Satellites monitoring Earth's magnetosphere* can provide warnings about communications blackouts and other effects of solar storms. For example, the European Space Agency (ESA) launched the four satellites of Cluster II into highly-elliptical polar orbits in 2000. The four Cluster satellites orbit the Earth in close formation with one another, and use their sensitive instruments to measure various phenomena arising out of the interaction between the solar wind and the Earth's magnetic field.

Since the 1960s, larger, more capable observatories employing increasingly advanced technologies have been launched to observe the

* **magnetosphere** the magnetic cavity that surrounds Earth or any other planet with a magnetic field. It is formed by the interaction of the solar wind with the planet's magnetic field

* **electromagnetic spectrum** the entire range of wavelengths of electromagnetic radiation

Sun and the rest of the universe over the entire electromagnetic spectrum*. The *Solar Dynamics Observatory* (*SDO*), for instance, uses its onboard instruments to investigate a variety of dynamic solar features. Launched in 2010, *SDO* is being used by scientists to understand the formation and structure of the solar magnetic field; how the Sun's magnetic field powers the solar wind; and how such solar phenomena effect the radiation arriving at Earth from the Sun. The *SDO* spacecraft is the first in a series of spacecraft from the National Aeronautics and Space Administration (NASA) for the Living With a Star (LWS) program. The LWS program is intended to focus on the particular aspects of the Sun that most directly affects the people and environment of Earth.

A well-known example of an optical telescope looking out into space (its instruments are much too sensitive to view the Sun) is the *Hubble Space Telescope.* The images that *Hubble* has returned have greatly expanded our knowledge of the universe. In the infrared portion of the electromagnetic spectrum, the *Herschel Space Telescope* has discovered large clouds of water with its highly sensitive instruments, both out in the Milky Way galaxy, as well as in our own solar system, in the form of a great cloud of water vapor around the planet Saturn. As a result of the data returned from these and many other scientific satellites, much more has been learned about a host of phenomena—everything from the Sun and how it affects Earth, to the origins of the universe.

▶ *See also* **Clarke, Arthur C. (Volume 1) • Navigation from Space (Volume 1) • Reconnaissance (Volume 1) • Remote Sensing Systems (Volume 1) • Satellite Industry (Volume 1) • Satellites, Future Designs (Volume 4) • Small Satellite Technology (Volume 1)**

Resources

Books and Articles

Bond, Peter *Jane's Space Systems & Industry 2010/2011.* Coulsdon, Surrey, UK: Jane's Information Group, 2011.

Eddy, John A. *The Sun, the Earth, and Near-Earth Space: A Guide to the Sun-Earth System.* Washington, D.C.: National Aeronautics and Space Administration, 2009.

Elbert, Bruce. *Introduction to Satellite Communication,* 3rd ed. Boston: Artech House, 2008.

Fortescue, Peter, John Stark, and Graham Swinerd, eds. *Spacecraft Systems Engineering.* 4th ed. Chichester, England: J. Wiley, 2011.

Gorn, Michael H. *Superstructures in Space: From Satellites to Space Stations.* New York: Merrell, 2008.

Johnson, Rebecca L. *Satellites.* Minneapolis, MN: Lerner Publications, 2006.

Peebles, Curtis. *The Corona Project: America's First Spy Satellites.* Annapolis, MD: Naval Institute Press, 1997.

Websites

Arthur C. Clarke Foundation. <http://www.clarkefoundation.org/acc/biography.php> (accessed October 2, 2011).

Global Positioning System. <http://www.gps.gov/> (accessed October 2, 2011).

Whalen, David J. *Communications Satellites: Making the Global Village Possible.* National Aeronautics and Space Administration. <http://www.hq.nasa.gov/office/pao/History/satcomhistory.html> (accessed October 2, 2011).

Search and Rescue

Commercial search and rescue missions to retrieve and repair valuable satellites may become commonplace in the twenty-first century. Telecommunications companies and other businesses typically spend between $50 and $300 million to manufacture and launch a new satellite. If all goes well, the spacecraft may function reliably for ten to twenty years. In the harsh environment of space, however, satellites may fail prematurely because of mechanical or electronic breakdowns, damage from solar flares*, or collisions with orbiting debris. Companies may reduce their economic losses from such perils by salvaging damaged or obsolete satellites at a cost lower than what they would pay for replacement spacecraft.

The National Aeronautics and Space Administration (NASA) successfully performed the first satellite search and rescue missions in 1984. In April of that year, astronauts on the space shuttle *Challenger* rendezvoused with the Solar Maximum Mission (SMM, or SolarMax) satellite, walked in space, and replaced electronics and other parts on the damaged spacecraft. They then released it back into orbit to continue its scientific mission of solar flare observations. Six months later, the crew of the space shuttle *Discovery* captured two commercial communications satellites and stowed them in the shuttle's cargo bay for return to Earth. Equipment malfunctions in February 1984 had left these satellites, the U.S. *Westar-6* and the Indonesian *Palapa-B2,* in improper orbits. Technicians on the ground repaired both satellites and successfully launched them back into orbit in April 1990: *Westar-6* was sold to China and became the satellite *AsiaSat* and *Palapa-B2* was sold back to Indonesia and renamed *Palapa-B2R.*

NASA reconsidered astronaut safety after the *Challenger* explosion on January 28, 1986. The agency decided to reduce the risk to astronauts by restricting most shuttle operations to scientific or military missions

* **solar flares** explosions on the Sun that release bursts of electromagnetic radiation, such as visible light, ultraviolet waves, and x rays, along with high speed protons and other particles

that required a human presence in space. The agency made a total of six exceptions to that general rule, only one of which involved a commercial satellite, the International Telecommunication Satellite Organization's Intelsat VI (F-3) in 1992. The other five repair missions involved repair and service missions to the Hubble Space Telescope in December 1993, February 1997, December 1999, March 2002, and May 2009. With the termination of the space shuttle program in July 2011, any additional search and rescue efforts through that avenue became impossible.

The loss of the space shuttle program as a possible mechanism for satellite servicing and repair has become especially troublesome given the huge growth in the satellite industry. Revenues generated by space commerce exceeded government expenditures for space exploration for the first time in 1996. Rapid growth of global telecommunications swelled space business revenues to $168.1 billion in 2010, an increase of five percent over the previous year, and a cumulative increase of 11.2 percent between 2005 and 2010. By June of 2011, there were 986 operational satellites in orbit, about one-third of which (365 satellites; 37 percent) were commercial communications satellites. The next largest categories of satellites were those designed for civil communications (108 satellites, or 11 percent), military communications (84; nine percent), military surveillance (89; 9 percent), and remote sensing (92; 9 percent). Insurance premiums on these satellites amounted to almost $600 million for 2010.

Satellite owners and insurance companies are therefore motivated to find new and creative ways to safeguard their business assets. In 1998, for example, insurers declared a loss on the *HGS-1* Asian television satellite, which had been stranded in a useless orbit after launch. Later, engineers at the Hughes Space and Communications Company (now, Boeing Satellite Development Center) found a way to boost the satellite on two looping orbits around the Moon, finally placing it in a useful parking orbit around Earth.

Repair and servicing of commercial satellites is likely to be a profitable venture in the long term. As of late 2011, however, no private company had made significant progress in developing a practical search-and-rescue system. By contrast, a number of governmental agencies have moved forward aggressively in the design and development of satellite repair and maintenance systems. For example, NASA launched an experimental humanoid robot called the Special Purpose Dexterous Manipulator ("Dextre," for short) to the International Space Station (ISS) on space shuttle flight STS-123 on March 11, 2008. The robot completed its first assignment, the unpacking of two boxes of materials for the ISS, on February 4, 2011. Dextre has the capability of repairing, upgrading, and servicing satellites working from the ISS platform. It was built for NASA by the Canadian firm of MacDonald Dettwiler, which also constructed the Canadarm and Canadarm2 mechanical arms used for loading and unloading on the ISS.

A project similar to the Dextre project has also been in development in Germany under the sponsorship of Deutsches Zentrum für Luft- und

Raumfahrt (German Center for Air- and Spaceflight; DLR), Germany's equivalent of NASA. DLR's humanoid robot "Justin" will be mounted on a spacecraft or satellite with the capability of traveling to other satellites that may need refueling, repair, maintenance, or other servicing. Ultimately, the robot will be able to act under its own control in space, although its current configuration requires instructions from Earth-bound controllers who wear a headpiece and exoskeleton that allow them to experience the same environment in which Justin finds itself. In this configuration, the controller can mimic the actions it wants the robot to take in space. Justin was developed at the Institute of Robotics and Mechatronics, a division of DLR. Other nations have or may become involved in the development of satellite maintenance and repair activities. The Japanese Space Agency, JAXA, for example, has announced that it plans to send a humanoid robot to the ISS in 2013. The Japanese explain that their robot will perform tasks beyond satellite and ISS repair, and will be designed to communicate and interact with human astronauts on board the ISS.

 See also **Satellites, Types of (Volume 1)** • **Servicing and Repair (Volume 1)** • **Tethers (Volume 4)**

Resources

Books and Articles

American Institute of Aeronautics and Astronautics. "Space Commercialization: An AIAA Position Paper Prepared by the Public Policy Committee." Reston, VA: Author, 1996.

Forward, Robert L., and Robert P. Hoyt. "Space Tethers." *Scientific American* 280, no. 2 (February 1999):86–87.

Gorn, Michael H. *Superstructures in Space: From Satellites to Space Stations.* New York: Merrell, 2008.

Johnson, Rebecca L. *Satellites.* Minneapolis, MN: Lerner Publications, 2006.

Lee, Wayne. *To Rise from Earth: An Easy-to-Understand Guide to Spaceflight.* New York: Facts on File, 2000.

Sellers, Jerry Jon. *Understanding Space: An Introduction to Astronautics.* Boston: McGraw-Hill Custom Pub., 2004.

Websites

Dextre, the Space Station's Robotic Arm, Will Try its Hand at Satellite Refueling. <http://www.popsci.com/technology/article/2011-06/dextre-space-stations-robotic-arm-will-try-its-hand-satellite-refueling> (accessed October 26, 2011).

Maintaining and Upgrading Hubble. National Aeronautics and Space Administration. <http://hubble.nasa.gov/missions/intro.php> (accessed October 26, 2011).

A Modular Architecture for an Interactive Real-Time Simulation and Training Environment for Satellite On-Orbit Servicing. <http://c4i.gmu.edu/events/conferences/2011/DS-RT/slides/Tuesday/AM/072-Modular%20Architecture%20for%20VR-OOS.pdf> (accessed October 26, 2011).

Robot Catches Two Balls at Once, Humans Wish They Could. <http://www.robotshop.com/blog/tag/justin> (accessed October 26, 2011).

Satelllite Servicing Capabilities Office. <http://ssco.gsfc.nasa.gov/index.html> (accessed October 26, 2011).

Whalen, David J. *Communications Satellites: Making the Global Village Possible.* National Aeronautics and Space Administration. <http://www.hq.nasa.gov/office/pao/History/satcomhistory.html> (accessed October 26, 2011).

Servicing and Repair

When the U.S. space shuttles became operational in the 1980s, access to space began to take on a whole new outlook. Space was about to become a place to carry out work as well as to explore. A key element in America's human-rated partially reusable space vehicle was the ability to provide access to space on a number of commercial fronts, including in-orbit satellite and spacecraft servicing and repair.

As of 2011, discussions of servicing and repair of space equipment is limited largely to the International Space Station (ISS) and the National Aeronautics and Space Administration's (NASA) Space Transportation system, better known as the space shuttle. The space shuttle operated from 1981 to 2011 and consisted of 135 flights, two of which ended in disaster. Servicing and repairs of other space objects, such as satellites, can be accomplished only by capturing those objects and bringing them to a shuttle or the ISS for attention, a very rare occurrence. Because so many satellites are now growing old, NASA is exploring methods for servicing and repairing those satellites while they are still in orbit, thus extending their lifetimes. One of the first devices constructed for this purpose is a humanoid robot nicknamed Dextre (Special Purpose Dexterous Manipulator; SPDM), built by the Canadian firm of MacDonald, Dettwiler and Associates. Dextre's first trial run on the ISS began on March 11, 2008, when it was launched aboard the space shuttle STS-123. Its first official assignment was the unpacking of two packages for use in the space station's Kounotori 2 module. Similar robots are expected to take over many of the most difficult and most dangerous space activities,

* **payload bay** the area in the shuttle or other spacecraft designed to carry cargo

* **low Earth orbit** an orbit between 300 and 800 kilometers (185 and 500 miles) above Earth's surface

such as repair projects that must be conducted outside of the space station, at some time in the future.

Until the space shuttle program ended in July 2011, most servicing and repair activities took place on this spacecraft. One of the workhorses onboard the shuttle was the robot arm called the Shuttle Remote Manipulator System (SRMS), or the Canadarm (Canadarm-1). The SRMS was capable of placing large items in or removing them from the shuttle's cargo bay. This 15-meter (50-foot) robot arm was used for a number of satellite repair and retrieval missions during the shuttle's first twenty years of operations. In 2001, an upgraded version of the device, Canadarm-2, was delivered to the ISS aboard STS-100. Canadarm-2 will be used for the same procedures on the space station as was its older cousin, Canadarm-1 for the shuttles.

During shuttle mission STS-41C (short for Space Transportation System 41C) in April 1984, the Long Duration Exposure Facility (LDEF) was deployed by the remote arm so that it could be left in orbit while experiments onboard were exposed to the emptiness of space. During that same mission, astronauts were able to retrieve the ailing Solar Maximum Mission (SMM, or SolarMax) satellite and repair it in the payload bay* of the shuttle. Later that year two communications satellites (the U.S. *Westar-6* and the Indonesian *Palapa-B2*) stranded in useless low Earth orbits* (LEOs) were successfully retrieved and brought back to Earth by the shuttle. Eventually, these two satellites, *Westar-6* and *Palapa-B2*, were successfully re-launched and deployed as the Chinese *AsiaSat* and Indonesian *Palapa-B2R*. In 1985, the Syncom IV-3 (or LEASAT-3) communications satellite, also stranded in a useless orbit, was retrieved, repaired, and again deployed by the STS-51D space shuttle crew in Earth orbit. It was initially deployed from the cargo bay but failed to properly sequence for its flight into its higher orbit

In the 1990s, there was a crucial repair mission carried out on the Hubble Space Telescope. Hubble had been deployed in 1990 with what was later discovered to be a flawed imaging system. In December 1993, astronauts carried out an emergency repair mission (Servicing Mission 1, SM-1) aboard the space shuttle *Endeavour* (STS-61) to correct Hubble's fault. During a series of space walks, the crew successfully repaired the imaging system, and the Hubble Space Telescope was able to continue its mission, making many outstanding astronomical discoveries over the next four years. The Hubble Space Telescope required four additional servicing and repair missions between 1997 and 2009. The second mission, Servicing Mission 2 (SM-2), was performed in February 1997 by the space shuttle *Discovery* crew. The spacewalking astronauts replaced several scientific instruments, including the Space Telescope Imaging Spectrograph and the Near Infrared Camera and Multi-Object Spectrometer, repaired the thermal insulation blanket protecting Hubble, and performed various other repairs on the telescope.

The Servicing Mission 3A (SM-3A) was conducted by the space shuttle *Discovery* crew in December 1999. The mission replaced Hubble's six

gyroscopes, along with its Fine Guidance Sensor, installed a computer and a Voltage Improvement Kit, and replaced the thermal insulation blankets around Hubble. The fourth mission, Servicing Mission 3B (SM-3B), was conducted by the *Columbia* crew (STS-109) in March 2002. The astronauts installed the Advanced Camera for Surveys, refilled the coolant within the Near Infrared Camera and Multi-Object Spectrometer, replaced Hubble's solar arrays, and made other repairs. The fifth and final servicing and repair mission for Hubble occurred in May 2009. The Servicing Mission 4

(SM-4) was performed by the STS-125 crew onboard the space shuttle *Atlantis*. After failing onboard Hubble just before the shuttle launch, the mission was delayed until a Science Instrument Command and Data Handling computer could be included on the mission as a replacement for the failed one. After launching, the astronauts also installed the new Wide Field Camera 3 and Cosmic Origins Spectrograph, along with repairing two instruments, the Advanced Camera for Surveys and the Space Telescope Imaging Spectrograph. The six nickel-hydrogen batteries used to provide electrical power on Hubble were also replaced. The Hubble Space Telescope is now expected to continue its important mission to explore the vastness of the universe, at least through 2014.

The space shuttle, however, was capable of achieving a maximum altitude of only 1,125 kilometers (700 miles) and was not designed to restart its main engines in order to attain escape velocity beyond LEO. Many satellites, such as communications satellites, are in geosynchronous orbit* 35,800 kilometers (22,300 miles) above Earth and require upper stages to boost them from LEO to their geosynchronous orbit and beyond to begin their operational missions. If a satellite failed to operate at this distance, it would be a total loss for its owners. In the early 1980s, business and government started to look at ways of solving this problem. One concept was a variant of a Martin Marietta Aerospace design of an Orbital Transfer Vehicle (OTV) for a 1980s U.S. space station concept. This vehicle would be able to take an astronaut to geosynchronous orbit to service satellites already in place. The OTV was renamed the Lunar Transfer Vehicle (LTV) in 1988 as lunar missions were preliminarily designed. However, the OTV/LTV concept was never developed for various reasons, including budget constraints. As newer, more powerful geosynchronous satellite systems are built, companies have designed satellites and their upper stages with preventive maintenance in mind.

Servicing and repair have also been an ongoing problem with the International Space Station. As with any large, complex device, parts break down or fail to function properly. One of the earliest examples of a potentially serious problem was the discovery of an air leak on the ISS in January 2004. At one point, almost five pounds of air was venting from the station each day. Astronauts Michael Foale and Aleksandr Kaleri were able to find and repair the leak, avoiding the possibility of having to shut the station down completely. Another potentially serious problem occurred in October 2007 during the deployment of two new solar arrays on the station. As the second solar array was being laid out, astronauts noticed first one tear, then a second tear, in its structure. They decided to try repairing those tears using the equipment that was available on hand (which did not include specialized equipment for such a job). Riding on the end of the station's Orbiter Boom Sensor System's arm astronaut Scott Parazynski was able to complete the necessary repairs. One of the most serious problems affecting the ISS was a computer failure on June 14, 2007. The problem resulted in a loss of power for its thrusters and failure of its oxygen generation, carbon dioxide scrubber, and other

* geosynchronous orbit the altitude above the surface of the Earth (22,300 miles up) where an orbiting object will be traveling around the Earth at the same speed that the Earth spins. This results in the orbiting object remaining over the same region of the Earth at all times, although it will progress north and south along a straight line (as seen from earth) over the course of a day, unless its orbit is directly over the equator, which results in a geostationary orbit

environmental control systems. A fix was found for the problem relatively quickly, and within two days, all systems were again operating normally.

 See also **Accessing Space (Volume 1) • Materials Testing in Space (Volume 2) • Robotics Technology (Volume 2) • Satellite Industry (Volume 1) • Satellites, Types of (Volume 1) • Search and Rescue (Volume 1) • Space Shuttle (Volume 3)**

Resources

Books and Articles

Gorn, Michael H. *Superstructures in Space: From Satellites to Space Stations.* New York: Merrell, 2008.

Johnson, Rebecca L. *Satellites.* Minneapolis, MN: Lerner Publications, 2006.

Lee, Wayne. *To Rise from Earth: An Easy-to-Understand Guide to Spaceflight.* New York: Facts on File, 2000.

Websites

The Hubble Space Telescope. Goddard Space Flight Center, National Aeronautics and Space Administration. <http://hubble.gsfc.nasa.gov/> (accessed October 25, 2011).

The International Space Station Solar Alpha Rotary Joint Anomaly Investigation. <http://ntrs.nasa.gov/archive/nasa/casi.ntrs.nasa.gov/20100003841_2010003501.pdf> (accessed October 25, 2011).

Maintaining and Upgrading Hubble. National Aeronautics and Space Administration. <http://hubble.nasa.gov/missions/intro.php> (accessed October 25, 2011).

Orbital Transfer Vehicle. Goddard Space Flight Center, National Aeronautics and Space Administration. <http://www.astronautix.com/craft/otv.htm> (accessed October 25, 2011).

Remote Manipulator System. Johnson Space Center, National Aeronautics and Space Administration. <http://prime.jsc.nasa.gov/ROV/rms.html> (accessed October 25, 2011).

Small Satellite Technology

The first satellites were very small. The Soviet Union's *Sputnik 1,* which opened the Space Age in 1957, weighed only 185 pounds on Earth (84 kilograms). The American response, *Explorer 1,* weighed 31 pounds (14 kilograms). These early small satellites proved that it was possible to put

Low launching costs and rapid development makes small satellites suitable for new applications that are not possible with larger spacecraft. This small 590-pound U.S./Argentina satellite, called *SAC-A* (Scientific Applications Satellite-S), was ejected from the space shuttle *Endeavour* (STS-88) in December 1998, and carried five technology experiments. *NASA.*

* radiation belts two wide bands of charged particles trapped in a planet's magnetic field

equipment in orbit and use it, providing the first opportunities for scientific observation outside Earth's atmosphere. *Explorer 1* provided the data that led to the identification of the Van Allen radiation belts* that surround Earth. Bell Labs' *Telstar 1,* about the size of a car tire, provided the first transatlantic television link, and *Pioneer 10,* weighing about 595 pounds (of 270 kilograms), was launched in 1972, and is believed to be the first satellite to leave the solar system. The last contact with *Pioneer 10* occurred on January 23, 2003. It has now traveled so far into deep space and its batteries are so weak (or dead) that no further contact with the satellite is likely to occur.

Almost all satellites are powered by sunlight, and small satellites, which intercept less of this resource, are limited in power as much as they are in size, mass, and budget. However, the modern revolution in digital electronics and portable computing technologies has enabled engineers to build satellites weighing just a few pounds that have capabilities rivaling those of older, larger satellites. Because launch costs have remained virtually constant since the beginning of the space age, small satellites and their lower costs are receiving renewed attention.

What Is a Small Satellite?

Small satellites (minisatellites, or minisats) are defined as those weighing less than 2,200 pounds (1,000 kilograms). Those below 220 pounds (1,000 kilograms) are referred to as microsatellites, those with a weight of less than 22 pounds (ten kilograms) are known as nanosatellites, and those under two pounds (one kilogram) are called picosatellites. However, the major difference between small and large satellites is not their weight but the way they are built. Small satellites are built by small, highly interactive teams that work with the satellite from conception through launch and operation. Large satellites are built in larger, more formally structured organizations. Small satellite teams typically have fewer than twenty members, whereas large satellites may be built by organizations with tens of thousands of people.

The small satellite team has the advantage of speed and efficiency, the ability to evaluate the implications of each design decision for the entire satellite, and the insight into all aspects of the satellite's design and application. The combination of low cost, rapid development, and low launch costs makes small satellites suitable for new applications that are not possible with larger spacecraft.

Students and hobbyists gain hands-on experience in space by building, launching, and operating satellites. Student-built satellites have hosted advanced communications experiments, astronomical and Earth-observing instruments, and video cameras that can be used to look at themselves and the satellites launched with them. Most amateur satellite activity focuses on building novel voice and digital communications links.

A very promising new application of small satellites is the inspection of larger satellites. A low-cost nanosatellite can observe the target spacecraft as it separates from its launch vehicle, deploys solar panels, and begins operations. Any problem during the initialization of operations, or later in the spacecraft's life, can be diagnosed by the escorting nanosatellite, which would have visible and infrared* cameras as well as radio-based diagnostics.

Developments in Small Satellite Technology

Small satellite manufacturers have experienced some notable successes in recent years. In 2011, for example, Surrey Satellite Technology produced and launched two satellites for the Nigerian government, NigeriaSat-2 and NigeraSat-X, to be used primarily for observation of that nation's physical features. The goal of the Nigerian mapping project is to enhance the nation's food supply by providing month-by-month reports on crop patterns, to provide general geographic and topographic information, to advance disaster management practices, and to advance the nation's space program. NigeriaSat-2 weighs about 600 pounds (270 kilograms), while the smaller NigeriaSat-X weighs about 220 pounds (100 kilograms). NigeriaSat-2 is expected to have an orbital lifetime of about 7 1/2 years, about 50 percent longer than its older cousin, NigeriaSat-1. Another example of small satellites launched in 2011 is RASAT (Radar And Sonar Alignment Target), designed and built by the Turkish Space Technologies Research Institute. The 200-pound (93-kilogram) satellite was the first to be built in Turkey. It will be used for four major purposes: mapping, disaster monitoring, environmental observations, and urban and regional planning. The satellite is expected to remain in orbit for about three years.

The Future of Small Satellites

Because small satellites require only a corner of a laboratory or manufacturing facility, plus some basic equipment for their construction, there are hundreds of small satellite developers around the world. By contrast, developers of large satellites include a few major corporations and government laboratories in the largest and wealthiest countries. The proliferation of developers and users of small spacecraft has unleashed the same creative forces that propelled the personal computer to its dominant position in the computer market.

Among the leading commercial developers of small satellites worldwide are AeroAstro (headquartered in Ashburn, Virginia, USA), CGS S.p.A. Compagnia Generale per lo Spazio (Milan, Italy), Lockheed Martin (USA), OHB System (Bremen, Germany), Quinetiq (Farnborough, UK), Space Systems Finland, and Surrey Satellite Technology (Guildford, UK). The development of small satellites has also become increasingly popular among universities, which can afford the low financial cost and small personnel investment for the production of such devices, especially nanosatellites and picosatellites that weigh only a few pounds.

The leading institution in this trend has been California Polytechnic State University (Cal Poly) at San Luis Obispo. Beginning in about 2004, researchers at Cal Poly designed a small satellite called the CubeSat that has become the standard of its type among academic institutions around the world. CubeSat is 1 meter (3.3 feet) in length on a side, with a mass of no more than about 1.3 kilograms (2.9 pounds). It contains technology for a wide variety of applications, including remote sensing, biological experiments, weather monitoring, and a variety of space research projects. As of late 2011, more than 60 universities worldwide were part of the CubeSat Project, coordinated by Cal Poly and the Space Systems Development Lab at Stanford University. The project is designed to help individual university groups to develop a CubeSat satellite, obtain necessary licenses and permits, make launch arrangements, transport the satellite to the launch site, and obtain real-time data on the launch itself.

In 2011, NASA also announced a program by which university groups could apply to the agency for inclusion in future NASA satellite launches for their own CubeSats. The selection of successful applicants in the NASA program was expected in early 2012.

The future of small satellites, and in large part the future of space exploration and application, will rely on the creativity of this diverse population of developers.

 See also **Communications Satellite Industry (Volume 1)** • **Satellite Industry (Volume 1)** • **Satellites, Future Designs (Volume 4)** • **Satellites, Types of (Volume 1)**

Resources

Books and Articles

Helvajian, Henry, and Siegfried W. Janson, eds. *Small Satellites: Past, Present, and Future..* El Segundo, CA: Aerospace Press 2009.

Sandau, Rainer, Hans-Peter Roser, and Arnoldo Valenzuela, eds. *Small Satellite Missions for Earth Observation: New Developments and Trends.* Berlin: Springer, 2010.

Thakker, Purvesh, and Wayne A. Shiroma, eds. *Emergence of Pico- and Nanosatellites for Atmospheric Research and Technology Testing.* Reston, VA: American Institute of Aeronautics and Astronautics, 2010.

Websites

CubeSat in the News. <http://www.cubesat.org/> (accessed October 15, 2011).

NASA's CubeSat Launch Initiative. <http://www.nasa.gov/directorates/heo/home/CubeSats_initiative.html> (accessed October 15, 2011).

Small Satellites Homepage. <http://centaur.sstl.co.uk/SSHP/> (accessed October 15, 2011).

Space Technology 5 (ST5). New Millennium Program. March 2006, <http://nmp.jpl.nasa.gov/st5/> (accessed August 5, 2009).

SSETI Express. European Space Agency. <http://www.esa.int/SPECIALS/sseti_express/index.html> (accessed August 31, 2009).

Tiny Satellites for Big Science. <http://www.astrobio.net/exclusive/3552/tiny-satellites-for-big-science> (accessed October 15, 2011).

Space Access	*See Accessing Space (Volume 1)*

Spaceports

Spaceports are facilities used to launch, and in some cases land, spacecraft (those vehicles that travel in space). Spaceports are similar to airports and seaports but have some unique features and requirements. They have to be able to support the testing, assembly, launching, and landing of large, powerful rockets and the satellites or other cargoes that they carry. There are only a handful of spaceports around the world, although, in the early part of the 2010s, more are being built as the demand for launches grow and the types of launch vehicles evolve.

Spaceport Components

The most familiar element of a spaceport is the launch pad. Originally developed on a patch of ground where rockets were hastily set up and launched, launch pads have evolved considerably as rockets became larger and more complex. Most launch pads have a tower, known as a gantry, which stands next to the rocket. Through the gantry, technicians have access to various levels of the rocket so they can check and repair systems, add propellant, and in the case of piloted rockets, provide a way for crews to get in and out.

Below the pad itself are pathways called flame trenches, which allow the hot exhaust from the rocket to move away from the pad at the time of the launch, so that it does not damage portions of the pad or the rocket itself. Some launch pads, such as the ones used in the past by the U.S. space shuttles, have water towers nearby that spray water onto the pad at launch. The water is designed to suppress the noise and vibration of the launch, which otherwise could reflect off the pad and damage the shuttle. As of the retirement of the U.S. space shuttle fleet in August 31, 2011, the launch pads used for shuttle flights are being converted for other uses.

The launch pad itself, though, is only a small part of a spaceport. Other facilities at spaceports include hangars on which sections of rockets are assembled before moving them to the launch pad. The Vehicle Assembly Building (VAB) at the Kennedy Space Center, built for the Apollo* program and used between 1981 and 2011 for the space shuttles, is one of the largest buildings in the world when measured by volume. Launch

* **Apollo** American program to land men on the Moon; *Apollo 11, Apollo 12, Apollo 14, Apollo 15, Apollo 16,* and *Apollo 17* delivered twelve men to the lunar surface (two per mission) between 1969 and 1972 and returned them safely back to Earth

Space shuttle *Columbia* (STS-1) is illuminated by Launch Pad A, Complex 39, at one of the best-known spaceports in the world, NASA's Kennedy Space Center, near Cape Canaveral, Florida. *NASA.* ▼

* **radar** a technique for detecting distant objects by emitting a pulse of electromagnetic radiation, usually microwaves or radio waves, and then recording echoes of the pulse off the distant objects

vehicles and their payloads are transported vertically from the assembly building to the launch site using large, slow-moving flatbed transporters. In Russia, launch vehicles are carried out to the pad horizontally on conventional rail lines. In some cases, rockets are assembled, stage by stage, at the launch site itself.

Spaceports also operate control centers where the progress of a countdown and launch is monitored. Nearby are radar* tracking systems that track both the rocket in flight as well as any planes or boats that may venture too close to the launch site. Spaceports usually notify pilots and ship captains of the regions of the ocean that will be off-limits during a launch because rocket stages or debris could fall there.

Spaceports of the World

United States One of the best-known spaceports in the world, the National Aeronautics and Space Administration's (NASA) John F. Kennedy Space Center, is located on Merritt Island, near Cape Canaveral, Florida. There are actually two separate spaceports there: NASA's Kennedy Space Center (KSC) and the U.S. Air Force's Cape Canaveral Air Force Station (CCAFS). NASA used KSC exclusively for launches of the space shuttle, from two launch complexes: 39A and 39B. These complexes, now that the space shuttles have been retired, are being configured for other uses in the future. The same pads were used to launch the Saturn 5 rockets for the Apollo Moon missions. CCAFS is home to a number of launch facilities for unmanned military and commercial rockets, including the Atlas and Delta. Other more recent rockets to be sent into space from these pads include those from private space transport company Space Exploration Technologies Corporation (SpaceX). One pad in the future will be exclusively used by the Spaceport Florida Authority, which is an aerospace economic development agency for the state of Florida.

There are several other spaceports in the United States. Vandenberg Air Force Base in southern California is used for launches of several types of unmanned rockets, including Delta boosters, for NASA, the U.S. military, and private companies. Vandenberg is used for launches into polar orbit because the only clear path for launches is south, over the Pacific Ocean.

The Kodiak Launch Complex, located on Kodiak Island in Alaska, was used for the first time for an unmanned orbital launch in 2001. It was built by the Alaska Aerospace Development Corporation at Narrow Cape on Kodiak Island, about 400 kilometers (240 miles) south of Anchorage. The Kodiak Launch Complex contains all-weather processing adaptable to all current small launch vehicles, and is one of the few commercial launch ranges in the United States not co-located with a federal facility.

Wallops Island Flight Facility (WFF), on Wallops Island, Virginia, has a launching pad for an expendable rocket, as well as runways for aircraft carrying the Pegasus small-winged rocket. It is located off the Eastern Shore of Virginia. Established in 1945 under NASA's predecessor, the

National Advisory Committee for Aeronautics, Wallops is one of the oldest launch sites in the world and supports scientific research and orbital and sub-orbital payloads for NASA. The Wallops Flight Facility focuses on providing fast, low cost, and highly flexible support for aerospace technology and science projects. The Virginia Commercial Space Flight Authority built the Mid-Atlantic Regional Spaceport (MARS) on the southern tip of Wallops Island in 1998. The first launch from this commercial site was made in 2006. Since June 30, 2011, four out of five launches have been successfully performed from this new spaceport.

The Omelek Island spaceport is located on Omelek island, which is part of the Kwajalein Atoll and part of the Republic of the Marshall Islands. The U.S. military leases the island as its Ronald Reagan Ballistic Missile Defense Test Site. On September 28, 2008, American space transport company Space Exploration Technologies Corporation (SpaceX) launched its first successful rocket, the Falcon 1, from Omelek Island. This fourth test of the rocket is considered the first successful launch of a privately-funded liquid-propelled orbital launch vehicle.

Europe   Outside of the United States there are several very active spaceports. Europe established a spaceport called the Guiana Space Center, near Kourou, French Guiana, on the northeast coast of South America in the late 1960s, for European launches. The European Space Agency and the commercial firm Arianespace use it for launches of the Ariane 5 boosters. The spaceport was deliberately built close to the equator at 5.3° north latitude to reduce the energy required for orbital plane-change maneuvers for missions to geostationary orbit.

Under the name ELA-1 (in French: l'Ensemble de Lancement Ariane 1; in English: Ariane Launch Area 1), the complex was used in the past to launch Ariane 1, 2, and 3 rockets. Currently, changed to the name ELV-1 (in French: l'Ensemble de Lancement Vega 1; in English: Vega Launch Area 1), the site is being rebuilt to support launches of the expendable Vega (European Advanced Generation Carrier Rocket). The Italian Space Agency and the European Space Agency are developing the Vega rocket. Completion of the launch complex is expected sometime in late 2011 or the following year. The ELA-2 Launch Complex was used for Ariane 4 rockets launched between 1988 and 2003. However, the Ariane 4 was retired in 2003 after the Ariane 5 was activated. As of 2008, the launch pad has been deactivated, though still usable. The ELA-3 Launch Complex was built specifically to serve the Ariane 5 heavy-lift vehicle, which began launches in 1996. It is designed to handle a launch rate of up to ten Ariane 5 missions per year.

The ELS (in French: l'Ensemble de Lancement Soyouz; in English: Soyuz Launch Area 1) Launch Complex is currently being built for the launching of Russian Soyuz-2 rockets. Build north of the Ariane 4 launch site, the first scheduled launch is October 20, 2011; however, many delays have occurred in the completion of the complex.

Russia Russia's primary launch site is called the Baikonur Cosmodrome, in Kazakhstan, formerly part of the Soviet Union. In reality, the Baikonur launch site is located more than 320 kilometers (200 miles) away from the town of that same name. Instead, the Baikonur Cosmodrome is situated north of the village of Tyuratam on the Syr Darya River (45.9° north latitude and 63.3° east longitude). The Baikonur name is a relic of Cold War deception. Despite the potential confusion, the Baikonur Cosmodrome is the site where *Sputnik 1,* Earth's first artificial satellite, was launched. A number of Russian rockets, including manned Soyuz missions, also use Baikonur. Unlike other spaceports, Baikonur is located in the middle of a continent, far from the ocean; spent rocket stages are dropped on desolate regions of Kazakhstan and Siberia rather than in the ocean.

Russia also operates spaceports in Plesetsk, in northern Russia, and Svobodny, which it uses for some unmanned flights and military missions. Russia launches all its human space missions as well as all geostationary, lunar, and planetary missions from the Baikonur Cosmodrome. Today, it is the only Russian site capable of launching the Proton launch vehicle and it is also used for manned and unmanned International Space Station missions. The Plesetsk Cosmodrome, Russia's northernmost launch complex, is used to launch satellites into high inclination, polar, and highly elliptical orbits. Another one of Russia's spaceports is the Kapustin Yar Cosmodrome in Astrakhan Oblast, between Volgograd and Astrakhan.

Beginning in 2006, the Yasny Cosmodrome became operational in Russia, located in the Orenburg Oblast, and 6 kilometers (3.7 miles) northwest of the village of Dombarovsky. The first three spacecraft launched from there used the Russian Dnepr-1 rocket. Their launch dates, names, and owners are: July 12, 2006, the *Genesis I* spacecraft by U.S.-based Bigelow Aerospace; on June 28, 2007, the *Genesis II* by Bigelow Aerospace; and on October 1, 2008, the THEOS satellite by Thailand-based Geo-Informatics and Space Technology Development Agency (GISTDA). The Russians will also begin launching some of its Soyuz rockets from the Guiana Space Center's ELS (l'Ensemble de Lancement Soyouz) launch site, sometime in the 2011-2012 timeframe. In addition, the Vostochny Cosmodrome in Siberia began construction in June 1, 2011, with an operational status tentatively scheduled for 2018. It is located at 51 degrees north latitude in the Amur Oblast.

Japan The Japanese use the Tanegashima Space Center as its largest space port. It contains two orbital launch complexes: Osaki and Yoshinobu. Located on an island 115 kilometers (71 miles) south of Kyushu, this 8.6-square-kilometer (3.3-square-mile) complex plays a central role in pre-launch countdown and post-launch tracking operations. On-site facilities include the Osaki Range, tracking and communication stations, several radar stations, and optical observation facilities. There are also related developmental facilities for firing of liquid-and solid-fuel rocket engines.

China   The Chinese have several spaceports—Jiuquan, Taiyuan, and Xichang—but the Xichang Satellite Launch Center, which is located within a military installation, supports all geostationary missions from its location in southern China. The Chinese are also building a fourth space port, its southernmost one, called the Wenchang Satellite Launch Center, on the northeastern coast of Hainan Island. A completion of the facility, with possibly three operational launch pads, is expected in 2014 or 2015; however, some launches have already taken place at the location.

Others   Other countries with national spaceports include:
- India: Satish Dhawan Space Centre
- Canada: John H. Chapman Space Centre
- South Korea: Naro Space Center
- Israel: Palmachim Air Base
- Brazil: Alcântara Launch Center

The Future

Like the rockets that use them, spaceports are evolving. As reusable launch vehicles, which launch and return, become more common, spaceports will have to support pre-launch preparations and post-landing operations. The Kennedy Space Center can handle both, as shown with thirty years of launching space shuttles into low-Earth orbit and processing them back to launch-ready status. Spaceports will also have to develop facilities to maintain these vehicles and prepare them for their next flights.

Currently, even the busiest spaceports, such as the Guiana Space Center and the Kennedy Space Center, handle only a couple dozen launches a year, which is near the maximum supportable with current technology. Past studies by NASA found that new technologies and an improved infrastructure would be needed to support higher flight rates. A greater demand for spaceflight should likely lead to the upgrading of existing spaceports and the creation of new spaceports. In addition, the development of single-stage reusable launch vehicles—which travel from the ground to space without dropping any stages along the way—would make it possible for spaceports to be located in many areas, not just near oceans. In the United States alone over a dozen states, including inland states such as Idaho and Oklahoma, have expressed an interest in developing spaceports for future reusable launch vehicles. The creation of these new spaceports could be a major step toward making space travel as routine as air travel. Some smaller spaceports that have been developed recently, or are still in the developmental stage, include the Mojave Air and Space Port, Spaceport America, and Spaceport Sweden.

Mojave Air and Space Port   One of the up-and-coming spaceports that has already seen manned missions to space is the Mojave Air and Space Port in Mojave, California, United States. It is unique in that the U.S. Federal Aviation Administration (FAA) gave the facility the first spaceport license (on June 17, 2004) for horizontal launches of reusable spacecraft. The Mojave Space Port was the site of the first flight of *SpaceShipOne,* which is a reusable spacecraft built by Scaled Composites. The spaceport has also been the scene of the first launch of a rocket-powered aircraft, the XCOR EZ-Rocket, which departed on December 3, 2005, with Dick Rutan at the wheel for its destination of California City, California. Twelve days later, the spaceport had its first arrival of a rocket-powered aircraft, when the XCOR EZ-Rocket arrived back from California City with Rick Searfoss as its pilot. On December 21, 2008, the first test flight of Scaled Composites carrier aircraft *WhiteKnightTwo* took off from Mojave. On May 26, 2010, the first ever successful flight of a vertical take-off, vertical landing (VTVL) vehicle was performed at Mojave by aerospace startup company Masten Space Systems, which is based in Mojave, California. The company is developing a line of VTVL spacecraft for unmanned suborbital research flights and, eventually, for unmanned orbital flights.

Spaceport America Still under development, Spaceport America (formerly the Southwest Regional Spaceport) is located near the White Sands Missile Range, north of Las Cruces, New Mexico and east of the city of Truth or Consequences, New Mexico. The operator of the facility, the New Mexico Spaceport Authority, received its FAA launch license (for both vertical and horizontal launches) in December 2008. Although construction (which began in April 2006) is not yet complete, several unmanned suborbital rocket launches have occurred on the site, including the vertical launch of the SpaceLoft XL rocket of Colorado-based UP Aerospace. The spaceport is expected to be fully operational by the end of 2011. The first commercial suborbital flight of Richard Branson's Virgin Galactic, with its spaceship *SpaceShipTwo* and its carrier aircraft *WhiteKnightTwo,* is expected to be launched from Spaceport America in the first quarter of 2013. Virgin Galactic will use Spaceport America's 3,000-meter (10,000-foot) runway to fly 2.5-hour commercial suborbital flights for paying customers to an altitude of 110 kilometers (70 miles), which is above the recognized boundary for outer space.

Spaceport Sweden Spaceport Sweden, which was founded in January 2007, is another spaceport that is under development, as of 2011. Being built by the Swedish Space Corporation, along with other Swedish organizations, the spaceport, located near the city of Kiruna, Sweden, is expected to be the European launch site for Virgin Galactic's suborbital flights into space starting sometime no earlier than 2012. The spaceport includes the Arena Arctica hangar facility, which is 5,000 square meters (53,820 square feet) in area, and a 2,500 kilometer (91,550-foot) runway.

▶ *See also* **Launch Services (Volume 1) • Launch Sites (Volume 3) • Launch Vehicles, Expendable (Volume 1) • Launch Vehicles, Reusable (Volume 1) • Reusable Launch Vehicles (Volume 4) • Rocket Engines (Volume 1) • Vehicle Assembly Building (Volume 3)**

Resources

Books and Articles

America's Spaceport: John F. Kennedy Space Center. Washington, D.C.: National Aeronautics and Space Administration, 2004.

Benson, Charles D., and William B. Flaherty. *Gateway to the Moon: Building the Kennedy Space Center Launch Complex.* Gainesville: University of Florida Press, 2001.

Lipartito, Kenneth, and Orville R. Butler. *A History of the Kennedy Space Center.* Gainesville: University Press of Florida, 2007.

West-Reynolds, David. *Kennedy Space Center: Gateway to Space.* Richmond Hills, Ontario, Canada: Firefly, 2006.

Websites

Alaska Aerospace Development Corporation. <http://www.akaerospace.com/> (accessed September 22, 2011).

Arianespace. <http://www.arianespace.com/index/index.asp> (accessed September 23, 2011).

Brazilian Space Agency. <http://www.aeb.gov.br/> (accessed September 7, 2011).

Canadian Space Agency. <http://www.asc-csa.gc.ca/eng/default.asp> (accessed September 7, 2011).

China National Space Administration. <http://www.cnsa.gov.cn/n615709/cindex.html> (accessed September 7, 2011).

European Space Agency. <http://www.esa.int/esaCP/index.html> (accessed September 7, 2011).

India Space Research Organisation. <http://www.isro.org/> (accessed September 7, 2011).

Israeli Space Agency. <http://www.most.gov.il/English/Units/Israel+Space+Agency/> (accessed September 7, 2011).

Italian Space Agency. <http://www.asi.it/en> (accessed September 7, 2011).

Japan Aerospace Exploration Agency. <http://www.jaxa.jp/index_e.htmln> (accessed September 7, 2011).

Korea Aerospace Research Institute. <http://www.kari.re.kr/english/> (accessed September 7, 2011).

National Aeronautics and Space Administration. <http://www.nasa.gov/> (accessed September 7, 2011).

* **ballistic** the path of an object in unpowered flight; the path of a spacecraft after the engines have shut down

Phase one of world's first commercial spaceport is now 90% completed — in time for first flights in 2013. MailOnline.co.uk. <http://www.daily-mail.co.uk/sciencetech/article-2034239/Phase-worlds-commercial-spaceport-90-cent-completed--time-flights-2013.html> (accessed September 22, 2011).

Russian Federal Space Agency. <http://www.roscosmos.ru/main.php?lang=en> (accessed September 7, 2011).

Spaceport Sweden. <http://www.spaceportsweden.com/> (accessed September 22, 2011).

United Kingdom Space Agency. <http://www.bis.gov.uk/ukspaceagency> (accessed September 14, 2011).

Virgin Galactic. <http://www.virgingalactic.com/> (accessed September 22, 2011).

Wallops Flight Facility. National Aeronautics and Space Administration. <http://www.nasa.gov/centers/wallops/home/index.html> (accessed September 22, 2011).

Spinoffs   *See Made with Space Technology (Volume 1)*

Sputnik

Sputnik (which means "traveling companion" or "satellite" in Russian) is the name given to a series of scientific research satellites launched by the Soviet Union during the period from 1957 to 1961. The satellites ranged in size and capability from the 83.6-kilogram (an equivalent weight on Earth of 184.3 pounds) *Sputnik 1,* which served only as a limited radio transmitter, to *Sputnik 10,* which had a mass of 4,695 kilograms (equivalent weight on Earth of 10,350 pounds). Together the Sputnik flights ushered in the space age and began the exploration of space by orbital satellites and humans. *Sputnik 1,* which launched on October 4, 1957, is the most famous in the series.

In August 1957, the Soviet Union conducted a successful test flight of a stage-and-a-half liquid-fueled intercontinental ballistic* missile (ICBM) called the R-7. Shortly thereafter Soviet scientists were quoted in the news media inside the Soviet Union saying that they were planning for the launch of an Earth satellite using a newly developed missile. Western observers scoffed at the accounts. In the late summer of 1957, Soviet scientists told a planning session of the International Geophysical Year celebrations that a scientific satellite was going to be placed into orbit, and they released to the press the radio frequency that the satellite would use

* **ionosphere** a charged particle region of several layers in the upper atmosphere created by radiation interacting with upper atmospheric gases

to transmit signals. Again, the statements were widely dismissed inside the United States as Soviet propaganda.

Late in the evening in the United States (Eastern Standard Time) on Friday, October 4, 1957, Radio Moscow announced that a small satellite designated *Sputnik 1* had been launched and had successfully achieved orbital flight around Earth. The U.S. Defense Department confirmed the fact shortly after the reports reached the West.

Sputnik 1 was the first artificial satellite to reach orbit. Launched from a secret rocket base in the Ural Mountains in Soviet central Asia, it had a mass of 83.6 kilograms (an equivalent weight on Earth of 184.3 pounds), was 0.58 meter (1.9 feet) wide, and carried four whip-style radio antennas that measured 1.5 to 2.9 meters (4.9 to 9.5 feet) in length. Aboard the tiny satellite were instruments capable of measuring the thickness and temperature of the high upper atmosphere and the composition of the ionosphere*, and the satellite was also capable of transmitting radio signals. The Soviet news agency *TASS* (in English: Telegraph Agency of the Soviet Union) released the final radio frequency of *Sputnik 1* and the timetables of its broadcasts, which were widely disseminated by news media worldwide. *Sputnik 1* transmitted for twenty-one days after reaching orbit and remained in orbit for ninety-six days. It burned up in the atmosphere on its 1,400th orbit of Earth.

Sputnik 2 was launched into orbit a month later on November 3, 1957. It was a much larger satellite, with a mass of 508 kilograms (an equivalent weight on Earth of 1,120 pounds), and contained the first living creature to be orbited, a dog named Laika. The dog, its capsule, and the upper part of the rocket that launched it remained attached in space for 103 days

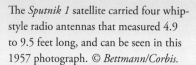

The *Sputnik 1* satellite carried four whip-style radio antennas that measured 4.9 to 9.5 feet long, and can be seen in this 1957 photograph. © *Bettmann/Corbis.*

before burning up after making 2,370 orbits. However, there was only enough oxygen, food, and water to keep Laika alive for a week. There were no provisions to either save the dog or return its capsule to Earth.

Sputnik 3 through *Sputnik 10* were research craft aimed at obtaining design data for the construction of a human-carrying spacecraft. *Sputnik 3* was launched on May 15, 1958, *Sputnik 4* on May 15, 1960, and *Sputnik 5* on August 19, 1960. *Sputnik 5* carried onboard two dogs (Belka and Strelka), along with a bunch of mice and rats, and a number of plants. All de-orbited and returned safely to Earth. Then, *Sputnik 6* was launched on December 1, 1960, *Sputnik 7* on February 4, 1961, *Sputnik 8* on February 12, 1961, *Sputnik 9* on March 9, 1961, and *Sputnik 10* on March 25, 1961.

Sputnik 10 was a full test version of the Vostok human-carrying space capsule, which carried the first human into space two weeks later on April 12, 1961. *Sputniks 5, 6, 9,* and *10* carried dogs. *Sputnik 10*'s canine passenger, Zvezdochka, was successfully recovered.

Sputnik designations, beginning with *Sputnik 11,* were unofficially given by the U.S. news media to a series of Soviet spacecraft but these craft were actually parts of later Soviet space programs such as Cosmos, Vostok, and Luna.

See also **Animals (Volume 3) • International Space Station (Volume 1 and volume 3) • Satellites, Types of (Volume 1) • Space Shuttle (Volume 3)**

Resources

Books and Articles

Brzezinski, Matthew. *Red Moon Rising: Sputnik and the Hidden Rivalries that Ignited the Space Age.* New York: Times Books, 2007.

Furniss, Tim. *A History of Space Exploration and Its Future.* London: Mercury, 2005.

Websites

Cowen, Ron. *Sputnik + 50: Remembering the Dawn of the Space Age.* CBS Interactive Business Network Resource Library. <http://findarticles.com/p/articles/mi_m1200/is_14_172/ai_n27405869/> (accessed October 13, 2011).

Sputnik. Russian Space Web. <http://www.russianspaceweb.com/sputnik.html> (accessed October 13, 2011).

Sputnik and the Dawn of the Space Age. National Aeronautics and Space Administration, NASA History Program. <http://history.nasa.gov/sputnik/> (accessed October 13, 2011).

T

Thompson, David
American Aeronautics Company Executive
1954–

David W. Thompson is the chairman of the Board of Directors and chief executive officer (CEO) of Orbital Sciences Corporation (OSC), a space technology and satellite services company he cofounded in 1982. These two positions he has held since 1982. From 1982 to 1999, Thompson was also the company's president. Before starting OSC, Thompson was a project manager and engineer who worked on advanced rocket engines at the National Aeronautic and Space Administration (NASA) facility the Marshall Space Flight Center, in Huntsville, Alabama. He also previously worked for Hughes Electronics Corporation and Charles Stark Draper Laboratory.

Thompson has a bachelor's degree in aeronautics and astronautics from Massachusetts Institute of Technology, a master's degree in aeronautics from California Institute of Technology, and a master's of business administration (MBA) degree from the Harvard Business School. As a graduate student, Thompson worked on the first Mars landing missions at NASA's Jet Propulsion Laboratory in Pasadena, California.

Thompson and his cofounders of OSC (Bruce Ferguson and Scott Webster) met at Harvard Business School, where they shared an interest in the commercial uses of space. Consequently, OSC was founded on the concept of commercial companies, not government agencies, being the driving force in the space industry. Whereas most established space companies' commercial businesses have evolved from government- or military-funded programs, OSC is devoted exclusively to the commercial aspects of the space industry.

OSC, commonly called Orbital, is one of the world's largest space-related companies, with about 3,700 employees (including over 1,800 scientists and engineers) worldwide. It is among the world's leading developer and manufacturer of small, affordable space and rocket systems. The company has its headquarters in Dulles, Virginia, and maintains major facilities in the United States, Canada, and several locations overseas. Thompson oversees the company's four major divisions: Launch Systems Group, Space Systems Group, Advanced Programs Group, and Technical Services Division.

The company's business activities involve such products as geosynchronous Earth orbit (GEO) and low Earth orbit (LEO) satellites, national security based spacecraft, launch vehicles, space robotics and software,

David Thompson, CEO of Orbital Sciences Corporation, cofounded his company with the vision of commercial companies forming the driving force in the space industry, instead of government agencies. *NASA.*

* **low Earth orbit** an orbit between 300 and 800 kilometers (185 and 500 miles) above Earth's surface

and deep-space planetary probes. For instance, under Thompson's direction, the Pegasus launch vehicle program saw its 500th mission in 2006. At the end of 2008, over 110 spacecraft had been delivered by OSC to customers around the world. One of these includes the Galaxy Evolution Explorer (GALEX), which was launched in 2003. Thompson is also directing OSC with its participation in NASA's Commercial Orbital Transportation Services (COTS) program, which will provide commercial deliveries to the International Space Station.

Thompson is a Fellow of the American Institute of Aeronautics and Astronautics, the American Astronautical Society, and the Royal Aeronautical Society. He is a member of the U.S. National Academy of Engineering and the International Academy of Astronautics. U.S. president George H.W. Bush awarded Thompson the National Medal of Technology, and the Smithsonian Institution awarded him the National Air and Space Museum Trophy.

▶ *See also* **Launch Vehicles, Expendable (Volume 1)** • **Launch Vehicles, Reusable (Volume 1)** • **Remote Sensing Systems (Volume 1)** • **Reusable Launch Vehicles (Volume 4)** • **Satellites, Types of (Volume 1)**

Resources

Books and Articles

Dorsey, Gary. *Silicon Sky: How One Small Start-Up Went over the Top to Beat the Big Boys into Satellite Heaven.* Reading, MA: Perseus Books, 1999.

Websites

Orbital Sciences Corporation. <http://www.orbital.com/> (accessed May 14, 2009).

Tourism

It is highly likely that the public will be traveling, touring, and living in space at some time in the twenty-first century. If history is repeated, the human expansion into space, on a large scale, is a foreseeable prospect for humankind. One possible scenario begins with 30-to-45-minute suborbital flights in the middle part of the 2010s, followed by orbital flights of one to three revolutions (ninety minutes to four-and-one-half hours) later that decade or the next. Surveys have shown that people would like to have a specific destination in space. That desire suggests a destination such as a resort hotel that can provide several days of accommodation in low Earth orbit*, and a hotel like this may possibly be available as early as in the 2020s. Beyond that, space hotels could be followed by orbiting lunar cruises with excursions to the Moon's surface in the 2040s or beyond. After suborbital rides become commonplace, a new aeroballistic* cargo and human transportation system could begin operation, leaving the transit time between major transportation hubs on Earth at an hour or less away.

Just exactly how and when these new modes of transportation and resorts will materialize is difficult to predict. However, there is an organized effort underway between the private and public sectors to assure that the correct ingredients and the proper catalysts* are brought together. This effort is multifaceted and includes the government, business, and the general public.

Introducing Space Tourism

Human space activity in the twentieth century was the exclusive domain of the Russian (formerly the Soviet Union) and U.S. governments. However, this situation has changed in the twenty-first century, at least to some degree. For instance, the Russian Federal Space Administration (Roscosmos, or RKA) has made available several seats per year on its Soyuz taxi flights to the International Space Station (ISS) to anyone who can mentally and physically qualify and pay the ticket price of $20 million or more. In April 2001, American businessperson Dennis Tito became the world's first space tourist by qualifying and paying the required fee for transportation and a week's stay at the station.

Mark Shuttleworth, a South African, became the second space tourist to the station in April 2002. The third space tourist, or spaceflight participant, whose trip (like all of them) was organized by Space Adventures Ltd. (Vienna, Virginia), was Gregory Olsen (from the United States) in 2005. Anousheh Ansari (Iran/U.S.) in 2006, Charles Simonyi (Hungary/U.S.) in 2007, and Richard Garriott (U.S.) followed Olsen in 2008. Simonyi became the first spaceflight participant to fly twice into space when he returned to space in March 26, 2009. Canadian businessperson Guy Laliberte was launched into space on September 30, 2009, for a 12-day visit to the International Space Station (ISS).

In 2010, the Russians temporarily halted the program for allowing space tourists to fly along with its cosmonauts in their Soyuz capsules. This was due to the size of the Expedition crews increasing from three to six beginning in 2011, along with the fact that the U.S. space shuttle program was ending in that same year. Russia hopes to reinstitute its spaceflight participant program in 2012 or 2013.

In the future, tourism in space will need a comprehensive array of services and facilities to support the continuing entrance of private citizens into space. The following are some of the activities currently active around the world.

Spaceports

The Mojave Air and Space Port, located in Mojave, California, became the first facility licensed by the Federal Aviation Administration (FAA) in the United States for horizontal launches of reusable spacecraft. It was

* **aeroballistic** describes the combined aerodynamics and ballistics of an object, such as a spacecraft, in flight

* **catalyst** a chemical compound that accelerates a chemical reaction without itself being used up; any process that acts to accelerate change in a system

certified as a spaceport by the FAA on June 17, 2004, and will remain certified through June 16, 2014. Most notably in the twenty-first century, the spaceport has become a home base for many small private companies developing spacecraft for private and governmental use. For instance, several teams for the Ansari X Prize (a competition sponsored by the X Prize Foundation that offered a $10 million prize for the first non-governmental organization to launch a reusable manned spacecraft into space twice within a two-week period), used the port for their suborbital flights. The winner of the Prize, Scaled Composites' *SpaceShipOne*, has gone on to team with Virgin Galactic, founded by Richard Branson, to pursue suborbital flights for space tourists. As of 2011, Orbital Sciences Corporation and Interorbital Systems also use Mojave for their developing suborbital spacecraft.

Spaceport America, formerly the Southwest Regional Spaceport, calls itself the "world's first purpose-built commercial spaceport." Located in New Mexico, near the White Sands Missile Range and about 30 miles (48 kilometers) east of Truth or Consequences, New Mexico, it is expected to be completed in 2011 or 2012. In December 2008, the operators of the facility, the New Mexico Spaceport Authority, received its FAA license, from the Office of Commercial Space Transportation, for vertical and horizontal launches of spacecraft. Thereafter, Virgin Galactic signed a 20-year lease for the spaceport's facilities. The spaceport will be the site from which Virgin Galactic will launch its suborbital flights for the paying public sometime around 2013.

Spaceport Sweden is a company that is planning to construct a European spaceport for suborbital and orbital spaceflights for space tourists. The company, a joint effort of the Swedish Space Corporation, IceHotel, Progressum, and LFV Group, plans to locate the spaceport in Kiruna, Sweden, a city in the northernmost part of the country, within Lapland province in Norrbotten County. Virgin Galactic, as of 2007, has an agreement with Spaceport Sweden that it would make its site the first one outside of the United States.

The Abu Dhabi Spaceport, a yet-to-be built spaceport in the United Arab Emirates (UAE), in Abu Dubai (the country's capital and largest city), is another location that Virgin Galactic has shown interest to eventually fly suborbital flights from in the future. UAE is on the central western coast of the Persian Gulf. Aabar Investments, a global investment company founded in 2005, secured a 32% interest in Virgin Galactic in 2009. Aabar is owned by several subsidiary companies of the Abu Dhabi government. Virgin Galactic and Aabar Investments has shown interest in building the Abu Dhabi Spaceport.

In 2011, XCOR Aerospace and Space Expedition Curacao (SXC) signed a lease contract for XCOR's Lynx suborbital spacecraft. The Caribbean island of Curacao now expects to begin construction of its own spaceport with an expected beginning date of operations in 2014. As of September 19, 2011, SXC had signed up 35 space tourists for the first flights of XCOR. SXC co-founder Michiel Mol stated, "A number of celebrities and notables have already signed up for this amazing experience, including Victoria's Secret model Doutzen Kroes, World Series Champion San Francisco Giants batting coach and Curacao native Hensley Meulens, and celebrity DJ Armin van Buuren, host of weekly radio show 'A State of Trance,' which attracts a reported 30 million listeners in 40 countries."

Suborbital Flights

In the future, civilians will have a chance to go into space onboard spacecraft operated by private enterprises such as Virgin Galactic. The company will be using nearly reusable launch vehicles developed by Scaled Composites. The reusable spacecraft called *SpaceShipTwo* is a spacecraft that will be launched with the help of the *WhiteKnightTwo* launching vehicle. The piloted *WhiteKnightTwo* will take *SpaceShipTwo,* with pilots and passengers, from launch to about 18 kilometers (11 miles) above Earth. The *SpaceShipTwo* then separates from the launch vehicle to complete its trip to space, about 100 kilometers (62 miles) above Earth. Test flights are scheduled for 2011 and 2012, with the first commercial flight scheduled sometime in 2013.

The Virgin Galactic/Scaled Composites partnership is one of several private companies attempting to provide private launch services to space with the use of RLVs. Other companies competing in this new market are Armadillo Aerospace and Blue Origin.

Armadillo Aerospace, which is based in Mesquite, Texas, was founded by John Carmack, who is the co-founder of id Software, a maker of computer games such as the series called *Doom* and *Quake.* The company, as of May 2010, is working with Space Adventures on a suborbital commercial rocket that will take tourists into space for about $102,000 per customer (priced as of 2010). The RLV spacecraft, tentatively called Black Armadillo, will be a vertical takeoff, vertical landing (VTVL) suborbital vehicle. No date has been set for the maiden voyage.

Blue Origin, which is headquartered in Kent, Washington, but with its main flight facility in Culberson County, Texas, was founded by Jeff Bezos, the founder of Amazon.com. In 2000, the company began work on its VTVL/RLV *New Shepard* spacecraft. The technology for the craft is based on the McDonnell Douglas DC-X. Unmanned test flights took place in 2011 with more flights planned for 2012 and beyond.

Modules, Space Stations, and Hotels in Space

Private company Bigelow Aerospace, headquartered in North Las Vegas, Nevada, United States, bought the commercial rights to the National Aeronautics and Space Administration (NASA) patents involving TransHab (for transit habitat). The TransHab module was a NASA-developed inflatable living space made of lightweight, flexible materials stronger than steel. Designed initially to fold inside the cargo bay of the space shuttle during transportation into space, the TransHab would, then, be inflated to its natural size when in orbit about Earth.

In July 2006, Bigelow launched its *Genesis I* module, which was followed up about eleven months later with its *Genesis II* module. As of October 1, 2011, both modules are in orbit about Earth, sending to mission control personnel in Las Vegas information on pressure, temperature, and radiation in and around their shells.

The company is expecting to launch its *BA 330* module (formerly called Nautilus space complex module) in 2014 or 2015. The space habitation module, expandable to 12,000 cubic feet (330 cubic meters), will directly support zero-gravity scientific and manufacturing research, along with indirect activities involving space tourism.

Bigelow is also designing a *BA 2100* module that would require a heavy-lift rocket to send it to low-Earth orbit. If manufactured, the module would be six times larger than the *BA 330*—at 77,690 cubic feet (2,200 cubic meters). In addition, Bigelow Aerospace is also developing a series of inflatable modules, under the name of Bigelow Next-Generation Commercial Space Station. The complex will include spacecraft modules, along with a central docking node, solar arrays, propulsion, and crew capsules. The company hopes to launch the components in 2014, with leasing options starting in 2015. So far, Bigelow has contracts with seven countries interested in leasing its space station modules.

NASA Helping Out Private Space Companies

NASA has been directed by the Barack Obama administration to help private industry ease itself into the ferrying of humans and cargo into low-Earth orbit, such as trips to and from the International Space Station (ISS). The Commercial Orbital Transportation Services (COTS) program by NASA helps to coordinate the delivery of humans and cargo to the ISS by private companies. With the termination of the space shuttle fleet, NASA needs a reliable means to get astronauts and cargo up into space so it can concentrate on the development of a deep-space propulsion system and capsule to get astronauts to asteroids, Mars, and other bodies in the inner solar system.

On December 23, 2008, NASA announced that Orbital Sciences Corporation (Orbital Sciences, based in Dulles, Virginia) and Space Exploration Technologies Corporation (SpaceX, headquartered in Hawthorne, California) were awarded contracts under a NASA Commercial Resupply Services (CRS) program. The contracts stipulate the delivery of a minimum of eight unmanned missions for Orbital Sciences (for $1.9 billion) and 12 unmanned missions for SpaceX (for $1.6 billion) between 2009 and 2016, to the International Space Station. Although both companies have encountered delays in their test flights, they are both proceeding with the hope of beginning cargo missions to the ISS in 2012.

Barriers and Obstacles to Space Travel and Tourism

Before space travel and tourism can be made economical, reliable, efficient, and safe for everybody, several obstacles must be overcome and many barriers will have to be removed.

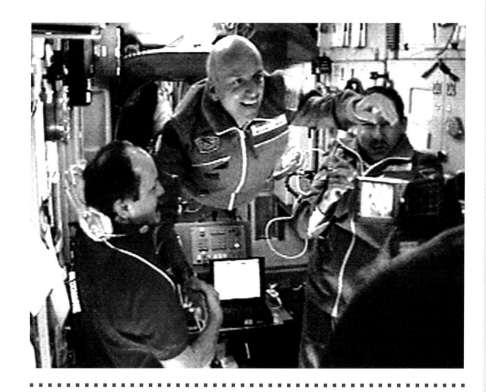

* **orbiter** spacecraft that uses engines and/or aerobraking, and is captured into circling a planet within orbits

Market Research and Development The space travel and tourism market, in the early 2010s, is attracting investors and businesspeople. A number of space travel and tourism market surveys and analytic studies have already been designed and performed by professionals in the market research field. In addition, ways to enhance the credibility of space tourism by piquing the interest of nontraditional space businesses, which stand to profit from its development, have already been developed.

Legislative Measures In 2004, President George W. Bush signed the Commercial Space Launch Amendments Act (H.R. 5382), which allows commercial tourism companies to send paying customers into space. With the signing of the bill, regulations were formulated over the next few years by the Federal Aviation administration (FAA) for space tourism companies and their paying customers. Currently, any company that expects to launch paying customers from U.S. spaceports onto a suborbital or orbital spacecraft must receive a license from the FAA Office of Commercial Space Transportation, such as those garnered by the Mojave Air and Space Port, in California, and Spaceport America, in New Mexico.

Technology and Operations There is a need to go far beyond space shuttle technology and operational capabilities. The costs per shuttle mission, depending on the annual budget and flight rate between its operational lifetime of 1981 and 2011, were between $500 million and $750 million per flight, and it took approximately six months to process orbiters* between flights. From these baseline parameters, it is essential

* **infrastructure** the physical structures, such as roads and bridges, necessary to the functioning of a complex system

to lower the unit cost and decrease the turnaround time between flights. Furthermore, reliability must be increased before space travel and tourism can become safe and affordable for the vast majority of the general public.

Medical Science There are volumes of recorded data about how a nearly physically perfect human specimen reacts to the space environment, but no information about people with common physical limitations and treatable maladies. For example, how would the medicines taken by a large percentage of the general public act on the human body in a state of extended weightlessness? Astronauts and cosmonauts are physiologically screened for their ability to react quickly and correctly under extreme pressure in emergencies. However, living in space by the first space tourists is, and will be in the near future, characterized by cramped living conditions, common hygienic and eating facilities, and semiprivate sleeping quarters. Such conditions are conducive to unrest and conflict among certain individuals, making screening of early space tourists for temperament and tolerance a must.

Regulatory Factors Methods must be devised through public and private sector efforts that will allow an orderly, safe, and reliable progression of certification and approval of a venture's equipment without the imposition of potentially crippling costs. Initially it will not be possible to match the safety and reliability levels of conventional aircraft that have evolved over time. Instead, a system is needed that will allow voluntary personal risk to be taken in excess of that involved in flying on modern aircraft while fully protecting the safety of third parties (people and property not affiliated with the operator and/or customer).

So far, the Commercial Space Launch Act (CSLA) of 1984, along with its 2004 Amendment, regulates all commercial launches, reentries, or landings conducted by U.S. companies. Under the CSLA, each launch or reentry within the United States must have a license. The FAA Office of Commercial Space Transportation issues these licenses. The FAA office conducts a policy review, a payload review, a safety evaluation, an environmental review, and a financial responsibility determination based on the data in the license application before issuing or refusing a license. The purpose of a launch license is to assure that "the public health and safety, safety of property, and the national security and foreign policy interests of the United States" are properly considered.

Legal Factors Just as there are laws for operating on Earth's land surface and oceans, there will be a need for laws for operating in space and on and around other celestial bodies. The United Nations treaty governing the use of space must be improved and expanded to take into account eventual space operations involving people and accommodating infrastructure*. From the navigational rules of space lanes to real estate claims for settlement or mining purposes, laws will have to be created by international legal bodies to provide order and justice on the final frontier.

* **reusable launch vehicles**
launch vehicles, such as the
space shuttle orbiter, designed to
be recovered and reused many
times

From the initial steps of establishing the principles for the exploration and use of outer space by nations and governments, space law in the twenty-first century encompasses rational and reasonable approaches to representing the demands of persons in virtually every part of the world for enhanced communications, education, entertainment, environmental, and transportation services. As space commerce grows, space law will continue to address the unique problems posed by commercial activities in space.

Finance and Insurance Perhaps the most prominent obstacle that must be overcome is the (sometimes) lack of financing available for private space ventures, particularly those involving new reusable launch vehicles* (RLVs). Several RLV development programs have been stalled because of an inability to find investors. Persuading investors to accept some front-end risk in return for the large rewards that should be realized in the years ahead is the main challenge. Legislation to ease the risk is one potential solution. Innovative methods for raising capital (e.g., tax-exempt bonds) and other ways to lower the risks to acceptable levels will have to come from the investment and insurance communities.

Space should be seen as another medium that will be developed for business and recreational purposes, contributing to the welfare and enjoyment of all the world's people. Before long, space will become an extension of Earth itself.

In 2011, the prospect of space tourists going up into space on a regular basis is producing a new avenue for the space insurance industry. With Virgin Galactic on the edge of providing sub-orbital flights as early as 2013, space insurance underwriter Lloyd's of London has been in negotiations with Virgin Galactic. The insurance will likely involve a combination of an aviation risk and a space risk while operating their flights both in Earth's atmosphere and above it. Talks with the pioneering space tourist company are also ongoing concerning third-party liability protection against claims from people living near the New Mexico launch site, such as claims from pollution caused by launches or from falling debris in the event of a crash over their homes. As space is no longer the sole domain of professional astronauts and scientists, and as more people get the chance to go into space, the space insurance industry will no doubt see new revenues coming its way.

The Future The exact size of this fledgling space tourism market, say, in twenty years, remains to be seen. The stated price of a trip to the International Space Station is 20 to 40 million U.S. dollars, while the price of a future suborbital flight starts at about $95,000, and goes upwards to $200,000. These prices make for an extremely small proportion of the population being able to experience space in these reusable spacecraft taking off from these spaceports over the first several years of actually sending tourists to space on a regular basis. However, as the space tourism industry grows—and as more and more people spend money to go into space—the price for such experiences will steadily decline. As with most industries, the most profits are garnered when the most people are allowed to participate in its products and services. To survive, these fledgling space tourism companies must quickly (but safely) expand in the number of flights offered each year and, thus, the number of people using their services. Space will be the final frontier for a large percentage of people only if the price is right.

 See also **Emerging Space Businesses (Volume 1) • Hotels (Volume 4) • Living in Space (Volume 3) • Space Tourism, Evolution of (Volume 4)**

Resources

Books and Articles

Haltermann, Robert L. *Going Public 2001: Moving Toward the Development of a Large Space Travel and Tourism Business.* Proceedings of the 3rd Space Travel and Tourism Conference. Washington DC: Space Transportation Association, 2001.

McCurdy, Howard E. *Space and the American Imagination.* Baltimore: Johns Hopkins University Press, 2011.

Parker, Martin, and David Bell, editors. *Space Travel and Culture: From Apollo to Space Tourism.* Malden, MA: Wiley-Blackwell/Sociological Review, 2009.

Pelt, Michael. *Space Tourism: Adventures in Earth's Orbit and Beyond.* New York: Springer, 2005.

Websites

Armadillo Aerospace. <http://www.armadilloaerospace.com/n.x/Armadillo/Home> (accessed October 3, 2011).

Blue Origin. <http://www.blueorigin.com/> (accessed October 3, 2011).

Boyle, Alan. *Private-spaceflight bill signed into law.* MSNBC. <http://www.msnbc.msn.com/id/6682611/> (accessed September 15, 2009).

Bonsor, Kevin. *How Space Tourism Works.* How Stuff Works. <http://science.howstuffworks.com/space-tourism.htm> (accessed September 28, 2011).

Civilians in space FAQs. CBC News. <http://www.cbc.ca/news/background/space/spacetourism.html> (accessed September 28, 2011).

Commercial Space Launch Amendments Act of 2004. National Aeronautics and Space Administration. <http://www.faa.gov/about/office_org/headquarters_offices/ast/media/PL108-492.pdf> (accessed October 3, 2011).

Deulgaonkar, Parag. *Virgin Galactic wants Abu Dhabi spaceport.* Emirates 24/7 News. <http://www.emirates247.com/news/virgin-galactic-wants-abu-dhabi-spaceport-2011-02-01-1.349881> (accessed October 3, 2011).

Future of Space Tourism: Who's Offering What. Space.com. <http://www.space.com/11477-space-tourism-options-private-spaceships.html> (accessed October 3, 2011).

Knapp, George. *I-Team: Bigelow Aerospace Begins Big Expansion.* 8NewsNow.com. <http://www.8newsnow.com/story/13967660/i-team-bigelow-aerospace-begins-big-expansion> (accessed September 29, 2011).

Lloyd's in talks to insure space flight. Lloyd's. March 30, 2006, <http://www.lloyds.com/News-and-Insight/News-and-Features/Archive/2006/03/Lloyds_in_talks_to_insure_space_flight> (accessed October 11, 2011).

Maliq, Tariq. *Virgin Galactic Unveils Suborbital Spaceliner Design.* Space.com. <http://www.space.com/news/080123-virgingalactic-ss2-design.html> (accessed September 28, 2011).

Office of Commercial Space Transportation. Federal Aviation Administration. <http://www.faa.gov/about/office_org/headquarters_offices/ast/> (accessed October 3, 2011).

Phase one of world's first commercial spaceport is now 90% completed — in time for first flights in 2013. MailOnline.co.uk. <http://www.dailymail.co.uk/sciencetech/article-2034239/Phase-worlds-commercial-spaceport-90-cent-completed--time-flights-2013.html> (accessed September 22, 2011).

RocketShip Adventures: XCOR Lynx Overview. Incredible Adventures. <http://www.incredible-adventures.com/xcor-lynx.html> (accessed September 30, 2011).

Russia to resume space tourism in 2012. RT.com <http://rt.com/news/sci-tech/russia-space-tourist-resume/> (accessed September 28, 2011).

Space Adventures. <http://www.spaceadventures.com> (accessed October 3, 2011).

Space Expedition Caracao. <http://www.spacexc.com/en/home/> (accessed October 4, 2011).

Virgin Galactic. <http://www.virgingalactic.com/> (accessed September 28, 2011).

"Virgin Galactic Unveils Suborbital Spaceliner Design." *Space.com.* <http://www.space.com/4869-virgin-galactic-unveils-suborbital-spaceliner-design.html> (accessed October 11, 2011).

XCOR Aerospace. <http://www.xcor.com/> (accessed October 3, 2011).

Wall, Mike. "First Space Tourist: How a U.S. Millionaire Bought a Ticket to Orbit." *Space.com.* <http://www.space.com/11492-space-tourism-pioneer-dennis-tito.html> (accessed October 3, 2011).

Toys

The term "toy" generally applies to any object used by children in play. However, there is a huge business in creating objects, usually in miniature, designed specifically for children's play. These toys often model adult culture and society, frequently with great accuracy. In the last half of the twentieth century, model National Aeronautics and Space Administration (NASA) spacecraft, toy ray guns with light and sound action, and spaceships and action figures related to popular films reflected a growth of interest in science fiction and space exploration. That interest in space-related toys continued into the twenty-first century.

In the early 2000s, there was a wide range of space toys available in the market. Children had access to transparent model Saturn V rockets, models of the International Space Station, models of the *Apollo 15* Lunar Lander with the Lunar Rover, models of the U.S. space shuttle, replicas of the Mars Exploration Rovers *Spirit* and *Opportunity,* and a complete Kennedy Space Center launch pad. The toy manufacturer Brio Corporation marketed a Space Discovery Set suitable for very young children that includes an astronaut, a launch vehicle, and a launch control center with a ground crew member. Lego offers numerous space-themed sets, for example the three toys in its Life on Mars series: the Excavation

Space-related toys, such as this Mattel Hot Wheels JPL Sojourner Mars Rover Action Pack set, can generate profits for toy manufacturers and publicize advancing space technology. © *AP Images.*

Searcher, the Mars Solar Explorer, and the Red Planet Protector. Action Products sells the Complete Space Explorer with models of the space shuttle, Apollo Lunar Lander, *Skylab* space station, and dozens of small action figures representing astronauts and ground crews.

The *Mars Pathfinder* mission, one of a series of robotic explorations of other planets, created many business opportunities for toy manufacturers. The Jet Propulsion Laboratory (JPL) and NASA have signed numerous licensing agreements for products related to this mission, including T-shirts, caps, and toys. One of the most interesting and ambitious toys was the Mattel Hot Wheels JPL Sojourner Mars Rover Action Pack set. This set includes toy models of the *Sojourner* rover, the *Mars Pathfinder* spacecraft, and a lander. Many of the *Sojourner* rover's unique attributes are included, such as its six-wheel independent suspension that allows it to navigate rough terrain. According to Joan Horvath, a business alliance manager with JPL's Technology Affiliates Program, these toy models help educate children and parents alike about the *Mars Pathfinder* mission (which lasted from 1996 to 1998) in the most user-friendly manner possible. Moreover, it made the business community aware of the many different aspects of the JPL's technology transfer programs.

The success of the Mattel Mars Pathfinder set led to another license agreement. JPL and Mattel teamed up for a toy version of NASA's *Galileo* spacecraft. The Hot Wheels Jupiter/Europa* Encounter Action Pack includes a highly detailed reproduction of the *Galileo* spacecraft, the *Galileo* descent probe, and of one of the ground-based antenna dishes. The *Galileo* mission lasted from 1989 to 2003.

The NASA *Mars Exploration Rover* (MER) mission consists of two rovers (*Spirit* and *Opportunity*) that are exploring the planet Mars. Here on Earth, identical miniatures of the rovers are being played with by children all over the world. In 2004, the Office of Technology Transfer, at the California Institute of Technology (in cooperation with the NASA Jet Propulsion Laboratory) licensed the image of the Mars exploration rovers to the Danish company Lego. When it was first offered for sale, the twin rovers, which contained 858 plastic parts, sold for about $90. The MER mission began in 2003 and was still underway in November 2011.

Toys in Space

Toys have also ventured into space. Carolyn Sumners of the Houston Museum of Natural Science in Houston, Texas, assembled a small group of toys to be flown on space shuttle mission STS-51D in April 1985. During the flight, crewmembers experimented with the toys, demonstrating the behavior of objects under conditions of apparent weightlessness. The first "Toys in Space" mission was so successful that a second group of toys was flown on the STS-54 mission in January 1993. During the second Toys in Space mission, astronauts John Casper and Susan Helms demonstrated how the behavior of several simple toys was quite different under microgravity* conditions.

* **Europa** one of the large satellites of Jupiter

* **microgravity** the condition experienced in freefall as a spacecraft orbits Earth or another body; commonly called weightlessness; only very small forces are perceived in freefall, on the order of one-millionth the force of gravity on Earth's surface

In June 2001, the Lego company teamed up with Space Media Inc. and RSC Energia to conduct the first experiment on the International Space Station using toys. The Life on Mars: Red Planet Protector was used to measure the mass of an object under zero-gravity conditions (those where astronauts do not perceive a presence of gravity while in free-fall, although it still exists). Cosmonauts Talgat Musabayev and Yuri Baturin demonstrated how an object's mass can be determined from oscillation frequency in a near weightless environment.

Educational toys related to space exploration can serve the dual roles of providing a good return on investment for toy manufacturers while providing a rich learning opportunity for children. The vision of Lego, sparking an interest in science and space, can provide a sound basis for socially conscious free enterprise. The cooperative model developed by Mattel and JPL has been mutually beneficial, serving as a strong profit center for Mattel while effectively publicizing the commitment of NASA and JPL to technology transfer.

Model Rockets

Many people young and old enjoy assembling and launching model rockets. A number of companies mass produce kits for this purpose; some leading manufacturers for the U.S. market are Aerotech, Estes Industries, and Quest. Other hobbyists create their rockets from scratch. Model rocketry enthusiasts have formed governing bodies, such as the National Association of Rocketry, to set safety standards and promote their hobby.

A typical model rocket is small and light, only a foot or so in length and weighing no more than a few ounces. These use black power rocket engines charges to reach heights of several hundred feet. Models of considerably more power are available, however, including multi-stage rockets. Skilled hobbyists can construct high power rockets capable of breaking the sound barrier and reaching altitudes of 10,000 feet. These rely on more advanced engines, using substances such as ammonium perchlorate composite propellant. Model rockets are typically equipped with a recovery system, such as a parachute, that allow them to return safely to the ground. Many can then be outfitted with a new engine and reused. Some models can be equipped with payloads for use during flight, such as cameras and altimeters.

 See also **Education (Volume 1) • Mars (Volume 2) • Microgravity (Volume 2) • Robotic Exploration of Space (Volume 2)**

Resources

Books and Articles

Bunte, Jim, et al. *Robots & Space Toys.* Iola, WI: Krause, 2000.

McCurdy, Howard. *Space and the American Imagination.* Baltimore: Johns Hopkins University Press, 2011.

Young, S. Mark. *America, Blast Off! Rockets, Robots, Ray Guns, and Rarities from the Golden Age of Space Toys.* New York: Dark House Books, 2001.

Wood, Lamont. *Out of Place in Time and Space.* Pompton, NJ: Career Press, 2011.

Websites

Kirby, Julia. *Seeing Robots Everywhere.* Harvard Business Review. <http://blogs.hbr.org/hbr/hbreditors/2011/11/seeing_robots_everywhere.html> (accessed November 6, 2011).

Mars Exploration Rovers. NASA Jet Propulsion Laboratory. <http://marsrover.nasa.gov/home/index.html> (accessed November 6, 2011).

Mars Rover Is Landing Again—In Toy Stores. MSNBC/Associated Press. <http://www.msnbc.msn.com/id/3995587/ns/technology_and_science-space/> (accessed November 6, 2011).

National Association of Rocketry. <http://www.nar.org/> (accessed November 11, 2011).

Smith, Ned. *New Robots for Kids Link Online and Offline Worlds.* BusinessNews. <http://www.businessnewsdaily.com/robots-toys-physical-virtual-seamless-experience-1915/> (accessed November 6, 2011).

U.S. Toy Industry Retail Sales Generated $21.87 Billion in 2010: Toy Industry Experiences 2 Percent Growth. NPD Group. January 27, 2011. <https://www.npd.com/press/releases/press_110127.html> (accessed November 6, 2011).

UV

United Space Alliance

In the early twenty-first century, thirty years after the successful maiden voyage of space shuttle *Columbia*, the National Aeronautics and Space Administration's (NASA's) Space Transportation System (STS)—commonly known as the space shuttle fleet—has been the workhorse of the U.S. manned space program. As of August 31, 2011, however, the fleet was retired. For those past three decades, the shuttle fleet was the only domestic U.S. transit system capable of supporting human spaceflight. Despite the shuttle's remarkable achievements and unique capabilities, those familiar with the shuttle program have come to realize that the shuttle failed to meet many of NASA's original objectives and expectations. However, a bright spot of the program was the privatization of the daily operations of the space shuttle program during its last decade of flight.

Higher Costs, Fewer Payloads Leads to Problems

At the beginning of the Space Shuttle program in the early 1980s, NASA expected the shuttle would fly some twenty-four times annually, launching astronauts, satellites, and other payloads* for the U.S. government as well as for other nations and private companies. By the mid–1990s, however, the space shuttle's average annual flight rate was a fraction of the predicted level and that reduced rate continued through 2011. In addition, fewer government payloads than expected had flown on the shuttle, and because of policy changes made after the seven astronauts were killed aboard space shuttle *Challenger* in 1986, commercial payloads had been effectively banned from the vehicle. Perhaps the greatest disappointment was that the costs of operating and refurbishing the shuttle were far higher than NASA's original projections.

Throughout the shuttle's history, NASA considered placing shuttle operations under private industry's control to reduce costs. That idea was continually rejected on grounds that NASA needed to maintain control of the shuttle for national security reasons. However, in 1995, then NASA administrator Daniel Goldin, who had previously spent twenty-five years at a private aerospace firm, asked a team of NASA, other governmental, and industry leaders to study shuttle operations management and propose a new, safe approach to reducing operations costs. The team found that shuttle operations tasks, as then assigned, were diffused* among many contractors and that no single entity was responsible for streamlining operations and reducing costs. After considering multiple management options, the commission recommended that NASA give a single,

* **payload** any cargo launched aboard a rocket that is destined for space, including communications satellites or modules, supplies, equipment, and astronauts; does not include the vehicle used to move the cargo or the propellant that powers the vehicle

* **diffuse** spread out; not concentrated

private contractor responsibility for shuttle operations that, at the time, were managed as twelve space shuttle program contracts. Goldin agreed, and NASA began soliciting bids from companies to take charge of shuttle operations.

United Space Alliance

Two companies that then held contracts to manage major elements of shuttle operations, Rockwell International and Lockheed Martin Corporation, recognized that failing to secure the prime contract under NASA's new management scheme would result in a substantial economic loss. As a result, the two companies decided to compete for the contract as a single entity. In August 1995, Rockwell and Lockheed Martin agreed to form a joint venture called United Space Alliance (USA). From the forty companies that responded to NASA's search for a shuttle prime contractor, NASA chose to award USA the contract. USA took over the individual Rockwell and Lockheed Martin contracts and on September 30, 1996, signed the Space Flight Operations Contract (SFOC), by which NASA designated USA as prime contractor for shuttle operations. That December, Rockwell sold its aerospace business to The Boeing Company, which took over Rockwell's share of USA.

A New Way of Doing Business

The SFOC was an unprecedented step for NASA. Never before had the space agency given so much authority and responsibility to a contractor for such a major program. Under the contract, which was set up for six years with options for two, two-year extensions, USA took over operations and maintenance of both ground and flight systems associated with the shuttle at NASA's two primary centers for human spaceflight activity. At the Johnson Space Center in Houston, Texas (the control and training center for shuttle missions), USA employees gained responsibility for flight operations, astronaut and flight controller training, mission control center management and operations, mission planning, flight design and analysis, and flight software development. Those at Florida's Kennedy Space Center (the shuttle's launch and processing site) took charge of vehicle testing and checkout, launch operations, procurement* and repair of shuttle hardware and ground support equipment, payload integration, and retrieval of the solid rocket boosters that were jettisoned into the Atlantic Ocean after launches. The SFOC also made USA responsible for training astronauts and planning for operations aboard the International Space Station.

The Good Stuff: Safety, Goals, and Efficiency

NASA emphasized that its expectations for USA under the contract included, in order of importance, maintaining safety of the shuttle system, supporting NASA's planned mission schedule, and reducing shuttle operating costs. In order to ensure that USA would meet these objectives, the SFOC was designed to reward USA on its quality of performance. The $7-billion contract—which could grow to a total of $12 billion with the extensions—was made

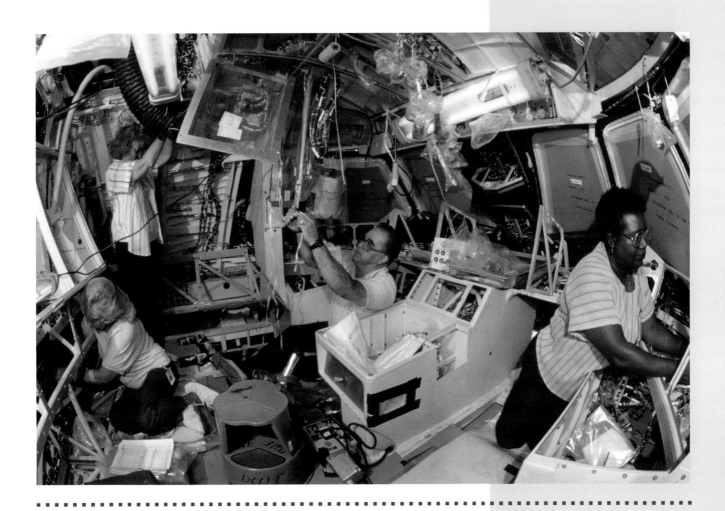

▲

Rockwell International workers prepare
the cockpit of space shuttle *Endeavour*
for extended flights. © *Roger Ressmeyer/
Corbis.*

contingent on the company's ability to meet safety standards, achieve mission and schedule objectives, and find more efficient ways to operate the shuttle program. Failure to meet these objectives could result in financial penalties.

The SFOC presented USA with many challenges. As the first and only company in the world to be fully responsible for maintaining and operating a reusable launch system, USA had to develop, from scratch, methods of fulfilling the basic contract requirements while finding ways to make operations less costly and more efficient. With accountability and quality of performance dominating the contract, USA was forced to accept a new way of earning a profit. Nonetheless, USA proved its ability to manage successfully the new type of contract and the responsibilities it brought. Since managing its first shuttle mission in November 1996, USA attempted to keep safety and reliability as top priorities. Both NASA and USA recognized that cost savings were realized through the SFOC, with USA reporting a reduction in operations costs of nearly $400 million between the fiscal years 1996 and 1998. NASA also pointed to more on-time launches and smoother prelaunch operations as indicators of USA's success in managing the shuttle program.

NASA's Space Transportation System (STS), commonly known as the space shuttle, is the only American transit system maintained by a national space agency that is capable of supporting human spaceflight. *NASA.*

Not Everything Turned Out Right

However, on February 1, 2003, the space shuttle *Columbia* was destroyed as it attempted to re-enter Earth's atmosphere as it completed its mission (STS–107) at the International Space Station. The destruction of the vehicle, and the loss of all seven astronauts onboard, was determined by an investigation board to be caused by a piece of foam (about the size of a suitcase) that broke away from the external tank during launch and hit the thermal protection system (TPS) on the leading edge of the shuttle's left wing. When *Columbia* de-orbited and re-entered Earth's atmosphere the intense heat from friction caused by its interaction with the atmosphere caused the wing to overheat and eventually tear apart—dooming the vehicle and its crew.

Thus, one major operational disaster occurred under the direction of NASA-contractor USA while under NASA's supervision. Since then,

USA has continued to add contracts for NASA's human spaceflight needs. In September 2006, NASA entered into the Space Program Operations Contract (SPOC) with USA, after recognizing the contribution that USA had performed in the past. The $1.1 billion contract ran through September 30, 2010, and covered space operations within the Space Shuttle Program. As with the previous SFOC, the new contract was a cost reimbursement contract, with provisions for performance and award fees. The contract allowed USA to complete the Space Shuttle Program and continue related support as it geared down that program. The contract included work and support for mission design and planning, astronaut and flight controller training, software development and integration, system integration, and flight operations. It also included vehicle processing, launch and recovery, flight crew equipment processing, and vehicle sustaining engineering.

Prospects for the Future

While USA took increasing responsibility for day-to-day shuttle operations, the SFOC made no provisions for ever giving USA ownership of the shuttle fleet. NASA maintained ownership and ultimate control of the space shuttle program to its retirement. With the government still in charge, NASA determined the nature of each shuttle mission and flew only government payloads. USA would have liked that to change, but it never did. The company's vision is still to pioneer human spaceflight as an affordable, viable business. Today, USA would like to see another manned spacecraft fully privatized—that is, given to the private sector to own, control, and fund—and commercialized, which would open manned spacecraft for use by paying customers from outside the U.S. government. USA also wishes to become increasingly involved in operations of the International Space Station as well as new space vehicles that NASA develops.

Many people—including NASA as well as USA officials—have stated in the past that privatization and commercialization of the shuttle would bring numerous benefits to NASA, private industry, and taxpayers alike. By either turning over this asset to private hands or opening its use to commercial customers—or both—NASA could cut costs, which in turn could translate into savings for taxpayers. These measures could also allow NASA to focus more attention on and apply some of the funds saved to activities such as exploring the solar system and universe. By fully owning, managing, and commercializing a spacecraft fleet, a private company potentially could realize revenues that far exceeded NASA's annual fiscal budget for operating the shuttle. As a result, the managing company could afford to conduct more missions and other space activities, in turn stimulating the growth of businesses whose satellites, experiments, or other hardware it launches.

USA's vision of complete spacecraft privatization did not become a reality with the space shuttle fleet. NASA did not relinquish control of its assets and functions. The U.S. space agency was primarily reluctant to give up control of the space shuttle fleet for reasons of national security and public safety. The agency was also aware that giving a single company full control of the shuttle (or any government sponsored space fleet) could be viewed by companies that manufacture, develop, and market other launch vehicles and spacecraft as a transfer to one company of government assets that were already paid for with public funds, which creates unfair competition. NASA, nonetheless, recognizes the benefits of privatization and thus intends, at the very least, to increase the private role in manned spacecraft management and operations in upcoming years. It is likely that, in any privatization scenario, the space agency will continue to maintain ownership and management of some launch infrastructure, play an active role in assuring safety of the program, and financially back the private company in the case of catastrophic disaster involving space vehicles.

As the U.S. space agency continues to look for the most competent and efficient company to assume the job of managing future manned space programs, USA must continue to perform at its best if it wishes to fulfill its vision of opening up the human spaceflight business. As NASA transitions from the Space Transportation System program (space shuttles) to deeper space missions, USA will likely continue to provide support to the U.S. manned space program through various service and support contracts.

For the 15 years that USA was in charge of the daily operations of the Space Shuttle Program, it controlled approximately 1.3 million spare parts for the space shuttle fleet program, along with the International Space Station Program. While working on the operations of the space shuttle fleet, according to its website, the company met 99% of the major mission objectives set by NASA on its two contracts. In addition, USA also achieved a 99% on-time delivery schedule with respect to products and a 99.5% schedule of hardware deliveries without deficiencies. The company also trained over 400 astronauts and astronaut candidates. During this time, USA won the George M. Low Award, which recognizes a company's commitment to "teamwork, safety, customer service, and technical and managerial excellence." USA continues to support the International Space Station program, which is expected to be funded through 2020, and possibly through 2028. In September 2011, NASA awarded USA a one-year extension to its SPOC so the company can continue to provide ground operations support at the Kennedy Space Center through September 30, 2012.

However, on the downside, USA was forced to lay off about 70% of its employees when NASA retired the space shuttle fleet in August 2011. As of August 12, 2011, according to KHOU.com in Houston, Texas, USA was able to retain only 3,100 employees out of 10,500 employees at the peak of its employment. Such a drastic decrease of the number of employees at USA will undoubtedly restrict the number and size of contracts that it can handle in the future. On the other hand, if USA can secure new contracts quickly, then former employees may still be available to work for the company. As primarily an engineering company, USA also has access to a large supply of engineering graduates coming out of the American college system.

In the future, USA is contributing to the processing, maintaining, and operating of reusable launch vehicles that are being developed for the next-generation of space vehicles for the United States. For instance, USA is a member of the Orion team, which is the program for the next-generational space capsule that will take U.S. astronauts on deep-space missions to the inner solar system. USA personnel will help to design and develop the Orion Multi-Purpose Crew Vehicle (MPCV).

In November 2010, USA was one of thirteen companies being considered for future awards for designing and developing heavy-lift launch vehicle systems and propulsion technologies for NASA's Space Launch System (SLS). USA is also involved with other activities for the Department of Defense (DOD) and various commercial and international customers. Some of the DOD activities include range management, safety and logistics, launch operations, engineering and fabrication, and maintenance, repair, and overhaul.

United Space Alliance is a good example of NASA allowing a private enterprise to successfully manage a large government contract related to space-based activities. Further private activities are in the works for the United States manned space program. In January 2010, U.S. President Barack Obama proposed a budget increase for NASA and announced plans to have the U.S. space agency encourage and oversee the advance of private launch capacity.

On April 15, 2010, President Obama made the following statement concerning the role of private enterprise and NASA's manned space program: "And in order to reach the space station, we will work with a growing array of private companies competing to make getting to space easier and more affordable. Now, I recognize that some have said it is unfeasible or unwise to work with the private sector in this way. I disagree. The truth is, NASA has always relied on private industry to help design and build the vehicles that carry astronauts to space, from the Mercury capsule that carried John Glenn into orbit nearly 50 years ago, to the space shuttle Discovery currently orbiting overhead. By buying the services of space transportation—rather than the vehicles themselves—we can continue to ensure rigorous safety standards are met. But we will also accelerate the pace of innovations as companies—from young startups to established leaders—compete to design and build and launch new means of carrying people and materials out of our atmosphere."

An example of private companies working with NASA for future spacecraft involves the NASA Commercial Orbital Transportation Services (COTS) contract. Two companies—Orbital Sciences Corporation (Orbital) and Space Exploration Technologies Corporation (SpaceX)—were awarded, on February 2008, contracts for a minimum of eight

Orbital missions and twelve SpaceX missions to the International Space Station. Orbital will launch its Cygnus spacecraft from a Taurus II rocket at a launch pad on Wallops Island (Virginia), while SpaceX will launch its Dragon spacecraft from its Falcon 9 rocket at Cape Canaveral Air Force Station (Florida). The first test flight for Orbital's Cygnus was tentatively scheduled for late 2011, while SpaceX's Dragon began its first test flight in December 2010, with a second plan scheduled for November 2011. Both spacecraft will be initially unmanned, only carrying cargo for resupply missions to the International Space Station, but eventually the two companies hope to modify them for manned missions.

Instead of NASA ferrying astronauts to and from the space station, these two privately run companies will hopefully take over such operations, which will allow NASA to concentrate on manned missions to asteroids, the Moon, Mars, and other possible locations in the inner solar system. In other future ventures, NASA will undoubtedly rely more and more on such companies as United Space Alliance for performing many of the activities associated with running a national manned space program.

▶ *See also* **Challenger (Volume 3) • Columbia (Volume 3) • Commercialization (Volume 1) • Emerging Space Businesses (Volume 1) • Launch Vehicles, Expendable (Volume 1) • Launch Vehicles, Reusable (Volume 1) • NASA (Volume 3) • Reusable Launch Vehicles (Volume 4) • Solid Rocket Boosters (Volume 3) • Space Shuttle (Volume 3)**

Resources

Books and Articles

Dittemore, Ronald D. *Concept of Privatization of the Space Shuttle Program.* Washington, D.C.: National Aeronautics and Space Administration, 2001.

Harris, Robert. *Space Enterprise: Living and Working Offworld in the 21st Century.* Berlin: Praxis, 2009.

McCurdy, Howard E. *Space and the American Imagination.* Baltimore: Johns Hopkins University Press, 2011.

Websites

Letter to the Honorable Ted Stevens. Congressional Budget Office. <http://www.cbo.gov/doc.cfm?index=4457&type=0> (accessed October 24, 2011).

National Aeronautics and Space Administration. <http://www.nasa.gov/> (accessed October 24, 2011).

NNJ06VA01C—Space Program Operations Contract (SPOC)—United Space Alliance. Johnson Space Center, National Aeronautics and

Space Administration. <http://www.nasa.gov/centers/johnson/news/contracts/NNJ06VA01C/NNJ06VA01C.html> (accessed October 24, 2011).

President Barack Obama on Space Exploration in the 21st Century. National Aeronautics and Space Administration. <http://www.nasa.gov/news/media/trans/obama_ksc_trans.html> (accessed October 24, 2011).

United Space Alliance. <http://www.unitedspacealliance.com/> (accessed October 24, 2011).

United Space Alliance lays off an estimated 500 Houston workers. KHOU.com. <http://www.khou.com/news/local/United-Space-Alliance-lays-off-an-estimated-500-Houston-workers-127603918.html> (accessed October 24, 2011).

USA Awarded contract Extension for KSC Operations. United Space Alliance. <http://www.unitedspacealliancenewsroom.com/usa-in-the-news/news-releases/usa-awarded-contract-extension-for-ksc-operations/> (accessed October 24, 2011).

▲

Jules Verne, author of *20,000 Leagues under the Sea* (1870), predicted many scientific advancements of the twentieth century and is considered one of the founding fathers of science fiction. *The Library of Congress.*

Verne, Jules

French Science Fiction Novelist
1828–1905

Jules Gabriel Verne, one of the founding fathers of science fiction, was born in Nantes, France, in 1828. He was the eldest son of a successful provincial lawyer. At twelve years of age, Verne ran off to be a cabin boy on a merchant ship, thinking he was going to have an adventure. However, his father caught up with the ship before it got very far and took Verne home to punish him. Verne promised in the future he would travel only in his imagination.

In 1847 Verne was sent to study law in Paris, France, and from 1848 until 1863 wrote opera librettos and plays as a hobby. He read incessantly and studied astronomy, geology, and engineering for many hours in Paris libraries. His first play was published in 1850, prompting his decision to discontinue his law studies. Displeased upon hearing this news, his father stopped paying his son's expenses in Paris. This forced Verne to earn money by selling his stories.

In 1862, at the age of thirty-four years, Verne sent a series of works called *Voyages Extraordinaire* to Pierre-Jules Hetzel, a writer and publisher of literature for children and young adults. Verne attained enough success with the first in the series, *Five Weeks in a Balloon,* published in 1863, for the Verne/Hetzel collaboration to continue throughout his entire career. Hetzel published Verne's stories in his periodical, *Magasin d'Education et de Recreation,* and later released them in book form.

Due to nineteenth-century interest in science and invention, Verne's work was received with enormous popular favor. He forecast with remarkable accuracy many scientific achievements of the twentieth century. He anticipated flights into outer space, automobiles, submarines, helicopters, atomic power, telephones, air conditioning, guided missiles, and motion pictures long before they were developed. In his novels, however, science and technology are not the heroes. Instead, his heroes are admirable men who master science and technology. His object was to write books from which the young could learn.

Among his most popular books are *Journey to the Center of the Earth* (1864), *From the Earth to the Moon* (1865), *20,000 Leagues under the Sea* (1870), *Mysterious Island* (1870), and *Around the World in Eighty Days* (1873). These five novels have remained in almost continuous print for well over a century. Verne also produced an illustrated geography of France, and his works have been the source of many films.

Because of the popularity of these and other novels, Verne became a wealthy man. In 1857, he married Honorine de Viane. In 1876, he bought a large yacht and sailed around Europe. This was the extent of his real-life adventuring, leaving the rest for his novels. He maintained a regular writing schedule of at least two volumes a year. Verne published sixty-five novels, thirty plays, librettos, geographies, occasional short stories, and essays.

The last novel he wrote before his death was *The Invasion of the Sea*. He died in the city of Amiens, France, in 1905. Verne was buried in the Madeleine Cemetery. One of his books, *Paris in the 20th Century*, was published in 1994, many years after his death, after it had been locked in a safe, and only discovered by his family in 1989. The book, written by Verne in 1863, talks about calculating machines, high-speed trains, a worldwide communications system, and many other technological marvels, which are common today in the twenty-first century.

 See also **Careers in Writing, Photography, and Film (Volume 1)** • **Literature (Volume 1)**

Resources

Books and Articles

Lottman, Herbert R. *Jules Verne: An Exploratory Biography.* New York: St. Martin's Press, 1996.

Jules-Verne, Jean. *Jules Verne: A Biography.* New York: Taplinger, 1976.

Verne, Jules. *Around the World in Eighty Days* (1873). New York: William Morrow &Company, 1988.

Verne, Jules. *20,000 Leagues under the Sea* (1870). New York: New York Scholastic Book Services, 1965.

Websites

From the Earth to the Moon Interactive. NASA Johnson Space Center. <http://er.jsc.nasa.gov/seh/index1.htm> (accessed November 17, 2011).

A Jules Verne Centennial: 1905–2005. Smithsonian Institution Libraries. <http://www.sil.si.edu/OnDisplay/JulesVerne100/> (accessed November 17, 2011).

X

X Prize

The X PRIZE Foundation is a nonprofit organization dedicated, as it states, to bringing "about radical breakthroughs for the benefit of humanity, thereby inspiring the formation of new industries and the revitalization of markets that are currently stuck due to existing failures or a commonly held belief that a solution is not possible." The Foundation creates and manages large "incentivized prize" competitions in various fledgling research and development projects. The general purpose of the X PRIZE competitions is to create exciting breakthroughs in various fields for the benefit of all humankind.

The X PRIZE competitions are grouped into the following areas: exploration (space and deep ocean), energy and environment, education and global development, and life sciences. The first X PRIZE competition involves space exploration and was inspired by past prizes awarded in marine and air travel. The mission of this prize is to change the way that people think about space. Rather than viewing spaceflight as the exclusive province of governments, the foundation's goal is to transform spaceflight into an enterprise in which the general public can directly participate, much in the way that people can fly on airplanes today.

Historical Prizes

In the early twenty-first century, millions of people fly on airplanes between cities around the world. With approximately six billion passengers flying annually in airplanes around the world, at any given time, hundreds of thousands of people are possibly airborne. However, it was not always this way. Only one hundred years ago, during the birth of aviation, flying in an airplane was a very expensive, risky, and infrequent activity, much the way spaceflight is in the early twenty-first century.

At the beginning of the twentieth century (1904–1930), one of the major activities that made aviation very popular, exciting, and affordable was a series of prizes or competitions. History has shown the amazing power of prizes to accelerate technological development. For example, in 1714, in response to a series of tragic maritime disasters, the British Parliament passed the Longitude Act, which provided a large financial prize for the demonstration of a marine clock that was sufficiently accurate to permit precise determination of a ship's longitude. Within twenty years of the announcement of the Longitude prize, a practical clock was demonstrated and marine navigation was revolutionized.

* **suborbital trajectory** the trajectory of a rocket or ballistic missile that has insufficient energy to reach orbit

In the twentieth century, the history of aviation contains hundreds of prizes that greatly advanced aircraft technology. One of the most significant prizes in the history of aviation (and the one from which the X PRIZE is modeled) was the Orteig Prize, an award for the first nonstop flight between New York and Paris, which was sponsored by Raymond Orteig (1870–1939), a wealthy hotel owner. Nine teams cumulatively spent $400,000, sixteen times the $25,000 purse, in pursuit of this prize. By offering a prize instead of backing one particular team or technology, Orteig automatically backed the winner. Had Orteig elected to back teams in order of their probability of success, as judged by the conventional wisdom of the day, he would have backed Charles Lindbergh last. However, Lindbergh achieved success, taking an unconventional single pilot/single engine approach. On May 20, 1927, Lindbergh flew his airplane, the *Spirit of St. Louis,* nonstop for thirty-three-and-one-half hours across the Atlantic Ocean from New York City (United States) to Paris (France), and won the $25,000 Orteig prize.

The First "X Prize"

The original X PRIZE was a competition that was created to inspire rocket scientists to build a new generation of spaceships designed to carry the average person into space on a suborbital flight to an altitude of 100 kilometers (62 miles). This flight is very similar to the flight made by U.S. astronaut Alan Shepard on May 5, 1961, on the Mercury Redstone rocket from Cape Canaveral, Florida. Shepard, who was the first American in space, did not actually go into orbit, as did the space shuttle when it flew into space, but instead flew a suborbital trajectory* that lasted about twenty minutes.

American entrepreneur Peter Diamandis, through the X PRIZE Foundation then headquartered in St. Louis, Missouri, offered a $10 million "X Prize" cash prize in 1996 to demonstrate the ability to fly a suborbital spacecraft into space. To win the prize, vehicles had to be privately financed and constructed, and competitors must demonstrate their ability to fly to an altitude of one hundred kilometers (62 miles) with three passengers. Furthermore, competitors had to prove their vehicle was reusable by flying it twice within a two-week period. The X PRIZE attracted twenty-six teams from seven countries. Later named the Ansari X PRIZE, a total of $100 million was invested by these teams. The name was changed on May 2004, when Iranian engineers and entrepreneurs Anousheh Ansari and Amir Ansari donated millions of dollars to the project.

On September 29, 2004, test pilot Michael "Mike" Melvill, of the Mojave Aerospace Ventures team, was aboard the reusable *SpaceShipOne* as the reusable carrier craft *WhiteKnightOne* lofted the spacecraft into the air. Once airborne, the two separated and Melvill piloted *SpaceShipOne* into outer space, accomplishing the first part (flight 16P) of the Ansari X PRIZE. Melvill flew to a height exceeding 100 kilometers (62 miles) with a 180-kilogram lead-ballast payload to simulate two human passengers. Then, on October 2004, pilot Brian Binnie completed the second half (flight 17P) of the competition by flying the same space plane, with assistance from its carrier aircraft, *WhiteKnightOne,* to outer space. These two successful flights won

Mojave Aerospace Ventures the X PRIZE on November 6, 2004. Mojave Aerospace Ventures is the name of the project (by Scaled Composites) vying for the X PRIZE. The organization backing the project was Tier One, along subgroup of Scaled Composites, with the parent company (Scaled Composites) manufacturing the aircraft and spacecraft.

In 2011, Scaled Composites, founded by Burt Rutan and financially backed by Paul Allen (a co-founder of Microsoft) is manufacturing a series of *SpaceShipTwo* spacecraft and *WhiteKnightTwo* carrier craft for Virgin Galactic. (Northrup Grumman now owns Scaled Composites.) Owned by Richard Branson, Virgin Galactic is developing a pioneering plan to send tourists into suborbital flights into space. A *WhiteKnightTwo* carrier craft will take a *SpaceShipTwo* spacecraft from launch to about 18 kilometers (11 miles) above Earth. At that time, the *SpaceShipTwo* spacecraft will continue under its own power while the carrier vehicle returns to Earth. The spacecraft then completes the trip to space, about 100 kilometers (62 miles) above Earth. Test flights took place in 2011 with more scheduled for 2012. The first commercial flight is scheduled for sometime in 2013. If the suborbital flights of *SpaceShipTwo* are successful, Virgin Galactic and Scaled Composites are preparing *SpaceShipThree*, an orbital spaceplane that will be able to re-enter Earth's atmosphere and land on a runway.

New Prizes

Since 2004, other X PRIZE competitions have been announced. For instance, on September 13, 2007, the X PRIZE Foundation announced the Google Lunar X PRIZE competition, sometimes also abbreviated GLXP and referred to as Moon 2.0. The challenge, this time, is to successfully launch, land, and operate a privately-funded lunar rover (for at least 500 meters, or, 1,640 feet) on the surface of the Moon, while transmitting images and/or data back to Earth. The first-prize winner would receive $20 million, while the second-prize winner would receive $5 million. As of October 28, 2011, 26 officially registered and still competing teams are still pursuing the prize money, which is available to be won until December 31, 2015. Some of the teams include Odyssey Moon, ARCA, Euroluna, Team Phoenicia, and Team Puli.

In 2009, the X PRIZE Foundation announced it would manage the Northrop Grumman Lunar Lander Challenge (formerly the X PRIZE Cup). Prize money came from the National Aeronautics and Space Administration (through its Centennial Challenges). The competition involved using privately funded rockets to take off and land vertically—what is called vertical takeoff/vertical landing (VTVL). These VTVL rockets, in order to win the prize, were to be able to produce sufficient velocity, from a fixed position, to achieve an orbit about the Moon. Armadillo Aerospace and Masten Space Systems claimed all of the prize money in November 2009.

The suborbital flights of the X PRIZE competition in space exploration are just the first step. The competition's goals are to bring about the creation of new generation of space-ships that will serve new markets such as space tourism and point-to-point package delivery (rocket mail). As X PRIZE teams gain experience and improve their technology, their ships will evolve from suborbital to orbital ships in the same fashion that one can draw the lineage from the Wright brothers' *Flyer* (which flew in 1903) to the Douglas DC-3 (of the 1930s and 1940s) and the Boeing 747 (introduced in the 1970s) aircraft, and eventually to the Boeing 787 aircraft (which began commercial service in October 2011).

The other X PRIZE Foundation competitions include the Archon Genomics X PRIZE (presented by Medco), the Progressive Insurance Automotive X PRIZE, and the Wendy Schmidt Oil Cleanup X CHALLENGE.

 See also **Launch Vehicles, Reusable (Volume 1) • Reusable Launch Vehicles (Volume 4) • Shepard, Alan (Volume 3) • Space Tourism, Evolution of (Volume 4) • Tourism (Volume 1)**

Resources

Books and Articles

Belfiore, Michael. *Rocketeers: How a Visionary Band of Business Leaders, Engineers, and Pilots is Boldly Privatizing Space.* New York: Smithsonian Books, 2007.

Dash, Joan. *The Longitude Prize.* New York: Farrar, Straus and Giroux, 2000.

Lindbergh, Charles A. *Spirit of St. Louis.* New York: Charles Scribner's Sons, 1953.

Linehan, Dan. *SpaceShipOne: An Illustrated History.* Minneapolis: Zenith Press, 2008.

Pelt, Michael. *Space Tourism: Adventures in Earth's Orbit and Beyond.* New York: Springer, 2005.

Websites

Armadillo Aerospace. <http://www.armadilloaerospace.com/n.x/Armadillo/Home> (accessed October 3, 2011).

Armadillo rocket takes $350,000 prize. MSNBC.com. <http://www.msnbc.msn.com/id/27368176/> (accessed October 28, 2011).

Google Lunar X PRIZE. <http://www.googlelunarxprize.org/> (accessed October 28, 2011).

Masten Space Systems. <http://masten-space.com/> (accessed October 7, 2011).

NASA and X Prize Announce Winners of Lunar Lander Challenge. National Aeronautics and Space Administration. <http://www.nasa.gov/home/hqnews/2009/nov/HQ_09-258-Lunar_Lander.html> (accessed October 28, 2011).

2009 Northrop Grumman Lunar Lander Challenge. X PRIZE Foundation. <http://space.xprize.org/lunar-lander-challenge> (accessed October 28, 2011).

Virgin Galactic. <http://www.virgingalactic.com/> (accessed October 28, 2011).

Virgin Galactic Unveils Suborbital Spaceliner Design. Space.com. <http://www.space.com/news/080123-virgingalactic-ss2-design.html> (accessed October 27, 2011).

X PRIZE Foundation. <http://www.xprize.org/> (accessed October 28, 2011).

Glossary

ablation removal of the outer layers of an object by erosion, melting, or vaporization

abort-to-orbit procedure emergency procedure planned for the space shuttle and other spacecraft if the spacecraft reaches a lower than planned orbit

accretion the growth of a star or planet through the accumulation of material from a companion star or the surrounding interstellar matter

adaptive optics the use of computers to adjust the shape of a telescope's optical system to compensate for gravity or temperature variations

aeroballistic describes the combined aerodynamics and ballistics of an object, such as a spacecraft, in flight

aerobraking the technique of using a planet's atmosphere to slow down an incoming spacecraft; its use requires the spacecraft to have a heat shield, because the friction that slows the craft is turned into intense heat

aerodynamic heating heating of the exterior skin of a spacecraft, aircraft, or other object moving at high speed through the atmosphere

Agena a multipurpose rocket designed to perform ascent, precision orbit injection, and missions from low Earth orbit (LEO) to interplanetary space; also served as a docking target for the Gemini spacecraft

algae simple photosynthetic organisms, often aquatic

alpha proton x-ray spectrometer analytical instrument that bombards a sample with alpha particles (consisting of two protons and two neutrons); the x rays are generated through the interaction of the alpha particles and the sample

altimeter an instrument designed to measure altitude above sea level

amplitude the height of a wave or other oscillation; the range or extent of a process or phenomenon

angular momentum the angular equivalent of linear momentum; the product of angular velocity and moment of inertia (moment of inertia = mass × radius²)

angular velocity the rotational speed of an object, usually measured in radians per second

anisotropy a quantity that is different when measured in different directions or along different axes

annular ring-like

anomalies phenomena that are different from what is expected

anorthosite a light-colored rock composed mainly of the mineral feldspar (an aluminum silicate); commonly occurs in the crusts of Earth and the Moon

antimatter matter composed of antiparticles, such as positrons and antiprotons, as opposed to normal matter composed of particles, such as electrons and protons

antipodal at the opposite pole; two points on a planet that are diametrically opposite

aperture an opening, door, or hatch

aphelion the point in an object's orbit that is farthest from the Sun

Apollo American program to land men on the Moon; *Apollo 11, Apollo 12, Apollo 14, Apollo 15, Apollo 16,* and *Apollo 17* delivered twelve men to the lunar surface (two per mission) between 1969 and 1972 and returned them safely back to Earth

asthenosphere the weaker portion of a planet's interior just below the rocky crust

astronomical unit the average distance between Earth and the Sun (152 million kilometers [93 million miles])

atmospheric probe a separate piece of a spacecraft that is launched from it and separately enters the atmosphere of a planet or other celestial body on a one-way trip, making measurements until it hits a surface, burns up, or otherwise ends its mission

atmospheric refraction the bending of sunlight or other light caused by the varying optical density of the atmosphere

atomic nucleus the protons and neutrons that make up the core of an atom

atrophy condition that involves withering, shrinking, or wasting away

auroras atmospheric phenomena consisting of glowing bands or sheets of light in the sky caused by high-speed charged particles striking atoms in Earth's upper atmosphere

ballistic the path of an object in unpowered flight; the path of a spacecraft after the engines have shut down

basalt a dark, volcanic rock with abundant iron and magnesium and relatively low silica common on all of the terrestrial planets

base-load the minimum amount of energy needed for a power grid

berth space the human accommodations needed by a space station, cargo ship, or other vessel

Big Bang name given by astronomers to the event marking the beginning of the universe when all matter and energy came into being

bioregenerative referring to a life support system in which biological processes are used; physiochemical and/or nonregenerative processes may also be used

bipolar outflow jets of material (gas and dust) flowing away from a central object (e.g., a protostar) in opposite directions

black holes objects so massive for their size that their gravitational pull prevents everything, even light, from escaping

bone mineral density the mass of minerals, mostly calcium, in a given volume of bone

breccia mixed rock composed of fragments of different rock types; formed by the shock and heat of meteorite impacts

bright rays lines of lighter material visible on the surface of a body and caused by relatively recent impacts

brown dwarf star-like object less massive than 0.08 times the mass of the Sun, which cannot undergo a thermonuclear process to generate its own luminosity

caldera the bowl-shaped crater at the top of a volcano caused by the collapse of the central part of the volcano

capsule a closed compartment designed to hold and protect humans, instruments, and/or equipment, as in a spacecraft

carbon-fiber composites combinations of carbon fibers with other materials such as resins or ceramics; carbon fiber composites are strong and lightweight

carbonaceous meteorites the rarest kind of meteorites, they contain a high percentage of carbon and carbon-rich compounds

carbonate a class of minerals, such as chalk and limestone, formed by carbon dioxide reacting in water

cartographic relating to the making of maps

Cassini-Huygens mission a robotic spacecraft mission to the planet Saturn that arrived in July 2004 and dropped its *Huygens* probe into Titan's atmosphere while the *Cassini* spacecraft studied the planet

catalyst a chemical compound that accelerates a chemical reaction without itself being used up; any process that acts to accelerate change in a system

cell culture a means of growing mammalian (including human) cells in the research laboratory under defined experimental conditions

cellular array the three-dimensional placement of cells within a tissue

centrifugal directed away from the center through spinning

centrifuge a device that uses centrifugal force caused by spinning to simulate gravity

Cepheid variables a class of variable stars whose luminosity is related to their period. Their periods can range from a few hours to about 100 days and the longer the period, the brighter the star

Čerenkov light light emitted by a charged particle moving through a medium, such as air or water, at a velocity greater than the phase velocity of light in that medium; usually a faint, eerie, bluish, optical glow

chassis frame on which a vehicle is constructed

chondrite meteorites a type of meteorite that contains spherical clumps of loosely consolidated minerals

cinder field an area dominated by volcanic rock, especially the cinders ejected from explosive volcanoes

Clarke orbit geostationary orbit; named after science fiction writer Arthur C. Clarke (1917–2008), who first realized the usefulness of this type of orbit for communication and weather satellites

coagulate to cause to come together into a coherent mass

commercial astronaut a person trained to go into space as part of a privately funded operation

communications infrastructure the physical structures that support a network of telephone, Internet, mobile phones, and other communication systems

convection currents mechanism by which thermal energy moves because its density differs from that of surrounding material. Convection current is the movement pattern of thermal energy transferring within a medium

convection the movement of heated fluid caused by a variation in density; hot fluid rises while cool fluid sinks

convective processes processes that are driven by the movement of heated fluids resulting from a variation in density

coronal holes large, dark holes seen when the Sun is viewed in x-ray or ultraviolet wavelengths; solar wind emanates from the coronal holes

coronal mass ejections large quantities of solar plasma and magnetic field launched from the Sun into space

cosmic microwave background ubiquitous, diffuse, nearly-uniform, thermal radiation created during the earliest hot phases of the universe

cosmic radiation high energy particles that enter Earth's atmosphere from outer space causing cascades of mesons and other particles

cosmology the scientific investigation of the universe: its structure, origin, evolution, and ultimate fate

cover glass a sheet of glass used to cover the solid state device in a solar cell

crash-lander or hard-lander; a spacecraft that collides with a planet or other celestial body, making no—or little—attempt to slow down; after collision, the spacecraft ceases to function because of the (intentional) catastrophic failure

crawler transporter large, tracked vehicle used to move the assembled Apollo/Saturn from the VAB to the launch pad

cryogenic related to extremely low temperatures; the temperature of liquid nitrogen or lower

crystal lattice the arrangement of atoms inside a crystal

crystallography the study of the internal structure of crystals

dark matter matter that interacts with ordinary matter by gravity but does not emit electromagnetic radiation; its composition is unknown

detruents microorganisms that act as decomposers in a controlled environmental life support system

diffraction the bending and subsequent spreading out of waves around an obstruction in their path

diffuse spread out; not concentrated

DNA deoxyribonucleic acid; the molecule used by all living organisms (and some viruses) on Earth to transmit genetic information

docking system mechanical and electronic devices that work jointly to bring together and physically link two spacecraft in space

downlink the radio dish and receiver through which a satellite or spacecraft transmits information back to Earth

drag a force that opposes the motion of an aircraft or spacecraft through the atmosphere

dunites rock type composed almost entirely of the mineral olivine, crystallized from magma beneath the Moon's surface

eccentric the term that describes how oval the orbit of an astronomical body is

ecliptic the plane of Earth's orbit

ejecta the pieces of material thrown off by a star when it explodes; or, material thrown out of an impact crater during its formation

electrolytes a substance that when dissolved in water creates an electrically conducting solution

electromagnetic spectrum the entire range of wavelengths of electromagnetic radiation

electron volts units of energy equal to the energy gained by an electron when it passes through a potential difference of 1 volt in a vacuum

electron a negatively charged subatomic particle

elliptical having an oval shape

encapsulation enclosing within a capsule

endocrine system in the body that creates and secretes substances called hormones into the blood

equatorial orbit an orbit parallel to a body's geographic equator

Europa one of the large satellites of Jupiter

eV an electron volt is the energy gained by an electron when moved across a potential of one volt. Ordinary molecules, such as air, have an energy of about 3×10^{-2} eV

event horizon the imaginary spherical shell surrounding a black hole that marks the boundary where no light nor matter can escape

expendable launch vehicles launch vehicles, such as a rocket, not intended to be reused

extrasolar planets planets orbiting stars other than the Sun

extravehicular activity a space walk conducted outside a spacecraft cabin, with the crew member protected from the environment by a pressurized space suit

extremophiles microorganisms that can survive in extreme environments such as high salinity or near boiling water

fairing a structure designed to provide low aerodynamic drag for an aircraft or spacecraft in flight

fault a fracture in rock in the upper crust of a planet along which there has been movement

feldspathic rock containing a high proportion of the mineral feldspar

fiber-optic cable a thin strand of ultrapure glass that carries information in the form of light (radiation), with the light turned on and off rapidly to represent the information sent

fission act of splitting a heavy atomic nucleus into two lighter ones, releasing tremendous energy

flares intense, sudden releases of energy

flyby flight path that takes the spacecraft close enough to a planet to obtain good observations; the spacecraft then continues on a path away from the planet but may make multiple passes

fracture any break in rock, from small "joints" that divide rocks into planar blocks (such as that seen in road cuts) to vast breaks in the crusts of unspecified movement

free radical a molecule with a high degree of chemical reactivity due to the presence of an unpaired electron

free fall the motion of a body acted on by no forces other than gravity, usually in orbit around Earth or another celestial body

frequencies the number of oscillations or vibrations per second of an electromagnetic wave or any wave

fusion the act of releasing nuclear energy by combining lighter elements such as hydrogen into heavier elements

G force the measure of acceleration (due to gravity at Earth's surface) of a body relative to free-fall (a body's g-force in free-fall is equal to one; an astronaut, for instance, accelerating five times that of free-fall has a g-force equal to five)

galaxy cluster a system typically consisting of hundreds to thousands of galaxies bound together by gravity

galaxy a system of as many as hundreds of billions, even trillions, of stars that have a common gravitational attraction

Galileo mission successful robot exploration of the outer solar system; this mission used gravity assists from Venus and Earth to reach Jupiter, where it dropped a probe into the atmosphere and studied the planet for nearly seven years

gamma rays a form of radiation with a shorter wavelength and more energy than x rays

Gemini the second series of American-piloted spacecraft, crewed by two astronauts; the Gemini missions were rehearsals of the space-flight techniques needed to go to the Moon

general relativity a branch of science first described by Albert Einstein showing the relationship between gravity and acceleration

geodetic survey determination of the exact position of points on Earth's surface and measurement of the size and shape of Earth and of Earth's gravitational and magnetic fields

geomagnetic field Earth's magnetic field; under the influence of solar wind, the magnetic field is compressed in the Sunward direction and stretched out in the downwind direction, creating the magnetosphere, a complex, teardrop-shaped cavity around Earth

geospatial relating to measurement of Earth's surface as well as positions on its surface

geostationary orbit a specific altitude of an equatorial orbit where the time required to circle the planet matches the time it takes the planet to rotate on its axis. An object in geostationary orbit will always remain over the same geographic location on the equator of the planet it orbits

geosynchronous orbit the altitude above the surface of the Earth (22,300 miles up) where an orbiting object will be traveling around the Earth at the same speed that the Earth spins. This results in the orbiting object remaining over the same region of the Earth at all times, although it will progress north and south along a straight line (as seen from Earth) over the course of a day, unless its orbit is directly over the equator, which results in a geostationary orbit

geosynchronous remaining fixed (as seen from Earth) in an orbit 35,786 kilometers (22,300 miles) above Earth's surface

gimbal motors motors that direct the nozzle of a rocket engine to provide steering

global positioning systems a system of satellites and receivers that provide direct determination of the geographical location of the receiver

globular clusters roughly spherical collections of hundreds of thousands of old stars found in galactic haloes

grand unified theory grand unified theory (GUT) states that, at a high enough energy level (about 10^{25} eV), the electromagnetic force, strong force, and weak force all merge into a single force

gravitational assist the technique of flying by a planet to use its energy to "catapult" a spacecraft on its way—this saves fuel and thus mass and cost of a mission; gravitational assists typically make the total mission duration longer, but they also make things possible that otherwise would not be possible

gravitational contraction the collapse of a cloud of gas and dust due to the mutual gravitational attraction of the parts of the cloud; a possible source of excess heat radiated by some Jovian planets

gravitational force the force of attraction between all masses, such as the attraction of Earth's mass for objects of mass in its vicinity

gravitational lenses two or more images of a distant object formed by the bending of light around an intervening massive object

gravity gradient the difference in the acceleration of gravity at different points on Earth and at different distances from Earth

gyroscope a spinning disk mounted so that its axis can turn freely and maintain a constant orientation in space

hard-lander spacecraft that collides with the planet or satellite, making no attempt to slow its descent; also called crash-landers

heliosphere the volume of space extending outward from the Sun that is dominated by solar wind; it ends where the solar wind transitions into the interstellar medium, somewhere between 40 and 100 astronomical units from the Sun

high-power klystron tubes a type of electron tube used to generate high frequency electromagnetic waves

hybrid rocket engine a rocket engine utilizing propellants that are in two different states of matter; one state is solid, while the other state is either gas or liquid

hydrazine a dangerous and corrosive compound of nitrogen and hydrogen commonly used in high powered rockets and jet engines

hydrothermal relating to high temperature water

hyperbaric chamber compartment where air pressure can be carefully controlled; used to gradually acclimate divers, astronauts, and others to changes in pressure and air composition

hypergolic fuels and oxidizers that ignite on contact with each other and need no ignition source

Imbrium Basin impact largest and latest of the giant impact events that formed the mare-filled basins on the lunar near side

impact craters bowl-shaped depressions on the surfaces of planets or satellites that result from the impact of space debris moving at high speeds

impact winter the period following a large asteroidal or cometary impact when the Sun is dimmed by stratospheric dust and the climate becomes cold worldwide

impact-melt molten material produced by the shock and heat transfer from an impacting asteroid or meteorite

in situ in the natural or original location

incandescence glowing due to high temperature

infrared radiation radiation whose wavelength is slightly longer than the wavelength of visible light

infrared portion of the electromagnetic spectrum with waves slightly longer than visible light

infrastructure the physical structures, such as roads and bridges, necessary to the functioning of a complex system

intercrater plains the oldest plains on Mercury that occur in the highlands and that formed during the period of heavy meteoroid bombardment

interferometers devices that use two or more telescopes to observe the same object at the same time in the same wavelength to increase angular resolution

interplanetary trajectories the solar orbits followed by spacecraft moving from one planet in the solar system to another

interstellar medium the gas and dust found in the space between the stars

interstellar between the stars

ion propulsion a propulsion system that uses charged particles accelerated by electric fields to provide thrust

ionosphere a charged particle region of several layers in the upper atmosphere created by radiation interacting with upper atmospheric gases

isotopic ratios the naturally occurring ratios between different isotopes of an element

James Webb Space Telescope (JWST) the telescope tentatively scheduled to be launched in 2018 that will partially replace the Hubble Space Telescope

jettison to eject, throw overboard, or get rid of

Jovian relating to the planet Jupiter

kinetic energy the energy an object has due to its motion

lander a spacecraft designed to travel from outer space onto the surface of a planet or other celestial body

laser-pulsing firing periodic pulses from a powerful laser at a surface and measuring the length of time for return in order to determine topography

light year the distance that light in a vacuum would travel in one year, or about 9.5 trillion kilometers (5.9 trillion miles)

lithosphere the rocky outer crust of a body

littoral the region along a coast or beach between high and low tides

lobate scarps a long sinuous cliff

low Earth orbit an orbit between 300 and 800 kilometers (185 and 500 miles) above Earth's surface

lunar maria the large, dark, lava-filled impact basins on the Moon thought by early astronomers to resemble seas

Lunar Orbiter a series of five unmanned missions in 1966 and 1967 that photographed much of the Moon at medium to high resolution from orbit

macromolecules large molecules such as proteins or DNA containing thousands or millions of individual atoms

magnetohydrodynamic waves a low frequency oscillation in a plasma in the presence of a magnetic field

magnetometer an instrument used to measure the strength and direction of a magnetic field

magnetosphere the magnetic cavity that surrounds Earth or any other planet, or other celestial body, with a magnetic field. It is formed by the interaction of the stellar wind with the body's magnetic field

malady a disorder or disease of the body

many-bodied problem in celestial mechanics, the problem of finding solutions to the equations for more than two orbiting bodies

mare dark-colored plains of solidified lava that mainly fill the large impact basins and other low-lying regions on the Moon

Mercury the first American piloted spacecraft, which carried a single astronaut into space; six Mercury missions took place between 1961 and 1963

mesons any of a family of subatomic particle that have masses between electrons and protons and that respond to the strong nuclear force; produced in the upper atmosphere by cosmic rays

meteor the physical manifestation of a meteoroid interacting with Earth's atmosphere; this includes visible light and radio frequency generation, and an ionized trail from which radar signals can be reflected. Also called a "shooting star"

meteorites any part of a meteoroid that survives passage through Earth's atmosphere

meteoroid a piece of interplanetary material smaller than an asteroid or comet

meteorology satellites satellites designed to take measurements of the atmosphere for determining weather and climate change

meteorology the study of atmospheric phenomena or weather

microgravity the condition experienced in free fall as a spacecraft orbits Earth or another body; commonly called weightlessness; only very small forces are perceived in free fall, on the order of one-millionth the force of gravity on Earth's surface

micrometeoroid flux the total mass of micrometeoroids falling into an atmosphere or on a surface per unit of time

micrometeoroid any meteoroid ranging in size from a speck of dust to a pebble

minerals crystalline arrangements of atoms and molecules of specified proportions that make up rocks

multi-bandgap photovoltaic cells photovoltaic cells designed to respond to several different wavelengths of electromagnetic radiation

muons the decay product of the mesons produced by cosmic rays; muons are about 100 times more massive than electrons but are still considered leptons that do not respond to the strong nuclear force

nebulae clouds of interstellar gas and/or dust

neutron star the dense core of matter composed almost entirely of neutrons that remain after a supernova explosion has ended the life of a massive star

neutron a subatomic particle with no electrical charge

nuclear black holes black holes that are in the centers of galaxies; they range in mass from a thousand to a billion times the mass of the Sun

nuclear fusion the combining of low-mass atoms to create heavier ones; the heavier atom's mass is slightly less than the sum of the mass of its constituents, with the remaining mass converted to energy

nucleon a proton or a neutron; one of the two particles found in a nucleus

occultations a phenomena that occurs when one astronomical object passes in front of another

orbit the circular or elliptical path of an object around a much larger object, governed by the gravitational field of the larger object

orbital velocity velocity at which an object needs to travel so that its flight path matches the curve of the planet it is circling; approximately eight kilometers (five miles) per second for low-altitude orbit around Earth

orbiter spacecraft that uses engines and/or aerobraking, and whose main function is to orbit about a planet or natural satellite

orthogonal composed of right angles or relating to right angles

oscillation energy that varies between alternate extremes with a definable period

osteoporosis the loss of bone density; can occur after extended stays in space

oxidizer a substance mixed with fuel to provide the oxygen needed for combustion

Paleozoic relating to the first appearance of animal life on Earth

parabolic trajectory trajectory followed by an object with velocity equal to escape velocity

parking orbit placing a spacecraft temporarily into Earth orbit, with the engines shut down, until it has been checked out or is in the correct location for the main burn that sends it away from Earth

payload bay the area in the shuttle or other spacecraft designed to carry cargo

payload fairing structure surrounding a payload; it is designed to reduce drag

payload operations experiments or procedures involving cargo or "payload" carried into orbit

payload specialists scientists or engineers selected by a company or a government employer for their expertise in conducting a specific experiment or commercial venture on a space shuttle mission

payload any cargo launched aboard a rocket that is destined for space, including communications satellites or modules, supplies, equipment, and astronauts; it does not include the vehicle used to move the cargo nor the propellant that powers the vehicle

perihelion the point in an object's orbit that is closest to the Sun

period of heavy meteoroid bombardment the earliest period in solar system history (more than 3.8 billion years ago) when the rate of meteoroid impact was very high compared to the present

perturbations term used in orbital mechanics to refer to changes in orbits due to "perturbing" forces, such as gravity

phased array a radar antenna design that allows rapid scanning of an area without the need to move the antenna; a computer controls the phase of each dipole in the antenna array

photometer instrument to measure intensity of light

photosynthesis a process performed by plants and algae whereby light is transformed into energy and sugars

photovoltaic pertaining to the direct generation of electricity from electromagnetic radiation (light)

plagioclase most common mineral of the light-colored lunar highlands

planetesimals objects in the early solar system that were the size of large asteroids or small moons, large enough to begin to gravitationally influence each other

point of presence an access point to the Internet with a unique Internet Protocol (IP) address; Internet service providers like AOL generally have multiple POPs on the Internet

polar orbits orbits that carry a satellite over the poles of a planet

porous allowing the passage of a fluid or gas through holes or passages in the substance

power law energy spectrum spectrum in which the distribution of energies appears to follow a power law

primary the body (planet) about which a satellite orbits

primordial swamp warm, wet conditions postulated to have occurred early in Earth's history as life was beginning to develop

procurement the process of obtaining

progenitor star the star that existed before a dramatic change, such as a supernova, occurred

prograde having the same general sense of motion or rotation as the rest of the solar system, that is, counterclockwise as seen from above Earth's north pole

prominences inactive "clouds" of solar material held above the solar surface by magnetic fields

proton a positively charged subatomic particle

pseudoscience a system of theories that assumes the form of science but fails to give reproducible results under conditions of controlled experiments

pyrotechnics fireworks display; the art of building fireworks

quasars luminous objects that appear star-like but are highly red-shifted and radiate more energy than an entire ordinary galaxy; likely powered by black holes in the centers of distant galaxies

quiescent inactive

radar altimetry using radar signals bounced off the surface of a planet to map its variations in elevation

radar images images made with radar illumination instead of visible light that show differences in radar brightness of the surface material or differences in brightness associated with surface slopes

radar a technique for detecting distant objects by emitting a pulse of electromagnetic radiation, usually microwaves or radio waves, and then recording echoes of the pulse off the distant objects

radiation belts two wide bands of charged particles trapped in a planet's magnetic field

radio lobes active galaxies show two regions of radio emission above and below the plane of the galaxy, and are thought to originate from powerful jets being emitted from the accretion disk surrounding the massive black hole at the center of active galaxies

radiogenic isotope techniques use of the ratio between various isotopes produced by radioactive decay to determine age or place of origin of an object in geology, archaeology, and other areas

radioisotope thermoelectric generator device using solid-state electronics and the heat produced by radioactive decay to generate electricity

radioisotope a naturally or artificially produced radioactive isotope of an element

range safety destruct systems system of procedures and equipment designed to safely abort a mission when a spacecraft malfunctioned, and destroy the rocket in such a way as to create no risk of injury or property damage

Ranger series of spacecraft sent to the Moon to investigate lunar landing sites; designed to hard-land on the lunar surface after sending back television pictures of the lunar surface; *Ranger 7, Ranger 8,* and *Ranger 9* (1964–1965) returned data

rarefaction decreased pressure and density in a material caused by the passage of a wave

reconnaissance a survey or preliminary exploration of a region of interest

regolith upper few meters of a body's surface, composed of inorganic matter, such as unconsolidated rocks and fine soil

relative zero velocity two objects having the same speed and direction of movement, usually so that spacecraft can rendezvous

relativistic time dilation effect predicted by the theory of relativity that causes clocks on objects in strong gravitational fields or moving near the speed of light to run slower when viewed by a stationary observer

remote manipulator system a system, such as the external Canadarm2 on the International Space Station, designed to be operated from a remote location inside the space station

remote sensing the act of observing from orbit what may be seen or sensed below on Earth or another planetary body

retrograde having the opposite general sense of motion or rotation as the rest of the solar system, clockwise as seen from above Earth's north pole

reusable launch vehicles launch vehicles, such as the space shuttle orbiter, designed to be recovered and reused many times

rift valley a linear depression in the surface, several hundred to thousand kilometers long, along which part of the surface has been stretched, faulted, and dropped down along many normal faults

rille lava channels in regions of maria, typically beginning at a volcanic vent and extending downslope into a smooth mare surface

rocket vehicle or device that is especially designed to travel through space, and is propelled by one or more engines

"rocky" planets nickname given to inner or solid-surface planets of the solar system, including Mercury, Venus, Mars, and Earth

rover vehicle used to move about on a surface

secondary crater crater formed by the impact of blocks of rock blasted out of the initial crater formed by an asteroid or large meteorite

sedentary lifestyle a lifestyle characterized by little movement or exercise

sedimentation process of depositing sediments, which results in a thick accumulation of rock debris eroded from high areas and deposited in low areas

semiconductor one of the groups of elements with properties intermediate between the metals and nonmetals

semimajor axis one half of the major axis of an ellipse, equal to the average distance of a planet from the Sun

shepherding small satellites exerting their gravitational influence to cause or maintain structure in the rings of the outer planets

shielding providing protection for humans and electronic equipment from cosmic rays, energetic particles from the Sun, and other radioactive materials

sidereal period the amount of time it takes for a celestial object to make one full orbit around the Sun, in relation to the distant "fixed" stars

sine wave a wave whose amplitude smoothly varies with time; a wave form that can be mathematically described by a sine function

smooth plains the youngest plains on Mercury with a relatively low impact crater abundance

soft-lander spacecraft that uses braking by engines or other techniques (e.g., parachutes, airbags) such that its landing is gentle enough that the spacecraft and its instruments are not

damaged, and observations at the surface can be made

solar arrays groups of solar cells or other solar power collectors arranged to capture energy from the Sun and use it to generate electrical power

solar corona the thin outer atmosphere of the Sun that gradually transitions into the solar wind

solar flares explosions on the Sun that release bursts of electromagnetic radiation, such as visible light, ultraviolet waves, and x rays, along with high speed protons and other particles

solar nebula the cloud of gas and dust out of which the solar system formed

solar prominence relatively cool material with temperatures typical of the solar photosphere or chromosphere suspended in the corona above the visible surface layers

solar radiation total energy of any wavelength and all charged particles emitted by the Sun

solar wind a continuous, but varying, stream of charged particles (mostly electrons and protons) generated by the Sun; it establishes and affects the interplanetary magnetic field; it also deforms the magnetic field about Earth and sends particles streaming toward Earth at its poles

sounding rocket a vehicle designed to fly straight up and then parachute back to Earth, usually designed to take measurements of the upper atmosphere

space station large orbital outpost equipped to support a human crew and designed to remain in orbit for an extended period; to date, only Earth-orbiting space stations have been launched

space-time in relativity, the four-dimensional space through which objects move and in which events happen

spacewalking moving around outside a spaceship or space station, also known as extravehicular activity

special theory of relativity the fundamental idea of Einstein's theories, which demonstrated that measurements of certain physical quantities such as mass, length, and time depended on the relative motion of the object and observer

spectra representations of the brightness of objects as a function of the wavelength of the emitted radiation

spectral lines the unique pattern of radiation at discrete wavelengths that many materials produce

spectrograph an instrument that can permanently record a spectra

spectrographic studies studies of the nature of matter and composition of substances by examining the radiation (light) they emit

spectrometers instruments with a scale for measuring the wavelength of light

spherules tiny glass spheres found in and among lunar rocks

spot beam technology narrow, pencil-like satellite beam that focuses highly radiated energy on a limited area of Earth's surface (about 160 to 800 kilometers [100 to 500 miles] in diameter) using steerable or directed antennas

stratigraphy the study of rock layers known as strata, especially the age and distribution of various kinds of sedimentary rocks

stratosphere a middle portion of a planet's atmosphere above the tropopause (the highest place where convection and "weather" occurs)

subduction the process by which one edge of a crustal plate is forced to move under another plate

sublimate to pass directly from a solid phase to a gas phase

suborbital trajectory the trajectory of a rocket or ballistic missile that has insufficient energy to reach orbit

subsolar point the point on a planet or other solar body that receives direct rays from the Sun

sunspots dark, cooler areas on the solar surface consisting of transient, concentrated magnetic fields

supernova an explosion ending the life of a massive star

supernovae ejecta the mix of gas enriched by heavy metals that is launched into space by a supernova explosion

superstrings supersymmetric strings are tiny, one-dimensional objects that are about 10^{-33} centimeters long, in a ten-dimensional space-time. Their different vibration modes and shapes account for the elementary particles scientists see in four-dimensional spacetime

Surveyor a series of spacecraft designed to soft-land robotic laboratories to analyze and photograph the lunar surface; *Surveyor 1, Surveyor 3,* and *Surveyor 5* through *Surveyor 7* landed between May 1966 and January 1968

synchrotron radiation the radiation from charged particles, such as electrons, moving at almost the speed of light inside giant magnetic accelerators of particles, called synchrotrons, on Earth, or from electrons moving at almost the speed of light from magnetic fields in space

synodic period the amount of time it takes for a celestial object to reappear at the same point in relation to two other bodies, as for instance when the Moon relative to the Sun as observed from Earth returns to the same illumination phase; the Moon's synodic period is greater than its sidereal period due to Earth's orbit around the Sun

synthesis the act of combining different things so as to form new and different products or ideas

technology transfer the acquisition by one country or firm of the capability to develop a particular technology through its interactions with the existing technological capability of another country or firm, rather than through its own research efforts

tectonism process of deformation in a planetary surface as a result of geological forces acting on the crust; includes faulting, folding, uplift, and down-warping of the surface and crust

telecommunications infrastructure the physical structures that support a network of telephone, Internet, mobile phones, and other communications systems

telescience the act of operation and monitoring of research equipment located in space by a scientist or engineer from their offices or laboratories on Earth

terrestrial planet one of the four rocky planets with high density orbiting close to the Sun, consisting of Mercury, Venus, Earth, and Mars

thermostabilized designed to maintain a constant temperature

thrust fault a fault where the block on one side of the fault plane has been thrust up and over the opposite block by horizontal compressive forces

torque measure of a force applied to a tool handle multiplied by the distance to the center of rotation

toxicological related to the study of the nature and effects on humans of poisons and the treatment of victims of poisoning

trajectories paths followed through space by missiles and spacecraft moving under the influence of gravity

transonic barrier the aerodynamic behavior of an aircraft changes dramatically as it moves near the speed of sound, and for early pioneers of transonic flight such changes were deemed dangerous, leading some to hypothesize there was a "sound barrier" where drag became infinite

transpiration process whereby water evaporates from the surface of leaves, allowing the plant to lose heat and to draw water up through the roots

transponder bandwidth-specific transmitter-receiver units

troctolite rock type composed of the minerals plagioclase and olivine, crystallized from magma

Tycho event the impact of a large meteoroid into the lunar surface as recently as 100 million years ago, leaving a distinct set of bright rays across the lunar surface including a ray through the *Apollo 17* landing site

ultramafic lavas dark, heavy lavas with a high percentage of magnesium and iron; usually found as boulders mixed in other lava rocks

ultraviolet radiation electromagnetic radiation with a shorter wavelength and higher energy than visible light

ultraviolet the portion of the electromagnetic spectrum with shorter wavelengths than visible light but longer wavelengths than x rays

uncompressed density the lower density a planet would have if it did not have the force of gravity compressing it

Universal time current time in Greenwich, England, which is recognized as the standard time that Earth's time zones are based on

vacuum an idealized region wherein air and all other molecules and atoms of matter have been removed, though such a complete absence of particles is never actually observed; within interstellar space density estimates range from about a hundred to a thousand atoms per cubic meter

Van Allen radiation belts   two belts of high energy charged particles captured from the solar wind by Earth's magnetic field

variable star   a star whose light output varies over time

vector sum   sum of two vector quantities taking both size and direction into consideration

velocity   speed and direction of a moving object; a vector quantity

visible spectrum   the part of the electromagnetic spectrum with wavelengths between 400 and 700 nanometers (where one nanometer is equal to one billionth of a meter); the part of the electromagnetic spectrum to which human eyes are sensitive

wavelength   the distance from crest to crest on a wave at an instant in time

x ray   form of high-energy radiation just beyond the ultraviolet portion of the electromagnetic spectrum

x-ray diffraction analysis   a method to determine the three-dimensional structure of molecules

Directory of Space Organizations

A

American Association of Amateur Astronomers (AAAA)
P.O. Box 7981
Dallas, TX 75209-0981
U.S.A.
Web site: http://www.astromax.com
Email: aaaa@astromax.com

American Astronautical Society (AAS)
6352 Rolling Mill Place, Suite 102
Springfield, VA 22152-0020
U.S.A.
Telephone: 703-866-0020
Fax: 703-866-3526
Web site: http://www.astronautical.org
Email: aas@astronautical.org

American Geophysical Union (AGU)
2000 Florida Avenue, N.W.
Washington, D.C. 20009-1277
U.S.A.
Telephone: 202-462-6900
Fax: 202-328-0566
Toll free: 800-966-2481 (North America only)
Web site: http://www.agu.org
Email: service@aqu.org

American Institute of Aeronautics and Astronautics (AIAA)
1801 Alexander Bell Drive,
Suite 500
Reston, VA 20191-4344
U.S.A.
Telephone: 703-264-7500
Fax: 703-264-7551
Toll free: 800-639-2422
Web site: http://www.aiaa.org

American Meteor Society (AMS)
121 EEE (Department of Astronomy and Astrophysics, Pennsylvania State University)
University Park, PA 16802
U.S.A.
Web site: http://www.amsmeteors.org

Association of Lunar and Planetary Observers (ALPO)
P.O. Box 13456
Springfield, IL 62791-3456
U.S.A.
Web site: http://www.alpo-astronomy.org

Astronaut Scholarship Foundation (ASF)
6225 Vectorspace Boulevard
Titusville, FL 32780
U.S.A.
Telephone: 321-455-7011
Fax: 321-264-9176
Web site: http://www.astronaut scholarship.org
Email: Linn@AstronautScholarship .org

Astronomical Society of the Pacific (ASP)
390 Ashton Avenue
San Francisco, CA 94112
U.S.A.
Telephone: 415-337-1100
Fax: 415-337-5205
Web site: http://www.astrosociety.org

Aviation and Space Education (AVSED), Federal Aviation Administration
800 Independence Avenue, S.W.
Washington, D.C. 20591
U.S.A.
Telephone: 781-238-7027
Web site: http://www.faa.gov/education

B

Brazilian Space Agency (AEB)
SPO, Area 5, Quadra 3, Bloco A
Brasilia, DF 70610 200
Brazil
Telephone: 61-3411-5159
Web site: http://www.aeb.gov.br

The British Planetary Society (BPS)
27-29 South Lambeth Road
London, SW8 1SZ
United Kingdom
Telephone: 20-7735-3160
Fax: 20-7587-5118
Web site: http://www.bis-space.com

Buzz Aldrin's ShareSpace Foundation
11901 Santa Monica Boulevard,
Suite 496
Los Angeles, CA 90025
U.S.A.
Web site: http://buzzaldrin.com/space-vision/sharespace-foundation

C

Canadian Space Agency (CSA)
6767 Route de l'Aéroport
Saint-Hubert, QC J3Y 8Y9
Canada
Telephone: 450-926-4800
Fax: 450-926-4352
Web site: http://www.asc-csa.gc.ca/eng/default.asp
Email: promo@asc-csa.gc.ca

Canadian Space Society (CSS)
65 Carl Hall Road, Box 1
(c/o Canadian Air and Space Museum)
Toronto, ON M3K 2E1
Canada
Web site: http://css.ca

Challenger Center for Space Science Education
300 North Lee Street, Suite 301
Alexandria, VA 22314
U.S.A.
Telephone: 703-683-9741
Web site: http://www.challenger.org
Email: info@challenger.org

China National Space Administration (CNSA)
Web site: http://www.cnsa.gov.cn/
n615709/cindex.html

E

European Space Agency (ESA)
8-10 rue, Mario Nikis
Paris, 75738 Cedex 15
France
Telephone: 5369-7654
Fax: 5369-7560
Web site: http://www.esa.int
Email: ContactESA@esa.int

F

Federation of Galaxy Explorers (FOGE)
6404 Ivy Lane, Suite 810
Greenbelt, MD 20770
U.S.A.
Fax: 240-764-1501
Toll free: 877-761-1266
Web site: http://www.foge.org
Email: info@foge.org

G

German Aerospace Center (DLR)
Linder Höhe
Cologne, 51147
Germany
Telephone: 2203-601-0
Fax: 2203-673-10
Web site: http://www.dlr.de/dlr/en

I

India Space Research Organisation (ISRO)
Antariksh Bhavan, New BEL Road
Bangalore, 560 231
India
Telephone: 234-152-75 or
221-722-96
Fax: 235-119-84
Web site: http://www.isro.org
Email: satish@isro.gov.in

International Astronautical Federation (IAF)
94 bis, Avenue de Suffren
Paris, 75015
France
Telephone: 4567-4260
Fax: 4273-2120
Web site: http://www.iafastro.com
Email: info@iafastro.org

International Astronomical Union (IAU)
98 bis, Boulevard Arago
Paris, 75014
France
Telephone: 4325-8358
Fax: 4325-2616
Web site: http://www.iau.org
Email: iau@iap.fr

International Space School Educational Trust (ISSET)
Carlton House, 5 Herbert Terrace
Penarth, CF64 2AH
United Kingdom
Telephone: 029-2071-0295
Web site: http://www.isset.org

Israeli Space Agency (ISA)
P.O. Box 49100 (Government Offices, Building 3, Hakirya Hamizrahit)
Jerusalem, 91490
Israel
Telephone: 5411-101
Fax: 5811-613
Web site: http://www.most.gov.il/
English/Units/Israel+Space+Agency

Italian Space Agency (ISA)
Viale Liegi, 26
Rome, 00198
Italy
Telephone: 06 8567.1
Web site: http://www.asi.it/en

J

Japan Aerospace Exploration Agency (JAXA)
Marunouchi Kitaguchi Building,
1-6-5 Marunouchi, Chiyoda-ku
Tokyo, 100-8260
Japan

Telephone: 6266-6400
Fax: 6266-6910
Web site: http://www.jaxa.jp/
index_e.html
Email: proffice@jaxa.jp

K

Korea Aerospace Research Institute (KARI)
115 Gwahangno (Yuseong)
Daejeon, Chungnam 305-333
Republic of Korea (South Korea)
Telephone: 42-860-2164
Fax: 42-860-2015
Web site: http://www.kari.re.kr/
english

M

Malaysian National Space Agency (ANGKASA)
Bangunan Komersil PjH, Tingkat
8, Lot 4C11, No 29, Persiaran
Perdana, Presint 4
Putrajaya, 62570
Malaysia
Telephone: 88888668
Fax: 88883480
Web site: http://www.angkasa
.gov.my

The Mars Society
11111 West 8th Avenue, Unit A
Lakewood, CA 80215
U.S.A.
Telephone: 303-980-0890
Web site: http://www.marssociety
.org
Email: info@marssociety.org

N

NASA Education
Web site: http://www.nasa.gov/audi
ence/forstudents/index.html
Email: education@nasa.gov

National Aeronautics and Space Administration (NASA)
300 E Street, N.W.
Washington, D.C. 20546
U.S.A.

Telephone: 202-358-0001
Fax: 202-358-4338
Web site: http://www.nasa.gov/
Email: public-inquiries@hq.nasa.gov

National Association of Rocketry (NAR)
P.O. Box 407
Marion, IA 52302
U.S.A.
Toll free: 800-262-4872
Web site: http://www.nar.org
Email: nar-hq@nar.org

National Center for Earth and Space Science Education (NCESSE)
P.O. Box 3806
Capitol Heights, MD 20791-3806
U.S.A.
Telephone: 301-395-0770
Web site: http://ncesse.org

National Center for Space Studies (CNES)
2 place, Maurice Quentin
Paris, 75039 CEDEX 01
France
Telephone: 4476-7500
Fax: 4476-7676
Web site: http://www.cnes.fr/web/
CNES-en/7114-home-cnes.php

National Space Agency of the Republic of Kazakhstan (Kazcosmos)
Web site: http://www.kazcosmos.kz/
index.php?lang=en

P

Pakistan Space and Upper Atmosphere Research Commission (SUPARCO, National Space Agency of Pakistan)
P.O. Box 8402 (Sector 28, Gulzar-e-Hijri, Off University Road)
Karachi, Sindh 75270
Pakistan
Telephone: 34690765
Fax: 34644928 or 34694941
Web site: http://www.suparco
.gov.pk
Email: am.pr@suparco.gov.pk

The Planetary Society
85 South Grand
Pasadena, CA 91105
U.S.A.
Telephone: 626-793-5100
Fax: 626-793-5528
Web site: http://www.planetary.org
Email: tps@planetary.org

R

Russian Federal Space Agency (RKA or Roscosmos)
42 Schepkina St.
Moscow, 107996 (GSP-6)
Russia
Telephone: 499-975-44-67
Fax: 495-688-9063 or
499-975-4467
Web site: http://www.roscosmos.ru/
main.php?lang=en

S

Sally Ride Science
9191 Towne Centre Drive, Suite L101
San Diego, CA 92122
U.S.A.
Fax: 858-638-1419
Toll free: 800-561-5161
Web site: https://www.sallyridesci
ence.com

SETI Institute
189 Bernardo Avenue,
Suite 100
Mountain View, CA 94043
U.S.A.
Telephone: 650-961-6633
Fax: 650-961-7099
Web site: http://www.seti.org

Smithsonian National Air and Space Museum
Independence Avenue at 6th Street, S.W.
Washington, D.C. 20560
U.S.A.
Telephone: 202-633-2214
Web site: http://www.nasm.si.edu
Email: NASM-VisitorServices@
si.edu

Space Camp
One Tranquility Base (U.S. Space and Rocket Center)
Huntsville, AL 35805
U.S.A.
Toll free: 800-637-7223
Web site: http://www.spacecamp
.com

The Space Foundation
4425 Arrowswest Drive
Colorado Springs, CO 80907
U.S.A.
Telephone: 719-576-8000
Fax: 719-576-8801
Web site: http://www.spacefounda
tion.org

Space Research Centre (SRC)
ul. Bartycka 18A
Warsaw, 00-716
Poland
Telephone: 4966-200
Fax: 8403-131
Web site: http://www.cbk
.waw.pl

Space Studies Institute (SSI)
1434 Flightline Street
Mojave, CA 93501
U.S.A.
Telephone: 661-750-2774
Web site: http://ssi.org
Email: info@ssi.org

Space Telescope Science Institute (STSI)
3700 San Martin Drive
Baltimore, MD 21218
U.S.A.
Telephone: 410-338-4700
Web site: http://www.stsci.edu
Email: help@stsci.edu

Space Tourism Society (STS)
3153 Purdue Avenue
Los Angeles, CA 90066
U.S.A.
Telephone: 310-313-6835
Fax: 310-313-0166
Web site: http://www.spacetourism
society.org
Email: jssdesign@aol.com

State Space Agency of Ukraine (SSAU)
Moskovska Str., 8
Kiev, 01010
Ukraine
Telephone: 281-62-00
Fax: 281-62-09
Web site: http://www.nkau.gov.ua/
nsau/nkau.nsf/indexE
Email: yd@nkau.gov.ua

Student Spaceflight Experiments Program (SSEP)
P.O. Box 3896 (c/o National Center for Earth and Space Science Education)
Capitol Heights, MD 20791-3806
U.S.A.
Telephone: 301-395-0770
Web site: http://ssep.ncesse.org
Email: ssep@ncesse.org

Students for the Exploration and Development of Space (SEDS)
77 Massachusetts Avenue (MIT Room W20-445)
Cambridge, MA 02139-4307
U.S.A.

Web site: http://seds.org
Email: chair@seds.org

Swedish National Space Board (SNSB)
Box 4006
Solna, 171-04
Sweden
Telephone: 627-64-80
Web site: http://www.snsb.se/en
Email: rymdstyrelsen@snsb.se

U

United Kingdom Space Agency (UKSA)
Polaris House, North Star Avenue
Swindon, Wiltshire SN2 1SZ
United Kingdom
Telephone: 020-7215-5000
Fax: 017-9341-8099
Web site: http://www.bis.gov.uk/
ukspaceagency

United Nations Office for Outer Space Affairs (UNOOSA)
Wagramerstrasse 5; Vienna
International Centre

Vienna, A-1220
Austria
Telephone: 260-60-4950
Fax: 260-60-5830
Web site: http://www.unoosa.org
Email: oosa@unvienna.org

Universities Space Research Association (USRA)
10211 Wincopin Circle,
Suite 500
Columbia, MD 21044-3432
U.S.A.
Telephone: 410-730-2656
Fax: 410-730-3496
Web site: http://www.usra.edu
Email: info@usra.edu

X

X Prize Foundation
5510 Lincoln Boulevard, Suite 100
Playa Vista, CA 90094-2034
U.S.A.
Telephone: 310-741-4880
Fax: 310-741-4974
Web site: http://www.xprize.org

Index

Overpopulation, **3:**406, **4:**64, 211, 282
Overwhelmingly Large Telescope, **2:**226
Oxygen, space suits, **3:**345, 349, 354, **4:**143, 156
Oxygen atmosphere in spacecraft, **3:**159, **263–265,** 334
  Apollo 1, **3:**14, 95–96, 154, 263
  Apollo 13, **1:**124, **3:**8, 96, 150, 311
  Soyuz 11, **3:**98
  space stations, **3:**263–264, *264*
Oxygen generation, space stations, **3:**101, 102
Oxygen production, on the Moon, **4:**63, 143, 157, 240, *241,* 289, 292
Ozone depletion, **4:**70, 222, 328

P

P4 (moon), **2:**137
Pacific Gas and Electric, **1:**120
Packet transmission (data), **4:**43, 102
Padalka, Gennady, **3:**181
Paints, innovations in, **1:**244, 245
Pakistan, **1:**192
Pal, George, **4:**28
Palapa-B2 (satellite), **1:**319, 323, **3:**222, 286
Pallene (moon), **2:**86, 300, 301
Palomar reflecting telescope, **2:**223
Parabolic maneuvers, **3:**232–233, 312, 420–421
Parallel universes
  science fiction, **4:**190, 314
  time travel theory, **4:**336
Paranel Observatory (Chile), **2:**225, 226
Parazynski, Scott F., **1:***164,* 325
Parker, Robert A., **3:**70
Parking orbit, **2:**280
Partial gravity environments, **3:**159, **4:**311
Partial Test Ban Treaty (1963), **4:**134
Particulate theory of light, **2:**76, **4:**73–74
Parts, components, and subsystems providers, **1:**17–18, 362
Patents, **1:**179, **4:**288
  habitat modules, **1:**345, **4:**343
  protection regulations, **1:**216
  Zubrin, Robert, **4:**361
Patsayev, Viktor, **3:**98, 154
PAVE PAWS radars, **3:**379
Payload for Antimatter Matter Exploration and Light-nuclei Astrophysics (PAMELA) (satellite), **4:**2

Payload Operations Control Centers, **1:**160
Payload specialists, **3:**46, **267–269**
Payloads, **3:269–272,** 331
  Getaway Special project, **3:**127–130
  nature of, launches, **4:**248
  tethers handling, **4:***332, 334*
Payloads and processing, **1:269–275,** *270*
  commercial payloads, **1:**216, 219–220, 357–358
  design and storage, **1:**271, *273,* 331–332
  launch preparation, **1:**272–274, **3:**195, 240, 397–398
  maximum capacity, **1:**197–198
  spacecraft buses, **2:**336, **3:**270
  specialist astronauts, **3:**27, 28
  See also Launch vehicles, expendable; Launch vehicles, reusable; Payload specialists
Peace, **4:**346
Pegaus rockets, **1:**12–13, 187, 299, 332, 342, **3:**292, **4:**128
Pendulum studies, **2:**106, *138,* 138–139
Pepsi Cola, **1:**7, *8*
Perfluorocarbons (PFCs), **4:**328
Periapsis, orbits, **2:**243
Perigee, orbits, **2:**243
Perihelion, orbits, **2:**243, 267
Perminov, Anatoly, **3:**165
Perseid meteor shower, **2:**192
Persian Gulf War (1990-91), **4:**182
Personal Satellite Assistant, **2:**291–292, **3:**84
Perspektivnaya Pilotiruemaya Transportnaya Sistema (spacecraft), **1:**4
Pešek, Ludek, **1:**30
Pettit, Don, **1:**178
Pharmaceutical research. *See* Drug research and design
Phenolic impregnated carbon ablator (PICA), **3:**146
Phobos (moon), **2:**89, 169, 313, *314,* 315, 316, **4:**171–172, 218, 242, 295
Phobos 2 (spacecraft and mission), **2:**172
Phobos-Grunt (spacecraft and mission), **2:**89, 177, 259, 285, **3:**164, **4:**171–172, 218
Phoebe (moon), **2:**299, 301
Phoenix Mars Lander (robotic probe), **2:**87, 88, 162, 163–164, 175–176,

258, 283, 284, **3:**84, **4:**15, 20–21, 144, 170–171, 240–241
Photoelectric effect, **2:**76, 77
Photography
  art and, **1:**29
  asteroids, **2:**282, **4:**7
  careers, **1:**82
  Hubble Space Telescope cameras, **2:**130–131, 133, 289–290
  Jupiter, **2:**282
  lunar, **2:**78, 79, *79,* 80, 164, 260–261, 282, **3:**372, 423–424
  Mars, **2:**81, *83,* 84, *161,* 162, 163, 170, 174, 282, 290, **4:**16, 168, 169, *173*
  Mercury, **2:***185–186,* 189, 282
  reconnaissance-related, **1:**11, 18, 277–279, 280, 313–314, **3:**394, **4:**176
  remote sensing, **1:**33, 88, *90,* 141–142, 178, 312, 314
  Saturn, **2:***298, 300,* 301
  Skylab, **3:**315, 317
  spaceflight, **1:**82
  telepresence, **4:**324
  Venus, **2:**81, 282
Photon pressure, **4:**139
Photons, **2:**54, 323, **4:**73–74, 190, 322
Photosynthesis, **4:**11–12, 13–14, 17, *79,* 201
Photovoltaic arrays, **1:**118, **4:**226, 268, 277–278, 279
Physical changes, bodies in space, **1:**19–20
Physical requirements for astronauts, **1:**59, 60, **3:**45, 244–245
Physics education
  needed, astronomy careers, **2:**27
  teacher training, **2:**30
Phytoplankton, **1:**287, 288, 290
Piazzi, Giuseppe, **2:**66
Picosatellites, **1:**327, 329, **4:**193, 194, 197, 198, 279
Pierce, John, **1:**94
Pilgrims (North America), **4:**237–238
Pilot astronauts, **3:**26, 27, 45, 228, 243, 308–309, 363–364, *364*
Pilot suits, **3:**345–346
Piloted guidance systems, **3:**132–133
Pioneer 1 (spacecraft), **1:**11
Pioneer 10 (spacecraft), **1:**327, **2:**82, 142, 261